AF725703

DRYING IN THE PROCESS INDUSTRY

DRYING IN THE PROCESS INDUSTRY

C.M. van 't Land

WILEY

A JOHN WILEY & SONS, INC., PUBLICATION

Copyright © 2012 by John Wiley & Sons. All rights reserved.

Published by John Wiley & Sons, Inc., Hoboken, New Jersey
Published simultaneously in Canada

No part of this publication may be reproduced, stored in a retrieval system, or transmitted in any form or by any means, electronic, mechanical, photocopying, recording, scanning, or otherwise, except as permitted under Section 107 or 108 of the 1976 United States Copyright Act, without either the prior written permission of the Publisher, or authorization through payment of the appropriate per-copy fee to the Copyright Clearance Center, Inc., 222 Rosewood Drive, Danvers, MA 01923, 978-750-8400, fax 978-750-4470, or on the web at www.copyright.com. Requests to the Publisher for permission should be addressed to the Permissions Department, John Wiley & Sons, Inc., 111 River Street, Hoboken, NJ 07030, 201-748-6011, fax 201-748-6008, or online at http://www.wiley.com/go/permission.

Limit of Liability/Disclaimer of Warranty: While the publisher and author have used their best efforts in preparing this book, they make no representations or warranties with respect to the accuracy or completeness of the contents of this book and specifically disclaim any implied warranties of merchantability or fitness for a particular purpose. No warranty may be created or extended by sales representatives or written sales materials. The advice and strategies contained herein may not be suitable for your situation. You should consult with a professional where appropriate. Neither the publisher nor author shall be liable for any loss of profit or any other commercial damages, including but not limited to special, incidental, consequential, or other damages.

For general information on our other products and services or for technical support, please contact our Customer Care Department within the United States at 877-762-2974, outside the United States at 317-572-3993 or fax 317-572-4002.

Wiley also publishes its books in a variety of electronic formats. Some content that appears in print may not be available in electronic formats. For more information about Wiley products, visit our web site at www.wiley.com

Library of Congress Cataloging-in-Publication Data

Land, C.M. van 't, 1937–
Drying in the process industry / C.M. van 't Land.
p. cm.
Includes bibliographical references and index.
ISBN 978-0-470-13117-6 (hardback)
1. Drying. 2. Drying apparatus. 3. Chemical processes. I. Title.
TP363.L229 2011
660'.28426–dc22

2011012195

10 9 8 7 6 5 4 3 2 1

CONTENTS

PREFACE

Drying is an important operation in the process industry. This book treats drying as a method for accomplishing liquid–solid separation by other than mechanical means. Usually, heat is supplied, leading to evaporation of a liquid (usually water), and this leaves a solid behind. Drying accomplishes the transformation of a process stream and, as such, often produces a salable product. As drying is an energy-intensive activity and dryers are expensive pieces of equipment, drying must be carried out as economically as possible.

This book is a follow-up to my earlier book, *Industrial Drying Equipment: Selection and Application*. In comparison to that book, the theoretical basis has been strengthened and the contents have been updated and extended.

The objective of this book is to assist the process development engineer, the process engineer, and the plant engineer in their selection of drying equipment. The theoretical background of drying and criteria to be observed when selecting drying equipment are discussed. Dryer descriptions and procedures for sizing them are treated. The subjects of product quality, process safety, process control, gas cleaning, and dryer feeding complete the book.

Acknowledgments

The writing of the earlier book was made possible by permission of Akzo Nobel Chemicals B.V., to whose management I am still grateful. The invaluable experience gained while in their employ was an important element in the design of that book.

Thanks are due a former colleague, Dave Buckland, who for the earlier book helped to convert my "Dutch English" into proper English and suggested a number of improvements to the contents. For the present book, the linguistic aspects of the modifications of and extensions to the earlier text were checked by the publisher, to whom I am grateful. Thanks are also due my former manager, Hans Postma, who read the manuscript of the earlier book on behalf of Akzo Nobel Chemicals B.V. and, in doing so, made useful suggestions.

Shortly after the earlier book appeared, I began to give seminars on drying in the process industry, mainly in Germany and The Netherlands. I am grateful for the information and suggestions given to me by participants in these seminars. The seminar interaction made clear in which direction industrial drying is going and provided useful contacts and material for the present book.

I began work as a consultant after my retirement. Thanks are due to the companies that I worked for, which thus helped me to extend my knowledge of industrial drying and keep it up to date. Particular appreciation is extended for the assistance given by:

M. Andreae-Jäckering, Altenburger Maschinen Jäckering GmbH
A. Bouwmeester, GMF-Gouda Processing Solutions
D.W. Dahlstrom, Alstom Power, Inc.
S. Gerl, Maschinenfabrik Gustav Eirich GmbH & Co. KG
A. Glockner, Glatt GmbH
A.K.E. Greune, Hazemag & EPR GmbH
W. Hinz, Buss-SMS-Canzler GmbH
W.J.L. Janssen, Deconsult
J. Schmid, FIMA Maschinenbau GmbH
H. Schneider, GoGaS Goch GmbH & Co. KG

I also thank the following companies, which most kindly provided data, drawings, and/or photographs:

Adolf Kühner AG, Birsfelden, Switzerland
Alstom Power, Inc., Warrenville, IL
Altenburger Maschinen Jäckering GmbH, Hamm, Germany
Andritz Fliessbettsysteme GmbH, Ravensburg, Germany
Andritz KMPT GmbH, Vierkirchen, Germany
Anhydro A/S, Søborg, Denmark
Bartec GmbH, Gotteszell, Germany
Bepex International LLC, Minneapolis, MN
Berthold Technologies GmbH & Co. KG, Bad Wildbad, Germany
Braunschweigische Maschinenbauanstalt AG, Braunschweig, Germany
Bucher Processtech AG, Niederweningen, Switzerland
Buss-SMS-Canzler GmbH, Butzbach, Germany
Carrier Vibrating Equipment, Inc., Louisville, KY
CPM Wolverine Proctor LLC, Horsham, PA
CPM Wolverine Proctor Ltd, Glasgow, UK
Deconsult, Heelsum, The Netherlands
FIMA Maschinenbau GmbH, Obersontheim, Germany
FLSmidth A/S, Valby, Denmark

Gala Industries, Inc., Eagle Rock, VA
GEA Barr-Rosin Ltd, Maidenhead, UK
GEA Pharma Systems nv, Wommelgem, Belgium
GEA Process Engineering A/S, Søborg, Denmark
GE General Eastern Instruments, Wilmington, MA
Glatt GmbH, Binzen, Germany
GMF-Gouda Processing Solutions, Waddinxveen, The Netherlands
GoGaS Goch GmbH & Co. KG, Dortmund, Germany
Grenzebach BSH GmbH, Bad Hersfeld, Germany
Hazemag & EPR GmbH, Dülmen, Germany
HERMETIC-Pumpen GmbH, Gundelfingen, Germany
Hosokawa Micron B.V., Doetinchem, The Netherlands
IMA Edwards Freeze Drying Solutions, Dongen, The Netherlands
Kidde Fenwal Inc., Ashland, MA
Kidde Products Limited, Colnbrook, UK
Komline-Sanderson Engineering Corporation, Peapack, NJ
Maschinenfabrik Gustav Eirich GmbH & Co. KG, Hardheim, Germany
Microdry Inc., Crestwood, KY
Mikropul GmbH, Cologne, Germany
Mitchell Dryers Ltd, Carlisle, UK
Nara Machinery Co., Ltd, Frechen, Germany
Oerlikon-Leybold Vacuum GmbH, Cologne, Germany
Patterson-Kelley/Harsco, East Stroudsburg, PA
Process Sensors Corp., Milford, MA
Rembe GmbH Safety + Control, Brilon, Germany
Rosenmund VTA AG, Liestal, Switzerland
SPX Flow Technology Danmark A/S, Søborg, Denmark
STALAM S.p.A., Nove, Italy
Strayfield Limited, Reading, UK
Streekmuseum voor Tholen en Sint-Philipsland "De Meestoof," Sint-Annaland, The Netherlands
Surface Measurement Systems Ltd, London, UK
Swenson Technology, Inc., Monee, IL
TREMA Verfahrenstechnik GmbH, Kemnath, Germany
Vaisala Oyj, Helsinki, Finland
3V Cogeim SRL, Dalmine, Italy
Vibra Maschinenfabrik Schultheis GmbH & Co., Offenbach am Main, Germany
Wyssmont Company, Inc., Fort Lee, NJ

I also wish to thank the following publishers, who most kindly provided permission to use material:

Access Intelligence, New York
Informations Chimie, Paris, France
The McGraw-Hill Companies, New York
Wiley-Blackwell, Oxford, UK

I am greatly indebted to my wife, Annechien, for her constant encouragement and patience.

C.M. van 't Land

1

INTRODUCTION

Drying can be defined as a unit operation in which a liquid–solid separation is accomplished by the supply of heat, with separation resulting from the evaporation of liquid. Although in the majority of cases water is the liquid being removed, solvent evaporation is also encountered. The definition may be extended to include the dehydration of food, feed, and salts, and the removal of hydroxyl groups from organic molecules.

This book is based on my personal experience gained in the selection and operation of drying equipment while employed by Akzo Nobel, a multinational company that manufactured, at that time, bulk and fine chemicals, pharmaceuticals, and coatings. Since 2000, I gained experience while working as an independent consultant.

Laboratory measurements and investigations concerning the drying of a product should be the first stage of the selection of a new dryer or the replacement of one. This aspect is discussed in Chapter 3. During the next stage, a person should seek the cooperation of a reputable dryer manufacturer. Close cooperation between the manufacturer and the potential user is essential, because one partner is knowledgeable about the equipment and the other person has expertise regarding the product. Since small-scale testing of drying equipment can be carried out, such testing can provide valuable insight into ultimate dryer selection. However, it is important that each partner have some insight into the other's field so that the user can develop value judgments on the equipment being recommended by the manufacturer. The size of the equipment must be checked, using various techniques (e.g., estimating methods, rules of thumb, rough-and-ready calculations). This book covers these techniques for each class of dryer.

Drying in the Process Industry, First Edition. C.M. van 't Land.
© 2012 John Wiley & Sons, Inc. Published 2012 by John Wiley & Sons, Inc.

Various reasons exist for drying materials to a specific level or range:

1. It is often necessary to obtain a free-flowing material that can be stored, packed, transported, or dosed.
2. Contractual limits exist for many products (e.g., salt, sand, yarn).
3. Statutory limits are in force for some materials (e.g., tobacco, flour).
4. A moisture content within a specified range may have to be obtained for quality control purposes. For many dried foods and feeds, too much moisture may adversely affect shelf life and nutritional value, whereas a moisture content too low, due to overdrying, may cause the loss of valuable nutrients. Moisture contents that are either too high or too low may render a product less enjoyable.
5. The feasibility of subsequent process steps sometimes requires that the moisture content be between specified limits, as in the milling of wheat or the pressing of pharmaceutical tablets. Another example is the low moisture content of rubber chemicals to be used in the vulcanization process of tires. Too much moisture causes the formation of blisters.
6. The onset of mildew and bacterial growth in such textiles as woolen cloth can be prevented by drying the cloth to a specific moisture content.
7. A drying step can be used as a shaping step. The manufacture of fluid cracking catalysts is an example. A spray-drying step produces hard and dry spheres of average diameter 80 μm. However, next, the spheres are leached with water to rcmovc sodium salts. That stcp is followcd by filtration and flash drying.

Typical dryer feeds are:

1. Objects (e.g., bricks)
2. Particulate materials (e.g., sodium sulfate crystals)
3. Filter and centrifuge cakes
4. Sheet material (e.g., paper for newspapers)
5. Pastes (e.g., dibenzoyl peroxide paste)
6. Liquids (i.e., solutions, emulsions, or suspensions)

Drying is an energy-intensive process. In general, heating and evaporation require large quantities of energy. An apple of mass 100 g hanging 4 m above the ground has a potential energy of approximately 4 J. Heating 1 kg of water from 15°C to 100°C requires 356,150 J. Evaporating 1 kg of water at 100°C and atmospheric pressure requires 2,285,000 J. Thus, in terms of energy, thermal effects are in general much more important than mechanical effects. This explains why the energy consumption in phase transformation and the heating in a drying operation exceeds the energy consumption of electromotors. In this respect, there is one more important aspect. The energy to evaporate 1 kmol of liquid is approximately constant for all liquids. Thus, it is possible to evaporate 18 kg of water (which has a kilomolecular weight of

18 kg·kmol^{-1}) with this heat of evaporation. However, it is also possible to evaporate 92 kg of toluene (which has a kilomolecular weight of 92 kg·kmol^{-1}) with this amount of heat. The explanation is that kilomoles of different substances contain the same number of molecules: $6.023 \cdot 10^{26}$ (Avogadro's number). Thus, on evaporating 1 kmol of a substance, the bonds between this number of molecules must be broken. The bonds between the molecules are relatively weak Van der Waals forces and are approximately equal. The evaporation of water occurs more frequently than the evaporation of organic liquids.

The energy consumption of the drying operation in the UK has been reviewed by Bahu and Kemp [1]:

- The energy consumption of drying is 8% of the industrial energy consumption. The industrial energy consumption comprises both processes and buildings.
- The annual water evaporation amounts to $2 \cdot 10^{10}$ kg. This is equivalent to 100-m water columns on 27.2 soccer fields (70·105 m^2). As the U.S. economy is about 5.5 times larger than the UK economy, the annual water evaporation in the United States due to drying could be $1.1 \cdot 10^{11}$ kg.
- In 1981, drying required $1.622 \cdot 10^{14}$ kJ. This figure was possibly 10 to 20% lower in 1991.
- $(1.622 \cdot 10^{14})/(2 \cdot 10^{10}) = 8110$ kJ per kilogram of evaporated water. This consumption figure includes electricity. Excluding electricity, the consumption figure is possibly 7000 kJ·kg^{-1}. Compared to the heat of evaporation of water at 0°C and atmospheric pressure (i.e., 2500 kJ·kg^{-1}), the consumption figure is quite high. In the chapters to come, the background of this state of affairs is discussed.
- Annual costs are determined by taking 32,000 kJ·nm^{-3} as the lower heating value of natural gas. The lower heating value is relevant if the heat of condensation of the water vapor in the combustion gases is not recovered. In the UK, an industrial price of €0.30 is typical:

$$\frac{1.622 \cdot 10^{14} \cdot 0.30}{32{,}000} = €1{,}520{,}625{,}000$$

These calculations illustrate that drying is an expensive means of accomplishing a liquid–solid separation; as a rule of thumb, 2 to 3 kg of steam is required for the evaporation of each kilogram of water. In a four-effect evaporation plant, approximately 4 kg of water can be evaporated with 1 kg of steam. Furthermore, performing a solid–liquid separation by means of a centrifuge or filter is usually much cheaper than using a dryer. Calculations concerning the energy required by the drying process begin with an assessment of the enthalpy difference between the process flows leaving the dryer and the process flow entering the dryer. Enthalpy differences are heat effects at constant pressure. In convective drying processes, the drying gas should be excluded from these calculations. Thus, the net heat is arrived at.

The heat required for drying can be supplied by the fundamentally different mechanisms of convection, conduction, and radiation:

1. *Convection.* A carrier gas (usually, air) supplies the heat for the evaporation of the liquid by the conversion of sensible heat into latent heat. The carrier gas subsequently entrains the volatile matter.
2. *Conduction.* The heat is supplied indirectly and the carrier gas serves only to remove the evaporated liquid. Typically, the airflow is approximately 10% of the airflow used in a convective process. Conduction of heat is the heat transport mechanism at contact drying.
3. *Radiation.* This type of drying can, in principle, be *nonpenetrating*, such as the drying of paint by infrared radiation, or *penetrating*, such as the drying of food or pharmaceuticals by dielectric drying. Dielectric drying (radio-frequency drying and microwave drying) is the only process in which heat is developed in the material being dried rather than having heat diffused into the material. Again a carrier gas is required to remove the evaporated liquid.

A combination of two mechanisms may be encountered in some dryer types.

The situation in the United States was analyzed by Strumillo and Lopez-Cacicedo [2], who found that 99% of dryer energy consumption could be attributed to six dryer types. In order of importance:

- Flash dryer
- Spray dryer
- Cylinder dryer for paper
- Convective rotary dryer
- Contact rotary dryer
- Fluid-bed dryer

This list illustrates that in terms of tonnage, convective drying is more important than conduction (contact) drying.

Dryer Types

A great variety of dryer types is commercially available. The reasons are as follows:

- Different products have very different drying times.
- The product quality often requires a certain dryer type or mode.
- It is often necessary to transport particulate material through a dryer.

A distinction should be made between free and bound moisture. Initially, *free water* is evaporated until the critical moisture content is reached. Free water's latent heat of evaporation is essentially equal to that of water on evaporating from a pool, with the heat transfer being the rate-determining step. Evaporation occurs at a constant rate if the heat supply is constant. Thus, as long as there is free water, the rate of evaporation is not a function of the water concentration. The order of the process is then zero. Drying to below the critical moisture content requires the evaporation of bound water, with the evaporation rate decreasing if the heat supply is kept constant. *Bound water* can be present in pores or crevices, can be physically absorbed, or can be present as water of hydration. The latent heat of evaporation of bound water is usually higher than that of free water; for example, the ratio of the latent heats of evaporation of water in wool containing 16 and 30% water by weight (the latter value is the critical moisture content) is approximately 1.1 : 1.

Temperature and Moisture Profiles

In this book we deal only with phenomena related to objects to be dried. Thus, transient temperature and moisture profiles in the product to be dried are not discussed.

Drying Systems

Unlike a centrifuge, for example, a dryer consists of a number of pieces of equipment grouped together in a subsystem. It is therefore more correct to refer to drying *systems*. Convective drying systems are often more extended than contact or radiation dryer systems. Drying is often the last process step, which is followed by a solids-handling system designed by mechanical engineers. In addition, being an energy-intensive process, drying is sometimes handled by energy specialists. It can therefore be considered a unit operation that falls at the interface of three disciplines: chemical, mechanical, and energy engineering.

In Chapter 2 it is recommended that the drying step not be considered in isolation but rather be reviewed in the context of the entire process. Upstream process modifications can have a great impact on the drying stage, whereas the method of drying is often of paramount importance to product quality.

Procedures for determining the optimum dryer to use are covered in Chapter 3. One scheme is presented for continuous dryers, with a separate scheme for batch dryers. Chapter 4 provides an introduction to convective drying, and Chapters 5 through 8 cover in detail the four main categories of convective dryers. In these chapters, the performance of dryers is analyzed, their literature data interpreted, and design methods are covered. The material that is presented permits an estimation of both fixed and relevant variable costs for convective dryers. In Chapter 9, miscellaneous continuous convective dryers and convective batch dryers are discussed, and atmospheric contact dryers are treated in Chapter 10. Vacuum drying, including freeze-drying, is covered in Chapter 11. Steam drying is treated in Chapter 12. Radiation drying (infrared, radio-frequency, and microwave drying) is dealt with in Chapter 13, and the important issues of product quality and safety are considered in Chapter 14. Fires and dust explosions are treated in the context of safety. Chapter 15 covers continuous solids- and gas-moisture measurement, dryer process control, and energy recovery.

Figure 1.1 Drying tower for madder roots. (Courtesy of Streekmuseum voor Tholen en Sint-Philipsland "De Meestoof", Sint-Annaland, The Netherlands.)

The separation of particulate solid material from spent drying gas by means of cyclones, fabric filters, scrubbers, and electrofilters are the topics in Chapter 16, and the selection of feeders for dryers is taken up in Chapter 17.

One hundred and fifty years ago, drying was often a very time-consuming process. We illustrate this by means of an example, the manufacture of a red dye from madder roots. Madder is a plant with long, thick roots that contain a red dye. From possibly 1400 until approximately 1900, this dye was manufactured industrially in Great Britain and The Netherlands. The roots were harvested and, as a first step, dried in a drying house (see Fig. 1.1). The roots were first laid on the lowest floor and were moved to higher floors as the drying proceeded. An oven at ground level heated the drying house. Control of the drying process was as follows:

- More or less intense fire
- Deposition of stones on the bottom ducts
- Degree of opening of the hatches at the top

The roots contained approximately 80% water by weight. A typical plant's annual capacity amounted to 100 metric tonnes, with a water evaporation of 400 metric tonnes. The installation of steam tubes around 1850 made possible a more reproducible drying process. The latter meant a switch from convective drying to contact drying. The first drying step was followed by a postdrying step on an oast and a milling step. The practice of manufacturing the red dye from madder was stopped around 1900 because in 1868, Gräbe and Liebermann discovered the synthesis of alizarine from anthracene, and alizarine could replace the red madder dye.

In general, contact drying in steam-heated rotary dryers began in 1830. The development of convective drying began in 1890 when cheap electromotors to drive air fans became available. Spray drying began between 1920 and 1930. Freeze-drying dates back to 1935, and microwave drying was introduced in 1955.

REFERENCES

[1] Bahu, R., Kemp, I. (1994). Chapter 6 (Drying) in *Separation Technology: The Next Ten Years,* edited by Garside, J., IChemE, Rugby, UK.

[2] Strumillo, C., Lopez-Cacicedo, C. (1991). Chapter 27 (Energy Aspects in Drying) in *Handbook of Industrial Drying,* edited by Mujumdar, A.S., Marcel Dekker, New York.

2

DRYING AS PART OF THE OVERALL PROCESS

In the early stages of investigating a drying problem, attention should be given to the entire manufacturing process. This holistic approach may yield one of the following conclusions:

1. The dried product can have a certain residual moisture content.
2. The dewatering step can be optimized.
3. It is possible to simplify the drying step via a process change.
4. The drying step can be combined with one or more other process steps.
5. It is possible to remove the water by a nonthermal method.
6. The drying step can be avoided by changing the process.
7. The product is not dried, whereas the process is not changed.

These seven options are examined below in greater detail.

2.1 RESIDUAL MOISTURE

To dry a product to a very low moisture content often requires a great deal of energy; however, it is sometimes sufficient to dry a product to a specific moisture content before selling it. This would reduce energy costs, and it would be advantageous that more product be sold at the same raw material cost. This option can be useful in combination with a reliable in-plant continuous moisture-monitoring system.

Drying in the Process Industry, First Edition. C.M. van 't Land.
© 2012 John Wiley & Sons, Inc. Published 2012 by John Wiley & Sons, Inc.

2.2 OPTIMIZATION OF THE DEWATERING STEP

Before drying, it is generally advantageous to remove as much water as possible by filtration or centrifugation. Centrifugation is in this respect in principle more effective than filtration, but it cannot always be used. Due to the centrifugal force, centrifuge cakes may become impermeable.

Example 2.1 The strong fiber Twaron (trade name of Teijin Twaron) is obtained by spinning a solution of the p-aramid polymer poly(*p*-phenyleneterephtaloylamide) (PPTA) in concentrated sulfuric acid. On spinning, the aramid molecules are arranged in parallel, which confers strength to the yarn through hydrogen bridges. The polymerization of terephtaloyldichloride and *p*-phenylenediamine to PPTA precedes this step. Prior to the dissolution in concentrated sulfuric acid, the polymer crumb is recovered from an aqueous slurry and dried. Initially, dewatering was carried out using a belt filter to produce an intermediate product containing 6.5 kg of water per kilogram of PPTA. In the 1990s the belt filter was replaced by a filter press, producing an intermediate product containing 2 kg of water per kilogram of PPTA.

Example 2.2 Another example of optimization of the dewatering step is that of the leaching of a cake in a liquid–solid separation system at an elevated temperature, which causes a reduction in the viscosity of the adhering liquid and hence leads to more efficient dewatering. This goal can be achieved by, for example, the use of steam in a leaching stage. A dramatic effect in the sugar industry has been described [1]: (1) leaching with cold water yields a sugar cake at 40°C containing about 2% water by weight; (2) treatment with steam results in a sugar cake at 80°C containing about 0.6% water by weight, with the additional benefit that further water loss occurs on the way to the dryer, so that the cake arrives at the dryer containing only 0.2 to 0.3% water by weight. Simons and Dahlstrom [2] reported moisture reductions by steam dewatering exceeding 60% for permeable filter cakes (a crystalline inorganic chemical with 50% by weight $> 200\ \mu m$ and a narrow size distribution). However, impermeable filter cakes cannot readily be dewatered further.

2.3 PROCESS CHANGES TO SIMPLIFY DRYING

Drying can often be simplified by increasing the particle size in the dryer feed. Various techniques, which are covered briefly below, can be used for particle-size enlargement. More detailed information may be found in standard textbooks on crystallization and precipitation (e.g., [3]).

The solubility of the material in the solvent affects the particle size. Materials that have moderate solubility in the solvent system being utilized (e.g., 1 to 30% by weight) are generally obtained in a coarse form with a weight-average particle size of 0.2 to 2 mm. This finding can be explained qualitatively since a small supersaturation/solubility ratio tends to lead to large crystals. For example, this behavior is found in sodium chloride, potassium chloride, and sugar.

Materials having a solubility of less than about 0.1% by weight tend to be obtained as small particles; for example, on precipitation, gypsum has a weight-average particle size in the range 1 to 100 μm. Particles in the size range 0.2 to 2 mm generally contain 1 to 5% moisture by weight on entering a dryer, whereas smaller particles may retain up to 30 to 40% by weight when discharged from a filter or centrifuge.

Particle size can be increased by changing the solubility of the dissolved material, by changing the solvent or pH, or by increasing the temperature, slurry density, or residence time of the crystallization process. Generally, a decrease in the system velocities of a crystallizer increases the average particle size.

Example 2.3 An organic acid is produced from an organic salt via acidification, which is followed immediately by precipitation. Process research showed that a good yield was obtained at pH 1.8; however, after filtration the precipitate had a moisture content of up to 40% by weight. On adjusting the pH to 2.3, the precipitate had a moisture content of 20 to 25% by weight, due to a different crystal modification; however, the yield was unsatisfactory. A plant design comprising two continuous-stirred tank reactors in series was chosen. The pH is adjusted to 2.3 in the first reactor, whereas pH 1.8 is selected for the second reactor. The bulk of the product is produced in the first reactor and has good filtration characteristics. The second reactor increases the yield while the good filtration characteristics are retained.

Example 2.4 Vacuum-pan salt is produced in multiple-effect evaporation plants. Modern salt plants contain crystallizers consisting of three main parts: vapor separator, heater, and pump. The three parts are connected by lines through which a salt slurry circulates. It is also possible to integrate these three parts into one piece of equipment. Plant measurements showed that the first type of crystallizer produces vacuum-pan salt having an average particle size of 450 μm, whereas the second type of crystallizer produces vacuum-pan salt having an average particle size of 650 μm. The difference is caused by the different pump tip velocities and velocities in the heater tubes, being 20 and 2 $m\cdot s^{-1}$ in the first case and 10 and 1 $m\cdot s^{-1}$ in the second case.

Combinations of more than one of the parameters cited above can also be used to achieve a desired particle-size distribution. Seeding the crystallizer contents can also increase the particle size. This procedure is applicable to systems that do not nucleate readily because of high viscosity, for example. Up to a certain level, supersaturation increases, at which point many nuclei may be produced. Seeding is practiced to prevent this, in sugar crystallization, for example. Seeding a crystallizer containing a material that nucleates readily (e.g., sodium chloride) can achieve a particle-size decrease.

Sometimes, because of product specification, it is not desirable to alter the average particle size; for example, rapid dissolution or proper dispersion of a product may require a small particle size. The average particle size and particle-size distribution are not the only factors that influence the moisture-retaining properties of materials; the particle shape (habit) and the specific area can also have a significant influence.

It is also possible to influence the process at the point where the particulate material is formed by replacing conventional crystallization (precipitation), liquid–solid separation, and drying by drum-drying or spray-drying systems. Two possible routes for the processing of clay tile-body suspension in water have been given [4]: (1) filter press–dryer–granulating unit–tile presses, and (2) spray dryer–tile presses. Spray drying and drum drying do not lead to the problem of having to dispose of an impure mother liquor.

The product as produced in a crystallizer or precipitator can be accepted, and an additional step can be introduced prior to liquid–solid separation and drying. Solids with a melting point of less than 100°C can be liquefied by the injection of live steam (i.e., the system is changed from a liquid–solid one to a liquid–liquid one). Subsequent cooling will lead to solidification and, if carried out correctly, can result in a particle-size increase. This process is termed *melt granulation*.

2.4 COMBINATION OF DRYING AND OTHER PROCESS STEPS

Since many possibilities exist in this category, drying can thus be combined with a chemical reaction, evaporation, mechanical liquid–solid separation, particle-size enlargement, and several other operations.

Example 2.5 The Solvay process is widely used for the manufacture of sodium carbonate (soda). Sodium bicarbonate is an intermediate particulate product separated from the mother liquor by rotary vacuum filters in which the crystals are also leached with water. Centrifuges are also used. The cake from the filters contains about 14% water by weight, whereas from the centrifuges it typically contains about 8%. Calcining of the bicarbonate to soda and drying take place in a single indirectly heated rotary-drum calciner, with the drying preceding the calcining.

$$2NaHCO_3 \rightarrow CO_2\uparrow + H_2O\uparrow + Na_2CO_3$$

Carbon dioxide is recycled after compression, and steam is generally used as the heating medium. The hot soda ash (ca. 200°C) is cooled, screened, and packaged or shipped in bulk. The product is called *light ash* because of its low bulk density.

Example 2.6 During the manufacture of potato chips, potatoes are peeled, sliced, and washed in 60°C water. The wet slices are then added to 160°C oil, and curing and drying occur in one step. Salt and spices are added before packaging.

Example 2.7 It is possible to combine liquid–solid separation, leaching, and drying in a single unit that functions batchwise. Such a step is used if, for example, it is desirable to protect the operators from dust. The slurry is pumped to the equipment and the dry solids are discharged at the end of the cycle.

Figure 2.1 Multifunction closed Nutsche vacuum filter–dryer. (Courtesy of Rosenmund VTA AG, Liestal, Switzerland.)

In the case of a filter–dryer, the equipment comprises a closed Nutsche-type vacuum filter with options (see Figs. 2.1 and 2.2). A typical cycle is made up of (1) feeding the slurry to the filter, (2) mother liquor removal, (3) leaching (displacement or reslurry), (4) smoothing and compressing the cake to remove liquid, (5) drying (indirect heat transfer, vacuum), and (6) discharge. For further treatment, see Section 11.1.

In the case of a centrifuge–dryer, the equipment consists of a closed filtering centrifuge with options (see Fig. 2.3). A typical cycle consists of (1) feeding the slurry to the centrifuge, (2) mother liquor removal, (3) leaching, (4) breaking up the centrifuge cake by nitrogen or air pressure, (5) convective drying by nitrogen or air circulation, and (6) discharge.

The filter–dryer and the centrifuge–dryer are described in more detail in Chapters 11 and 9, respectively. Possible applications of these pieces of equipment include the automation of existing processes, processing light-sensitive materials, solvent recovery, and handling toxic materials.

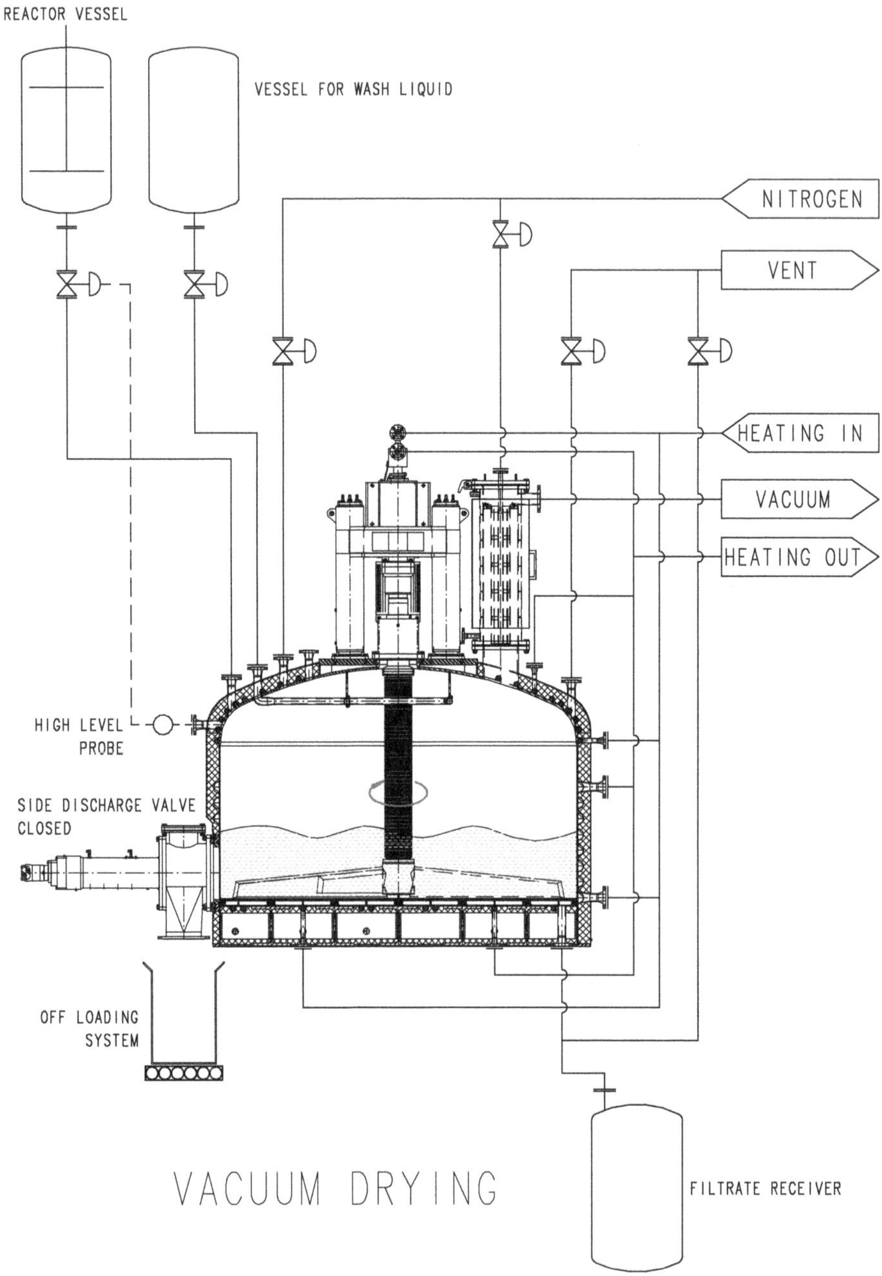

Figure 2.2 Vacuum drying step in a closed filter–dryer. (Courtesy of Rosenmund VTA AG, Liestal, Switzerland.)

Figure 2.3 Centrifuge–dryer. (Courtesy of FIMA Maschinenbau GmbH, Obersontheim, Germany.)

Reference to spray drying and drum drying was made in Section 2.3. With the use of this equipment, evaporation, crystallization, liquid–solid separation, and drying can be combined. The combination is also possible in a thin-film evaporator (see Fig. 2.4). The choice of such a system may not be energetically preferable; however, it can be the best choice if conventional crystallization, liquid–solid separation, and drying are complicated or impossible, or if the resulting product has favorable properties (e.g., the rapid dissolution of spray-dried coffee powder, in which the formed particles appear as hollow spheres).

2.5 NONTHERMAL DRYING

As a general rule, moisture that can be removed mechanically should not be removed thermally. The mechanical removal of moisture is not considered in this section, but physical absorption and chemical reaction, the two principal nonthermal methods for moisture removal, are.

Example 2.8 An inorganic hydrate containing a small percentage of moisture leaves a centrifuge. The material can be admixed in a screw conveyor with a small amount of a lower hydrate that picks up the water and is itself converted to the higher hydrate.

$$x \cdot \mathrm{aH_2O} + b\mathrm{H_2O} \rightarrow x \cdot (a + b)\mathrm{H_2O}$$

The nonthermal drying of centrifuged soda decahydrate by admixing it with a small amount of anhydrous soda is an application. x stands for a chemical compound.

Figure 2.4 Vertical thin-film dryer. (Courtesy of Buss-SMS-Canzler GmbH, Butzbach, Germany.)

Example 2.9 In casting operations for certain pottery objects, such as teapots, gypsum molds are used to dewater a clay stream (slip) in order to produce an article in the correct physical form. The slip typically contains 30 to 40% water by weight. Additives allow the concentrated slurry to flow. A porous gypsum mold absorbs the water and must be dried before it can be reused. After 10 to 20 cycles, the pores of the mold are plugged and it cannot be regenerated further.

Example 2.10 Williams-Gardner has described the use of starch molds for the shaping of confections [4]. Some of the water in the syrup containing dissolved material is absorbed to accomplish product shaping.

2.6 PROCESS CHANGES TO AVOID DRYING

Process change can also enable a product to be obtained in a form that makes drying superfluous.

Example 2.11 Acetylene is manufactured by the reaction of calcium carbide and water:

$$CaC_2 + 2H_2O \rightarrow Ca(OH)_2 + C_2H_2\uparrow \quad -134\,\text{kJ}\cdot\text{mol}^{-1}$$

Originally, an excess of water was used to control the reaction and to suspend the lime, the hydrated lime slurry being sent to a second plant for use there. Because of changes in the latter plant, the practice of pumping the slurry had to be discontinued. Liquid–solid separation and drying of the lime were considered; however, a better solution was to switch from a wet to a dry manufacturing process. This process uses about 1 kg of water per kilogram of carbide, and the heat of reaction is dissipated by vaporization of the water. The hydrated lime still contains about 5% water by weight but can be packaged and sold directly to the building trade, for example. The rates of water addition and mixing are critical and must be controlled carefully; nevertheless, the process is now used widely.

Example 2.12 A method to circumvent the drying of aliphatic diacyl peroxides, which are solid at room temperature and fusible below 120°C, has been described [5, 6]. These compounds are manufactured by the addition of acid chloride to an aqueous solution of sodium peroxide:

$$2R\text{—}\overset{\overset{\displaystyle O}{\|}}{C}\text{—}Cl + Na_2O_2 \rightarrow R\text{—}\overset{\overset{\displaystyle O}{\|}}{C}\text{—}O\text{—}O\text{—}\overset{\overset{\displaystyle O}{\|}}{C}\text{—}R + 2NaCl$$

The reaction proceeds at room temperature, with the chemical reaction followed by precipitation. Since the products are almost insoluble in the aqueous phase, the resulting particles are small. Conventionally, the solid is separated from the liquor by filtration, then leached and dried in tray dryers. The alternative process route is to inject live steam into the reaction slurry as it is pumped through a line, the system temperature being elevated to a value slightly higher than the melting point of the peroxide. The liquid–solid system is thus converted into a liquid–liquid system that can be separated in a disk centrifuge. The organic phase is solidified rapidly on a cooling belt. To avoid decomposition, the peroxide is held at the elevated temperature for just a few seconds. Although the short process time is an important advantage, the principal benefit is that the product obtained is in a purer form than when obtained by the conventional process, because leaching of the filter cake is difficult. The solid-handling characteristics are also improved, with the product being obtained as a flake rather than as a powder.

Example 2.13 A similar process improvement for the manufacture of dialkyl peroxydicarbonates has been described [7]. An ester of chloroformic acid reacts with an aqueous solution of sodium peroxide at an ambient or slightly elevated temperature:

$$2\mathrm{R{-}O{-}\overset{\overset{\displaystyle O}{\|}}{C}{-}Cl} + \mathrm{Na_2O_2} \rightarrow \mathrm{R{-}O{-}\overset{\overset{\displaystyle O}{\|}}{C}{-}O{-}O{-}\overset{\overset{\displaystyle O}{\|}}{C}{-}O{-}R} + 2\mathrm{NaCl}$$

(where R is a C_6 to C_{18} linear, cyclic, or branched alkyl group). It is also possible to use a mixture of esters. The patent concerns principally the manufacture of both dimyristyl and dicetyl peroxydicarbonate. Once again live steam is injected into the reaction slurry to liquefy the product, which is separated from the aqueous phase by means of a disk centrifuge. The product is subsequently solidified using a flaker. The particulate product contains little dust.

Figure 2.5 Centrifuging and drying can be combined in this piece of equipment. (Courtesy of Gala Industries, Eagle Rock, VA.)

2.7 NO DRYING

The process is left as is except for the drying stage, which is simply not carried out. On analyzing the full picture, it sometimes becomes evident that the manufacturer of a product takes pains to dry the product knowing full well that upon its receipt the customer dissolves or suspends the particles in water. If, for example, packaging, transporting, and unloading of the product is not hindered, a delivery of wet cake is simpler. In general, this can be realized for materials that are not soluble in the adhering liquid, because soluble products may present caking problems.

Example 2.14 Dibenzoyl peroxide is a solid product prepared by reaction and crystallization in an aqueous phase. It was marketed originally as a dry powder obtained through liquid–solid separation followed by drying. But in this form the dry powder has unfavorable safety characteristics (low impact resistance), and severe in-plant decompositions have been experienced during its manufacture. A reevaluation of the market led to customer acceptance of a wet-cake form and mixtures of the dry product and inert fillers (phlegmatization). In the latter case the wet cake was, prior to drying, admixed with these inert fillers.

Example 2.15 High-pressure polyethylene is extruded and cut into pellets in water. The water–pellet slurry can be transported to a dewatering screen. The wet pellets are subjected to centrifugal action, and they can also be dried with warm air simultaneously and then bagged for sale (see Fig. 2.5).

REFERENCES

[1] Våhl, L. (1957). The drying of particulate material, elucidated with sugar and starch as examples. *De Ingenieur*, *69*, 77–86 (in German).

[2] Simons, C.S., Dahlstrom, D.A. (1966). Steam dewatering of filter cakes. *Chemical Engineering Progress*, *62*, 75–81.

[3] Mullin, J.W. (2001). *Crystallization*, Butterworth-Heinemann, Oxford, UK.

[4] Williams-Gardner, A. (1976). *Industrial Drying*, George Godwin, London, pp. 4–5.

[5] van Holten, J., Ribbens, C. (1971). Improvements in or relating to the purification of organic peroxides. British Patent 1 239 088.

[6] van Holten, J., Ribbens, C. (1971). Apparatus useful in the purification of organic peroxides. British Patent 1 239 089.

[7] Appel, H., Brossmann, G. (1984). Process for the continuous manufacture of dialkyl peroxy-di-carbonates. European Patent 0 049 740 (in German).

3

PROCEDURES FOR CHOOSING A DRYER

A large variety of drying equipment is currently available from manufacturers. In this chapter the screening procedures that offer a preliminary choice for a specific drying duty are described. In subsequent chapters we provide more details (e.g., dimensions and energy consumption) on the main classes of dryers.

The selection schemes described here are for batch and continuous dryers. They do not cover every possible type of dryer, but many of the industrially important systems are considered.

Production capacities exceeding 100 $kg \cdot h^{-1}$ often require a continuous dryer, but the choice between batch and continuous dryers also depends on the nature of the equipment preceding and following the dryer. Table 3.1 outlines the data that have to be collected before the process of selecting a dryer system can be begun, and Table 3.2 lists some of the criteria for evaluating dried material.

3.1 SELECTION SCHEMES

Figures 3.1 and 3.2 provide, for a particulate material, step-by-step procedures for the selection of a batch dryer and a continuous dryer, respectively, and the information presented in each subsection supplements the respective chart. Batch dryers are discussed first.

Adapted and reprinted by special permission from *Chemical Engineering*, March 5, 1984. Copyright © 2011 by Access Intelligence, New York, NY 10038.

Drying in the Process Industry, First Edition. C.M. van 't Land.
© 2012 John Wiley & Sons, Inc. Published 2012 by John Wiley & Sons, Inc.

Table 3.1 Data To Be Assessed Before Attempting Dryer Selection[a]

Production capacity ($kg \cdot h^{-1}$)	Experience already gained
Initial moisture content	Moisture isotherms
Particle-size distribution	Solubility of the solid in the liquid
Drying curve	Solution vapor pressure
Maximum allowable product temperature	Contamination by the drying gas
Explosion characteristics (vapor–air and dust–air)	Corrosion aspects
Toxicological properties	Physical data on the relevant materials

[a]Methods of determining the numerical values of the various criteria must be agreed upon.

Batch Dryers

Vacuum Dryers Vacuum drying is usually carried out batchwise. The subject receives attention in Chapter 11, in which vacuum pumps are treated as well.

If the maximum product temperature is lower than or equal to 30°C, it is worthwhile looking at a vacuum dryer. A good driving force for evaporation can be created while keeping the temperature low. The vacuum tray dryer is the simplest, but the product must usually be sieved to break down any agglomerates (the breakdown may be aided mechanically).

The capacity of the vacuum tray dryer is rather low. It may be economical to consider an agitated vacuum dryer (Fig. 3.3), in which the contents are moved mechanically. Such dryers are widely used. If the product is oxidized by air during drying, consider either vacuum drying or inert-gas drying.

If either the product or the liquid removed is toxic, the equipment must be kept closed as much as possible. Again, a vacuum dryer can render good service. (In addition, dust formation is avoided.)

Fluid-Bed Batch Dryers If the average particle size is about 0.1 mm or larger, fluid-bed drying (Fig. 3.4) may be considered. (If smaller particles must be dealt with, the equipment required to handle them may be too large to be feasible.) Inert gas may be used if there is the possibility of fire or explosion of either the vapor or dust in the air. If such a dryer is being considered, it is easy to carry out tests in a small fluid-bed dryer.

Table 3.2 Some Criteria for Judging a Dried Particulate Material[a]

Moisture content	Odor, taste
Particle-size distribution	Appearance
Bulk density	Dispersibility
Hardness	Dissolution or rewetting behavior
Dust content	Assay
Flow characteristics	Caking tendency
Color	Segregation of originally dissolved components (food)

[a]Methods of determining the numerical values of the various criteria must be agreed upon.

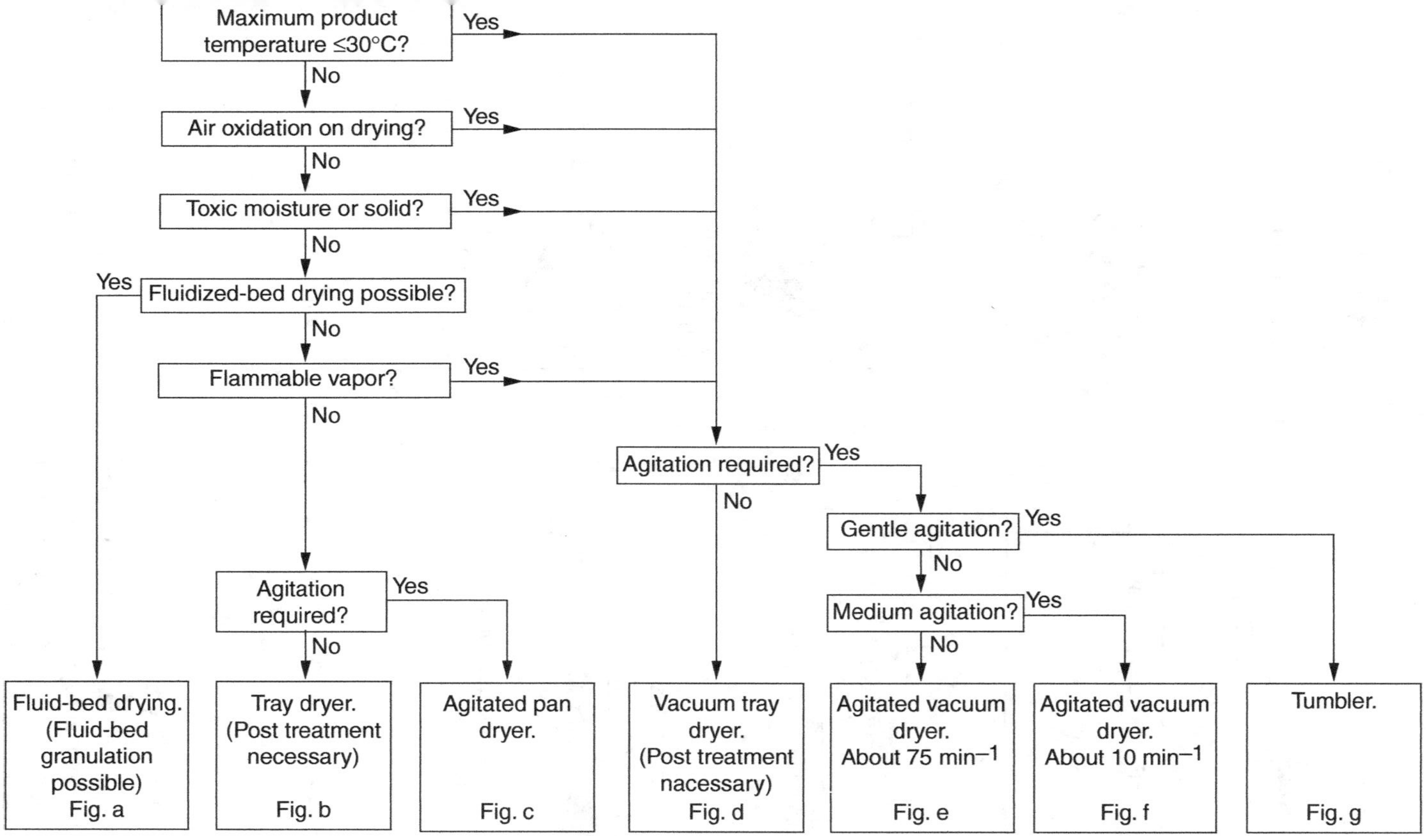

Figure 3.1 Decision tree for the selection of a batch dryer suitable for any particular process need, together with sketches of the various dryers suggested. (*Continued on next page.*)

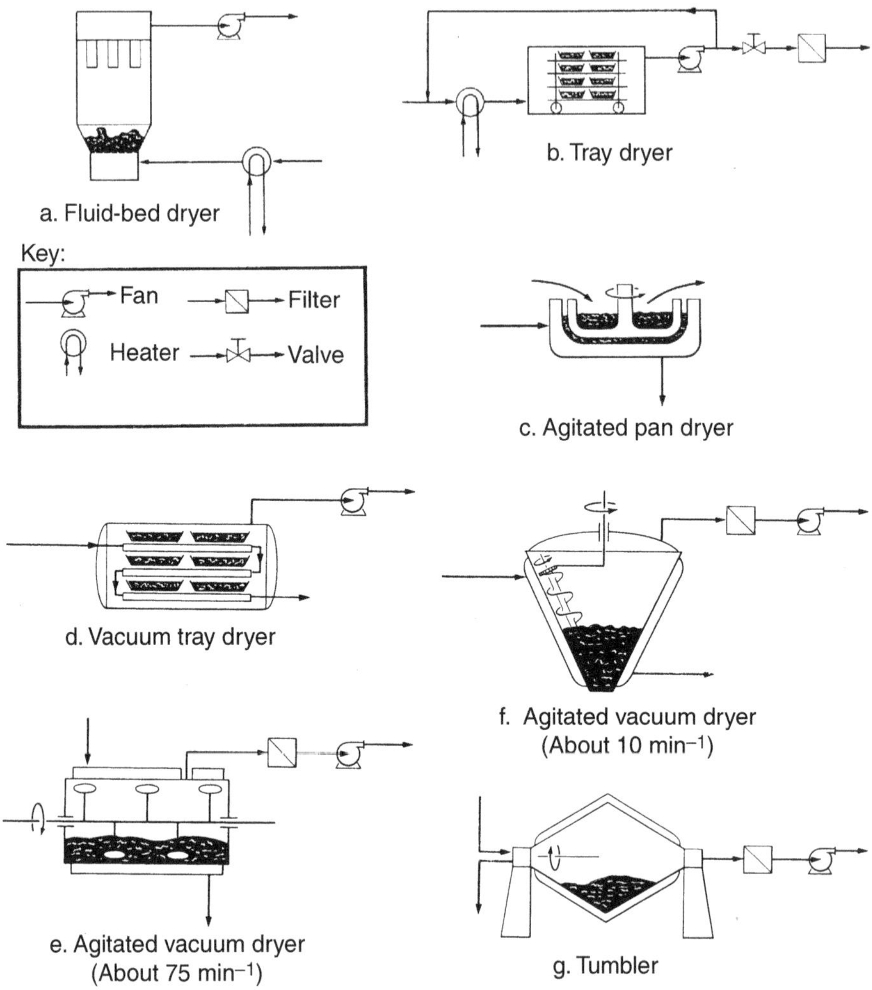

Figure 3.1 (*Continued*)

Other Batch Dryers As Figure 3.1 shows, the remaining possibilities regarding batch drying are the tray dryer and the agitated pan dryer.

Attention is focused next on continuous dryers.

Continuous Dryers

Solvent Evaporation If a solvent must be evaporated and then recovered, it is often not optimal to choose a convection dryer. Since solvent must be condensed from a large carrier-gas flow, the condenser and other equipment become rather large. A plate dryer could be a good option. It is a contact dryer with stationary plates in

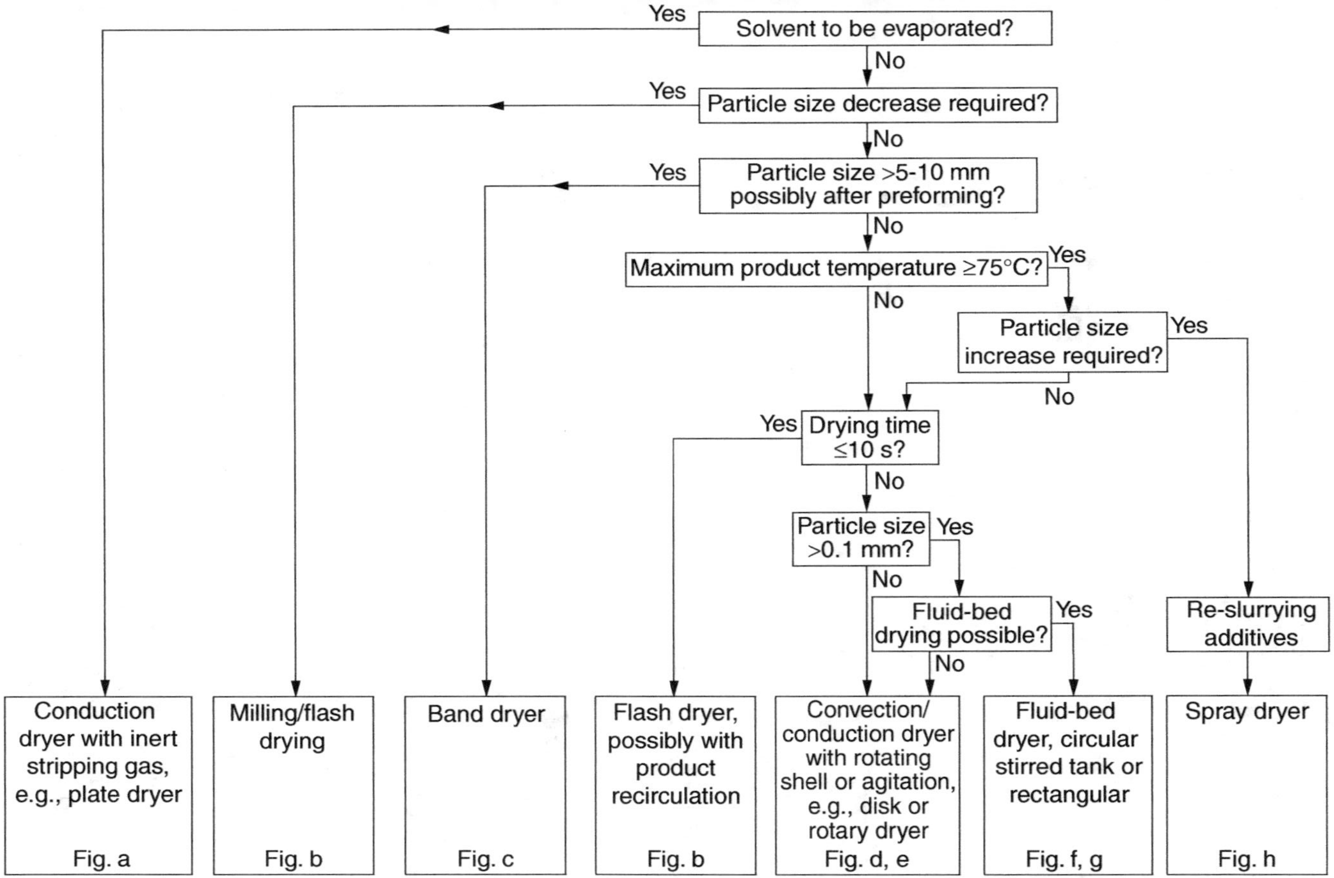

Figure 3.2 Decision tree for the selection of a continuous dryer, leading the user to one of the dryers shown in the sketches. (*Continued on next page.*)

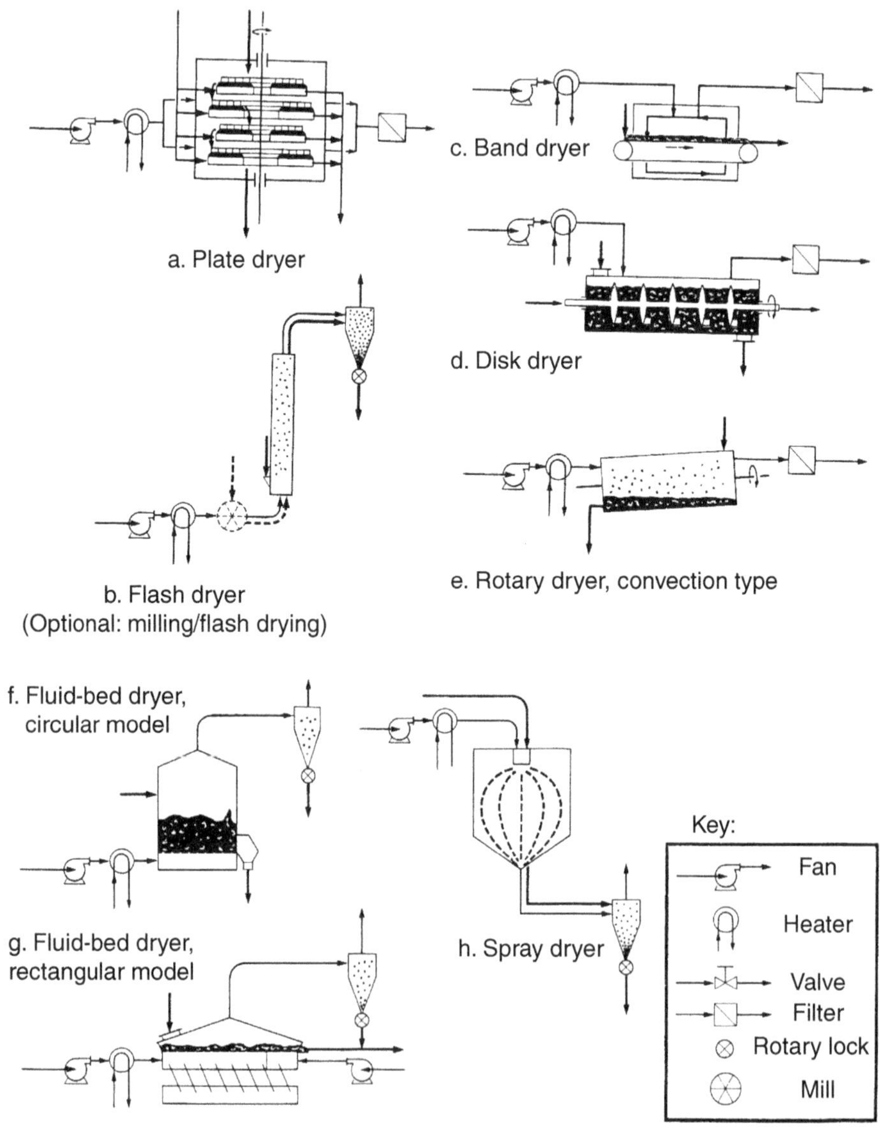

Figure 3.2 (*Continued*)

which the material is transported from the top to the bottom by means of rotating rakes. The heating medium circulates through the plates. This dryer is discussed in Chapter 10. The plate dryer's capacity, however, is limited. Andritz Fliessbett Systeme GmbH in Ravensburg, Germany reports the successful installation of large rectangular fluid-bed dryers in which high-density polyethylene is dried by evaporating hexane by means of warm nitrogen. The fluid-bed dryer is suitable for this application, as it does not contain moving parts.

Figure 3.3 Vacuum dryer for drying ion-exchanging resins contaminated by radioactivity. (Courtesy of Hosokawa Micron B.V., Doetinchem, The Netherlands.)

Milling and Drying If it is necessary to decrease particle size in addition to drying, the two operations may be combined advantageously. This dryer is also chosen if it is desired to prevent particle agglomeration. The wet particulate solid is transported to a mill by warm or hot gas. Gas and particulate solid leave the mill, fly through a line, and are separated. The comminution often greatly helps the drying by exposing internal moisture. This type of drying is encountered in cases where the fineness is of great importance to the application. Examples are cases where a rapid and complete dispersion (or dissolution) or a high level of activity ($m^2 \cdot g^{-1}$) are being sought. This dryer is discussed in Chapter 9.

Band (Belt) Dryers A band dryer (Fig. 3.5) is preferable if the particles are rather coarse (i.e., over 5 to 10 mm). The particles are spread evenly onto a perforated belt which is moving slowly (e.g., with a velocity of 5 $mm \cdot s^{-1}$). The belt moves into a drying cabinet and warm gas passes upward or downward through the layer. This type of dryer is chosen when it is not possible or desirable to suspend the particles in the drying gas. The dryer must offer a residence time (e.g., 15 min), because bound moisture must diffuse through the pellet.

The performance of such a dryer can be predicted from a determination of the drying curve on a small scale, employing realistic conditions (pellet characteristics, layer thickness, and drying-gas parameters). Many band dryers are used to dry preformed particles. The wet particulate material is mixed with additives, granulated, and dried. One reason to do this is that direct drying of the wet material may yield a dusty material, whereas the granules are less dusty. Band dryers are also used in food applications (e.g., diced carrots). The dryer is discussed further in Chapter 9.

Figure 3.4 Batch type of fluid-bed dryer. (Courtesy of Glatt GmbH, Binzen, Germany.)

Flash Dryers The flash dryer is the workhorse of industry (Fig. 3.6). However, because drying must take place within a couple of seconds, the removal of bound moisture is difficult. Since the dryer is essentially a vertical line, drying and vertical transport can be combined.

Miscellaneous Continuous Dryers Jobs that cannot be handled by a flash dryer or a fluid-bed dryer can often be accomplished in a direct-heat rotary dryer or in a conduction dryer, such as a steam-tube rotary dryer (for dusty products such as silica). Direct-heat rotary dryers are often employed for the drying of inorganic products, such as sand and limestone. Further continuous industrial dryers are treated in Chapters 9 and 10.

Figure 3.5 Band dryer uses a slowly moving perforated belt.

Fluid-Bed Continuous Dryers Use of a fluid-bed dryer is a possibility if the particle size exceeds 0.1 mm. A round piece of equipment holding a thick product layer is one option. The holdup must be large, as the composition of the dryer contents equals the outlet composition. The thick product layer means that considerable fan power is needed to push the drying gas through it. Because caking will not occur

Figure 3.6 Flash dryer, one of the most common types used in industry. (Courtesy of Grenzebach BSH GmbH, Bad Hersfeld, Germany.)

Figure 3.7 Continuous stationary fluid-bed dryer with in-bed heat exchanger for poly(vinyl chloride). (Courtesy of Andritz Fliessbett Systeme GmbH, Ravensburg, Germany.)

easily, the construction can be stationary. Such constructions can allow high drying-gas temperatures—up to 500 to 600°C.

A rectangular dryer will permit plug flow. Figure 3.7 shows a stationary type with a thick product layer; a shallow layer may require vibration for transport and to prevent caking. However, a vibrated construction (Fig. 3.8) cannot withstand high temperatures. A realistic maximum drying-gas inlet temperature is 300°C. Moreover,

Figure 3.8 Continuous vibrated fluid-bed dryer for detergents. (Courtesy of Andritz Fliessbett Systeme GmbH, Ravensburg, Germany.)

Figure 3.9 Spray dryer often produces a spherical product. (Courtesy of GEA Process Engineering A/S, Søborg, Denmark.)

the hot drying gas must pass through the flexible devices that couple moving and stationary parts; generally, these cannot withstand high temperatures either.

Spray Dryers In this scheme, a spray dryer (Fig. 3.9) can be used if the aim is the conversion of a fine material (e.g., 15 μm) into a coarser material of spherical form (e.g., 150 μm). The material thus obtained is free-flowing and less dusty. However, this is an expensive particle-size enlargement method in terms of both investment and energy.

The wet filter cake is reslurried (e.g., to 40% solids by weight), additives are introduced, and the mixture is fed to the spray dryer, where the liquid is evaporated. To keep the size of the equipment reasonable, a minimum inlet-gas temperature is required (perhaps 200°C) to produce a solids outlet temperature that exceeds 75°C.

3.2 PROCESSING LIQUIDS, SLURRIES, AND PASTES

Up to this new section we have been considering primarily the drying of masses of wet particulates, such as filter cakes. Next, we look at points to consider when choosing equipment for continuous and batchwise processing of liquids, slurries, and pastes with the aim of obtaining a dry particulate material.

Spray Dryers

Spray dryers be chosen if the isolation of a solid from a solution or slurry via conventional crystallization and liquid–solid separation is either impossible or too complicated. Sometimes, spray drying is chosen if the characteristic spherical particle shape is desired (as in making ceramic powder). Typically, the average particle size (on a weight basis) falls in the range 50 to 200 μm. This dryer's short residence time is an advantage when drying heat-sensitive products (e.g., milk powder, organic salts).

Continuous Fluid-Bed Dryers/Granulators

Continuous drying and granulation are often accomplished in a circular-type fluid-bed dryer operating with a thick product layer. The feed is either a solution or a slurry passed into the product layer. Hot or warm air flows through the fluidized product layer and through heat transfer causes the water in the feed to evaporate. If the feed is a liquid (the processing of an aqueous solution of calcium chloride into particles in the size range 0.5 to 1.55 mm is a typical example), granulation is the result of both nucleation and growth. If the feed is a suspension of a particulate material in water in which the solid is insoluble, granulation is the result of agglomeration of the original particles.

Film Drum Dryers

Film drum dryers usually consist of two counterrotating heated rolls. The material to be processed is fed into a space between the drums. As with spray dryers, they may be chosen in cases where crystallization and liquid–solid separation are not feasible. This type of dryer can be used for pastes, and it can be placed under vacuum for heat-sensitive products. Examples of its use include the flaking of mashed potatoes and the drying of other instant foods. These dryers are discussed in Chapter 10.

Cylindrical Scraped-Surface Evaporator/Crystallizer/Dryer

Cylindrical scraped-surface dryers resemble the equipment used for other thin-film techniques. They can be arranged either horizontally or vertically. Horizontal dryers are more common than vertical dryers. The fluid to be dried passes through in plug flow. Between the feed point (for a solution or suspension) and the product outlet (for a more or less free-flowing powder), there is a zone where much power is consumed (per unit wall area), as a viscous paste is converted into particles. In principle, the equipment is suitable for pastes as well as for liquids and slurries. Vacuum may be applied for heat-sensitive products, and solvent recovery is possible. These dryers are treated in Chapter 10. They are also used to dry filter or centrifuge cakes of materials that tend to cake.

Figure 3.10 Agitated vacuum dryer for batch drying of liquids, slurries, and pastes.

Agitated Pan or Vacuum Dryers

Agitated pan or vacuum dryers (Fig. 3.10) should be considered for batchwise drying of liquids, slurries, or pastes. In both of these dryers, solvent recovery is possible. For liquids and slurries, as the drying proceeds, the power consumption rises to a maximum and then decreases. (As noted above, a liquid or slurry is converted into a more or less free-flowing powder via a viscous paste.) Hence, the motor, gearing, and stirrer must be sized adequately for maximum power consumption.

3.3 SPECIAL DRYING TECHNIQUES

Infrared Drying

Infrared drying is generally confined to surface drying, as of newly painted automobile bodies. For details, see Chapter 13.

Dielectric Drying

Dielectric drying is able to raise the heat where there is moisture. It is thus used in those cases where it is time consuming to apply heat by conventional means: by convection and conduction. One application is for relatively large and expensive articles that contain little moisture. Others are for drying pharmaceutical products under vacuum and in the drying of pastas [1]. Dielectric drying receives attention in Chapter 13.

Freeze-Drying

In the freeze-drying technique, ice is sublimed from a frozen product while under vacuum. Drying temperatures typically range from –30°C to –10°C, so the process is ideal for heat-sensitive products such as those of a biological nature. Freeze-drying is treated in Chapter 11.

Steam Drying

The material to be dried is contacted with superheated steam in the steam drying technique. Heat is transferred from the superheated steam to the material being processed, and this heat is used to evaporate water from the material. The steam generated can often be used for other purposes, such as in the concentration of diluted solutions. Steam drying is dealt with in Chapter 12.

3.4 SOME ADDITIONAL COMMENTS

Convective, Conduction, and Dielectric Dryers Compared

Convective, conduction, and dielectric dryers are compared in Table 3.3. The criteria provided in Table 3.3 are discussed below.

For large capacities, convective dryers are often chosen. The areas for heat transfer and the heat transfer coefficients of convective dryers are much larger than the areas and coefficients of conduction dryers. Next, there is usually practically no limit to scaling up a convective dryer. Furthermore, it is a drawback of conduction dryers that the square meters of wall heating area per cubic meter of dryer content decreases on scaling up. It is inversely proportional to a characteristic linear dimension. This is illustrated in Figure 3.11.

The construction of convective dryers is generally simpler than the construction of conduction dryers; hence, as a rule, convective dryers are cheaper. However, convection dryers (e.g., flash and spray dryers) often require relatively large solid–gas separation equipment, and this leads to additional costs.

Table 3.3 Qualitative Comparison of Three Different Dryer Types[a]

	Convective	Conduction	Dielectric
Large capacities	+	|	–
Fixed costs	+	|	–
Variable costs	|	+	–
Product temperature	+	|	–
Product quality	|	|	+
Scaling up (see Fig. 3.11)	+	|	–
High temperatures	+	–	|
Environment	–	|	+
Dust	–	+	+

[a] +, favorable; –, less favorable; |, neutral.

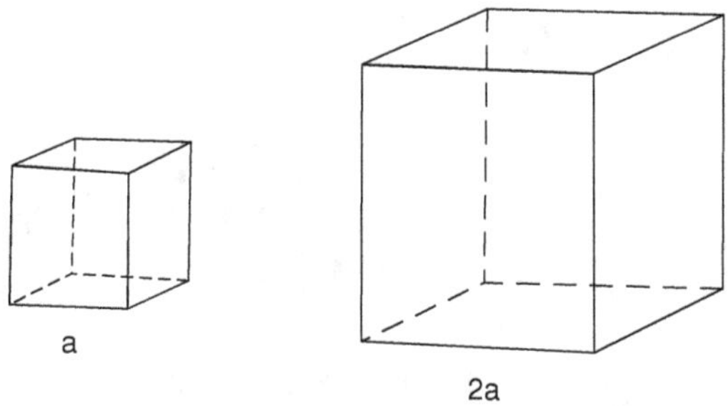

			Factor
Edge	a	2a	2
Area	$6a^2$	$24a^2$	4
Volume	a^3	$8a^3$	8
$m^2 \cdot m^{-3}$	6/a	3/a	1/2

Figure 3.11 $m^2 \cdot m^{-3}$ decreases on scaling up.

An auxiliary material, generally air, is used in drying convectively. Conduction drying does not require such material. Having to use an auxiliary material implies worse thermal efficiency. The reason is that such a material leaves the drying process at a higher temperature than that at which it entered. Heat recovery from this flow is not always easy, as will be elaborated when we discuss energy recovery. Thus, in general, the variable costs of drying convectively are higher than the variable costs of drying conductively. When heating conductively, the heating medium is returned to the heat source (e.g., a boiler house). The thermal efficiency of convection dryers increases with increasing inlet temperature of the drying gas. There is no such effect with conduction dryers.

A different aspect concerns the relationship between product temperature and heating-medium temperature. In convection dryers, as long as there is free moisture, the product adopts the adiabatic saturation temperature. But in conduction dryers, product in contact with hot metal will, as long as there is free moisture, attain a higher temperature, that is, the boiling temperature at the prevailing pressure.

The next criterion is product quality. Both time and temperature are inportant in regard to thermal degradation. The residence time in conduction dryers can be up to several hours; the residence time in flash and spray dryers can be short: 10 s or less. In special cases, dielectric drying benefits the quality. Scaling-up aspects were mentioned when large capacities were discussed.

In convective drying, the drying gas temperature can be quite high; temperatures as high as 600°C are used in drying inorganic materials. For quality or process safety reasons, when drying organic materials convectively, the drying gas temperature usually does not exceed 200°C. When conduction dryers are heated by means of steam, the steam condensation temperature usually does not exceed 200°C. This temperature requires a steam pressure of 15 bar absolute. When conduction dryers are heated indirectly by means of thermal oil, the maximum temperature is 350°C. On using a molten salt, a temperature as high as 500°C can be chosen.

Continuous convective drying often implies large airflows. When returned to the environment, the quality of this air is usually worse than the quality of the inlet

Figure 3.12 Krauss Maffei plate dryer is useful for the thermal processing of friable powders. (Courtesy of Andritz KMPT GmbH, Vierkirchen, Germany.)

air. Therefore, convective drying does not imply a bonus for the environment. Both conduction drying and dielectric drying are, in principle, better in this respect. Dust is more likely to be raised at convective drying than at both conduction and dielectric drying.

In the classification we have used, the plate dryer (Fig. 3.12) is a hybrid type, although the main contribution is by conduction. This type of dryer can be useful for solvent recovery or dusty products. The raking counteracts caking.

Combining Dryer Types

A combination of two different types of dryers is sometimes optimum. One example is for the spray drying of milk, followed by posttreatment of the powder in a fluid-bed dryer.

Fires and Dust Explosions

If there is the possibility of a fire or a dust explosion in a convection dryer, special precautions must be taken. These may involve lowering the drying air inlet temperature, use of inert gas, provision of explosion panels, triggered injection of suppressing agents, and the like. Process safety aspects are treated in Chapter 14.

Vacuum Drying

Continuous vacuum drying is not in common use; it is difficult to feed and extract the material under vacuum. However, there are some special cases where a liquid, slurry, or pumpable paste is fed and a dried solid (with relatively good handling characteristics) is removed.

3.5 TESTING ON SMALL-SCALE DRYERS

Laboratory measurements and investigations concerning the drying of a product should be the first stage of the selection of a new dryer or the replacement of a dryer. This aspect will receive attention in Chapter 4.

For testing, it is usually wise to seek the cooperation of a limited number of reputable drying-equipment manufacturers. It is often possible to ship samples of the material to be dried and have them tested in the manufacturer's small-scale units. This is relatively simple and straightforward, but there are risks:

1. The material may have changed in character between the time the wet product was removed from the process stream and the time of drying.
2. Because the quantity of sample shipped is limited, it is not possible to check long-run performance. For some desired process results, such as proper transport of the material through the dryer, and the absence of caking or dusting, it is essential to carry out experiments over a long period. A run lasting 4 h is a start.
3. The performance of the dryer is not checked in relationship to the process-plant infrastructure: for example, the behavior of the dried material in a plant's solids-handling equipment.

Consequently, if the experiments done by the equipment supplier are successful, it is good practice to install a small-scale dryer in the process plant, where it can be fed from a sidestream. If a new product or process is being dealt with, the dryer should be in the pilot plant. If there is no pilot plant, representative material must be made or obtained and then dried at the equipment manufacturer's facility. If you are interested in seeing full-scale equipment in operation, the manufacturer may be able to arrange an introduction to a facility using such dryers.

Before purchase, it is necessary to obtain guarantees from the dryer manufacturer and to agree on the methods of analysis that will determine whether or not the guarantees are met. Obtaining these guarantees means that the dryer manufacturer believes that the dryer will work. Even more important than these guarantees is the future user's conviction that the dryer will function satisfactorily.

3.6 EXAMPLES OF DRYER SELECTION

Example 3.1 In an existing plant, the rotary dryers being used for salt (NaCl) dated back to the 1930s. These continuously functioning dryers were worn out and needed replacement. After a successful drying test at a manufacturer's facility, it was decided to install a flash dryer. The choice was based on the following (see Fig. 3.2):

1. Water was to be evaporated.
2. There was no need to change the particle size.
3. The average particle size was about 0.4 mm.
4. The inorganic salt was not temperature-sensitive.
5. Only surface moisture was to be removed.

Space economy and the possibility of combining drying and vertical transport were additional aspects. The flash dryer was installed but was not successful, owing to the formation of a fine product mist from the equipment. Two reasons for this were suggested: (1) the high velocities in the flash dryer abrade the crystal, producing fine particles, and (2) the rapid evaporation of the surface moisture causes nucleation therein (the surface moisture being a saturated solution of salt in water).

The problem disappeared when the flash dryer was replaced by a troughlike vibrated fluid-bed dryer having a shallow product layer. Residence time is now many minutes instead of a few seconds, and velocities are much lower. Prior to this installation, a small-scale vibrated fluid-bed dryer was installed in the process plant and fed from a sidestream.

Example 3.2 A plant was built to produce about 40 $kg \cdot h^{-1}$ (dry basis) of a solid organic peroxide. There was a drying step in the manufacturing process at which water (about 25% by weight of the dryer feed) was evaporated. The relatively low capacity called for a batch dryer. On consulting Figure 3.1, a fluid-bed dryer was selected, for these reasons:

1. The maximum product temperature was 40°C.
2. Oxidation by air did not occur.
3. The solid was not very toxic.
4. Laboratory tests showed the feasibility of fluid-bed drying. The average particle size was about 500 μm.

Organic peroxides are compounds containing a relatively weak –O–O– bond, which is the cause of their thermal instability. Organic peroxides can decompose at relatively low temperatures to form free radicals. They can burn and give rise to a dust explosion, as they are organic compounds. For these reasons it was decided to use recycled nitrogen in the convective drying step. The nitrogen is warmed, indirectly, by warm water to 40°C. The diameter of the supporting plate is 0.92 m.

The process is controlled by measuring the temperature of the leaving gas. When it reaches a specified temperature, the warm water supply is switched off, and the batch is cooled by nitrogen for 10 min.

Example 3.3 A case of drying of organotin compounds is described by Schaake and Stigter [2]. Many of these compounds are toxic. Between 10 and 30% of moisture (a mixture of various solvents, which may include water) is to be evaporated. The maximum allowable product temperature is between 50 and 90°C. Batch size is 1500 to 2000 kg, and the bulk density of the dry product is about 500 kg·m^{-3}.

Originally, a tumbler dryer was used, but it was later replaced (to increase capacity) by a conduction dryer, comprising a conical mixer with a screw-type mixing element rotating along the circumference of the cone and around its own axis. Both dryers were of stainless steel. Their capacities were:

	Tumbler	Conical Mixer
Drying time (h)	25–40	10–30
Volume (L)	6800	4000

Hourly production is about 100 kg, so the choice of a batch dryer seems logical. A vacuum dryer was selected because of the toxicity of many of the products and the flammability of the vapors. Vacuum used in the conical mixer is between 30 and 50 torr. The reasons for replacing the tumbler dryer by the vacuum dryer with medium agitation were:

1. Shorter drying time because of better heat transfer (no crusts)
2. Fixed provision for charging and discharging
3. Homogeneous product: no manual cleaning because of the absence of crusts
4. Low maintenance costs: no need to convey utilities from stationary parts to rotary parts (and vice versa), as in the tumbler.

According to the authors, by replacing the tumbler, the drying cost per unit of product was reduced by 40%.

Example 3.4 A plant was designed for the production of various organic chemicals via reaction and crystallization in methanol. The capacity of the continuous plant is much larger than 100 kg·h^{-1}. A peeler centrifuge was planned for the liquid–solid separation, and experiments showed that the centrifuge cake contained about 5% by weight of methanol. Figure 3.2 suggests a plate dryer for drying these chemicals. Small-scale experiments were successful, and installation of an Andritz–KMPT plate dryer in the plant also became a success. Warm water is circulated through the plates, and nitrogen is used to entrain the evaporated methanol.

REFERENCES

[1] Hubble, P.E. (1982). Consider microwave drying. *Chemical Engineering*, *89*, 125–127.

[2] Schaake, P., Stigter, H. (1975). The vacuum drying of sensitive specialty chemicals, *Aufbereitungstechnik*, *16*, 27–28 (in German).

4

CONVECTIVE DRYING

Convective drying is accomplished by contacting a process stream with a warm- or hot-gas stream. The heat required for evaporation of the moisture is provided by cooling of the gas stream; the gas stream is also used as a carrier for the removal of the evaporated moisture.

Convective dryers can be divided conveniently into three classes:

1. Dryers in which all the product is entrained by the gas stream (flash and spray)
2. Dryers in which part of the product is entrained (fluid bed, direct rotary, and Hazemag Rapid)
3. Dryers in which a small but not negligible part of the product is entrained (band, plate, and tray)

The four continuous convective dryers that are most important to industry—fluid-bed, direct-heat rotary, flash (pneumatic-conveyor), and spray dryers—are discussed in Chapters 5 to 8. Each chapter starts with a general description of the equipment and provides details of the suitability of the dryer for various drying functions. Theoretical aspects are also covered: for example, minimum fluidization velocity, a literature design model for direct-heat rotary dryers, rapid drying of wet particles in flash dryers, and particle-size distribution on spray drying. The theoretical aspects are then followed by examples concerning dryer design. Further continuous convective dryers and several convective batch dryers are discussed in Chapter 9. Thus, Chapters 5 to 9 should provide convenient details for choosing among the main types of convective dryers, whereas in this chapter we concentrate on aspects that are common for convective dryers.

Drying in the Process Industry, First Edition. C.M. van 't Land.
© 2012 John Wiley & Sons, Inc. Published 2012 by John Wiley & Sons, Inc.

4.1 COMMON ASPECTS OF CONTINUOUS CONVECTIVE DRYERS

Modes of Flow

Four types of flow are identified:

1. Countercurrent flow
2. Concurrent or parallel flow
3. A mixture of countercurrent and cocurrent flow, often called *mixed flow*
4. Cross-flow

Countercurrent flow exposes the dried material to the highest temperature. This mode of flow permits a very low final moisture content but also creates the risk of overheating the processed material. Parallel flow exposes the feed to the highest temperature, leading to a high initial drying rate. The fast initial drying basically occurs at the adiabatic saturation temperature; however, evaporation that is too rapid may damage the product or hinder subsequent drying due to case hardening.

Spray and rotary dryers can accommodate the first three modes. The first two modes are frequently adopted for rotary dryers. Cocurrent flow and mixed flow are often used in spray dryers. Flash and Hazemag Rapid dryers have parallel flow, whereas fluid-bed dryers and band dryers have cross-flow.

The flow mode does not strongly influence operational efficiency since the bulk of the drying takes place when the solid is at the adiabatic saturation temperature.

Air-Inlet and Air-Outlet Temperatures

Generally, the air-inlet temperature varies between 100 and 800°C, whereas the corresponding air-outlet temperature varies only between 50 and 150°C. Relatively low drying-gas temperatures (e.g., up to 200°C) are used to dry many organic chemicals. Higher temperatures are usually not recommendable for product quality or process safety reasons. However, relatively low drying-gas temperatures (e.g., up to 300°C) are also used for coarse inorganic crystals that have between 1 and 5% by weight of initial moisture. If higher drying-gas temperatures are used, heating the dry solid flow consumes too much energy. Wet inorganic materials can often be dried with high air-inlet temperatures.

It will now be made plausible that, on convective drying, air-outlet temperatures are approximately 100°C. Assume that bone-dry air, on convective drying, cools down from 200°C to 45°C. The evaporated moisture is water. The air-outlet temperature is, at this point, an assumption that will be checked. It is also assumed that the moisture is at 45°C (i.e., the water phase transition takes place at 45°C). The air temperature drop is $200 - 45 = 155$ K. The air water absorption is $155 \cdot 1.02/2394 = 0.066$ kg of water per kilogram of dry air; 1.02 is the air specific heat in $kJ \cdot kg^{-1} \cdot K^{-1}$, and 2394 is the water heat of evaporation at 45°C in $kJ \cdot kg^{-1}$. Now, the following table is applicable.

	kg	% by Weight	mol	% by Moles	Partial Pressure ($N{\cdot}m^{-2}$)
H_2O	0.066	6.2	3.67	9.6	9,725
Air	+ 1.000	+ 93.8	+ 34.48	+ 90.4	+ 91,575
	1.066	100.0	38.15	100.0	101,300

The air molecular weight was taken to be 29 $g{\cdot}mol^{-1}$. Dalton's law says that if a gaseous mixture contains 9.6% by moles of a component, the component also exerts 9.6% of the system pressure. The system pressure is assumed to be atmospheric: 101,300 $N{\cdot}m^{-2}$. Thus, in the example, the water partial pressure should equal the saturated water vapor pressure at 45°C. The latter value is 9541 $N{\cdot}m^{-2}$ and is sufficiently close to 9725 $N{\cdot}m^{-2}$. The conclusion is that the original assumption of a cooling down of the air to 45°C is correct. The adiabatic saturation temperature of bone-dry air of 200°C was calculated in this example. This temperature could also have been found using Mollier's diagram, the Grosvenor chart, or a humidity chart. The humidity chart is discussed in this chapter.

If, on convective drying with bone-dry air having an air-inlet temperature of 200°C, the air would leave at 45°C, there would be no driving force to extract bound water from the product to be dried. Thus, the air-outlet temperature will have to be higher than 45°C: for example, 65°C. If the calculation in the example would have been carried out for higher air-inlet temperatures, the air-outlet temperature would approach 100°C.

4.2 SATURATED WATER VAPOR PRESSURE

The saturated water vapor pressure is a function of the temperature and can be calculated using the *Antoine equation* [1]:

$$^{10}\log\, p_g^* = A - \frac{B}{T + C - 273.15} \tag{4.1}$$

where

p_g^* is expressed in bars absolute (1 bar = 10^5 $N{\cdot}m^{-2}$)
$A = 5.11564$
$B = 1687.537$
$C = 230.17$
T is the temperature in K

This equation is valid in the range (bar/K) 0.01/273.20 to 16/473.20.

Example 4.1 Calculate the saturated water vapor pressure at 50°C.

$$^{10}\log\, p_g^* = 5.11564 - \frac{1687.537}{323 + 230.17 - 273.15} = -0.9108$$

$$p_g^* = 0.123\ \text{bar}$$

4.3 WET-BULB TEMPERATURE

The wet-bulb temperature of a gas stream is measured to determine the moisture content of the gas stream. This temperature is discussed here for the air–water system. A thermometer surrounded by wet cotton will indicate this temperature if it is held in an airflow. This temperature is lower than the temperature of the airflow, and the temperature difference is caused by evaporation. In this situation, the evaporation hardly affects the composition of the airflow. This is a case of combined heat and mass transfer. The temperature difference between the airflow and the piece of cotton enables the transport of heat from the air to the wet cotton. The transport of heat is necessary to enable mass transport by evaporation from the wet cotton to the airflow. The situation is illustrated in Figure 4.1. The temperature difference is maximum for bone-dry air, becomes smaller if the air already contains some water vapor, and becomes nil for saturated air. An equation that enables the calculation of the water concentration in air will be derived. The derivation may be found in Beek et al. [2]. The equation can be used for air temperatures up to approximately 200°C, but not beyond, because the derivation has been simplified. The first equation is

$$\phi_h'' = \phi_{\text{mol}}'' \cdot \Delta H$$

The second equation is the equation for the heat flux:

$$\phi_h'' = \alpha(T_g - T_W)$$

The equation for the mass flux is the third equation:

$$\phi_{\text{mol}}'' = k(c_g^* - c_g) = \frac{k}{R \cdot T}(p_g^* - p_g)$$

T is the arithmetic average of T_W and T_g. This is the simplification mentioned earlier. On combining the three formal equations, we obtain

$$\frac{p_g^* - p_g}{T_g - T_W} = \frac{R \cdot T}{\Delta H} \cdot \frac{\alpha}{k} \tag{4.2}$$

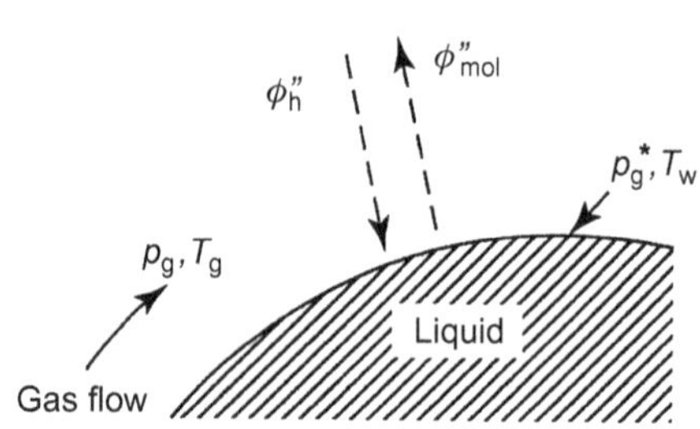

Figure 4.1 Evaporation from a wet surface.

In general, the heat transfer correlation for this situation is

$$\text{Nu} = C\cdot\text{Re}^m\cdot\text{Pr}^{1/3}$$

Also, in general, the mass transfer correlation for this situation is

$$\text{Sh} = C\cdot\text{Re}^m\cdot\text{Sc}^{1/3}$$

Because of the analogy between heat and mass transfer, the coefficient and the exponents on the right-hand side of the latter two equations are equal. Dividing the first correlation by the second correlation gives

$$\frac{\alpha}{k} = c_p\cdot\rho\left(\frac{\text{Sc}}{\text{Pr}}\right)^{2/3} = c_p\cdot\rho(\text{Le})^{2/3}$$

Substitution of the expression for α/k in equation (4.2) gives

$$\frac{p_g^* - p_g}{T_g - T_w} = \frac{R\cdot T\cdot c_p\cdot\rho(\text{Le})^{2/3}}{\Delta H} \tag{4.3}$$

Le is the Lewis number and is the quotient of Sc and Pr. $(\text{Le})^{2/3}$ for the air–water system is 0.95. c_p and ρ are taken at temperature T. p_g^* is the saturated water vapor pressure at temperature T_W. On measuring T_W and T_g, p_g can be calculated.

Example 4.2 The wet-bulb temperature of 40°C air is 23°C. What is the relative humidity of the air at 40°C? What is the water content of this air in grams per kilogram of dry air?

Use Table E4.2 for the calculations. The numbers are calculated using the Antoine equation [equation (4.1)].

Table E4.2

Temperature (°C)	Water Vapor Pressure ($\text{N}\cdot\text{m}^{-2}$)
0	601.4
5	861.5
10	1215.7
15	1691.6
20	2322.9
25	3150.4
30	4222.9
35	5598.3
40	7344.6

We begin with equation (4.3):

$$\frac{p_g^* - p_g}{T_g - T_W} = \frac{R \cdot T \cdot c_p \cdot \rho}{\Delta H} \cdot \text{Le}^{2/3}$$

where

$$\begin{aligned}
p_g^* &= 2819\ \text{N}\cdot\text{m}^{-2}\ \text{(saturated water vapor pressure at 23°C)}\\
T_g &= 40°\text{C} = 313\ \text{K}\\
T_W &= 23°\text{C} = 296\ \text{K}\\
R &= 8314\ \text{J}\cdot\text{kmol}^{-1}\cdot\text{K}^{-1}\\
T &= \frac{23+40}{2} = 31.5°\text{C} = 304.5\ \text{K}\\
c_p &= 1010\ \text{J}\cdot\text{kg}^{-1}\cdot\text{K}^{-1}\\
\rho &= 1.30 \cdot \frac{273}{273+31.5} = 1.17\ \text{kg}\cdot\text{m}^{-3}\\
\Delta H &= 2{,}450{,}000 \cdot 18\ \text{J}\cdot\text{kmol}^{-1}\\
\text{Le}^{2/3} &= 0.95
\end{aligned}$$

It follows that $p_g = 1723\ \text{N}\cdot\text{m}^{-2}$. This pressure is 23.4% of the saturated water vapor pressure at 40°C. Thus, the relative humidity is 23.4%.

	Partial Pressure ($\text{N}\cdot\text{m}^{-2}$)	% by Moles	$\text{g}\cdot\text{mol}^{-1}$	% by Weight
Water	1,723	1.70	18	1.06
Air	+ 99,577	+ 98.30	29	+ 98.94
	101,300	100.00		100.00

This air contains 10.7 g of water per kilogram of dry air.

4.4 ADIABATIC SATURATION TEMPERATURE

The adiabatic saturation temperature T_s of drying air is the temperature to which the air cools down when it is used to evaporate free water. The calculation of T_s is based on an energy balance for a unit mass of the gas. The transport rates of heat and mass do not play a role. The amount of water evaporated per kilogram of air is

$$\frac{c_p(T_g - T_s)}{\Delta H} \qquad \text{kmol}\cdot\text{kg}^{-1}$$

The number of kilomoles of water per kilogram of air can also be given as

$$\frac{p_g^* - p_g}{\rho \cdot R \cdot T} \qquad \text{kmol}\cdot\text{kg}^{-1}$$

This follows from application of the universal gas law. The temperature T is the arithmetic average of T_g and T_s, and ρ is the arithmetic average of the gas specific masses at T_g and T_s. On equating the two expressions, we obtain

$$\frac{p_g^* - p_g}{T_g - T_s} = \frac{R \cdot T \cdot c_p \cdot \rho}{\Delta H} \tag{4.4}$$

Equation (4.4) resembles equation (4.3). Equation (4.3) contains $\mathrm{Le}^{2/3}$, whereas it is absent in equation (4.4). $\mathrm{Le}^{2/3}$ is 0.95 for the air–water system. Thus, in the practice of industrial drying, the wet-bulb temperature and the adiabatic saturation temperature are almost equal. However, the two temperatures are of a physically different nature.

4.5 HUMIDITY CHART

Figure 4.2 is a humidity chart for the air–water system. It is also called a *psychrometric chart*. It can be used if free water is evaporated. To start, it is assumed that ambient air is at 20°C and has a relative humidity of 50%. This air quality is indicated by point A in the diagram. The air is heated indirectly by, for example, warm water to 70°C. This step can be represented by line AB. This air could be used to dry a pharmaceutical product in a batch fluid-bed dryer. At the right ordinate an absolute humidity of 0.006 kg of water per kilogram of dry air can be read for both points A and B. On going from A to B, the temperature is raised while moisture is neither added to nor removed from the air. Point B reflects the quality of the air prior to the drying step. As far as the removal of free moisture is concerned, the drying step is characterized by a straight sloping line, intersecting the curved line of 100% relative

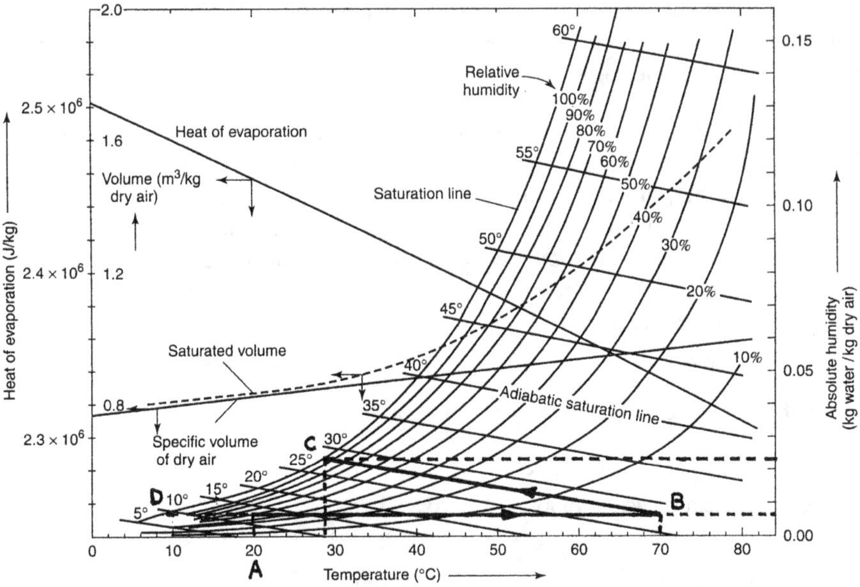

Figure 4.2 Humidity chart. (From [2].)

humidity in point C. The adiabatic saturation temperature of 28°C can be read on the abcissa. The straight sloping lines are isenthalpic lines, meaning that on going from B to C, for example, the enthalpy per kilogram of dry air (mixed with water vapor) is constant. The reference temperature for the enthalpy is 0°C. The drying step in a convective dryer is an adiabatic process, meaning that, ideally, heat exchange with the surroundings does not occur. The step from B to C is both isenthalpic and adiabatic. The energy present at 70°C per kilogram of dry air is equal to the energy present at 28°C per kilogram of dry air; however, at 28°C the energy is at a lower level. The presence of water vapor is accounted for in both situations.

On going from B to C, the moisture content of the air increases from 0.006 $kg \cdot kg^{-1}$ to 0.023 $kg \cdot kg^{-1}$. These absolute humidities can be converted into water vapor partial pressures. (See the example in Section 4.3, where water vapor partial pressure is converted into absolute humidity.) The conception partial pressure is the pressure the water vapor would exert if it would fill the space alone.

If, at point A, the temperature would be decreased while moisture was neither added nor removed, condensation at D would occur at 9°C. The latter temperature is the dew point corresponding to both A and B.

If free water is evaporated from a solution, the drying step ends before the line of 100% relative humidity is reached. For example, the water vapor pressure of saturated brine is 75% of the corresponding water vapor pressure between 0 and 100°C. If salt crystals were dried with air at 70°C having an absolute humidity of 0.006 $kg \cdot kg^{-1}$, the sloping line would stop at a relative humidity of 75% and the adiabatic saturation temperature would be 32°C. Diagrams giving further information, including enthalpy data, are the Mollier H/x diagram (H is the enthalpy and x is the absolute humidity in $kg \cdot kg^{-1}$) and the Grosvenor chart.

The humidity chart given is applicable for atmospheric pressure (i.e., 101,300 $N \cdot m^{-2}$). A different chart is applicable for drying at Mexico City, for example, which is 2240 m above sea level.

Question: Can 1 kg of dry air at 40°C absorb more or less water at saturation at Mexico City than at Amsterdam? *Hint:* Check the calculation in Section 4.1. *Further hint:* Consider that ambient air that is used for a compressed-air system is dried. Otherwise, water would condense in the system. Thus, on compressing, the amount of water vapor that can be combined with 1 kg of dry air decreases. On moving in the other direction (i.e., on decreasing the pressure) the reverse occurs.

Mollier H/x diagrams and Grosvenor charts also exist for other systems (e.g., nitrogen–methanol).

Example 4.3 Check the enthalpy at both points B and C in Figure 4.2.

Specific heat of air: 1.0 $kJ \cdot kg^{-1} \cdot K^{-1}$

Specific heat of water vapor: 1.9 $kJ \cdot kg^{-1} \cdot K^{-1}$

Water heat of evaporation at 0°C: 2500 $kJ \cdot kg^{-1}$

Point B: $1 \cdot 1 \cdot 70 + 0.006(2500 + 70 \cdot 1.9) = 85.8\ kJ \cdot kg^{-1}$

Point C: $1 \cdot 1 \cdot 28 + 0.023(2500 + 28 \cdot 1.9) = 86.7\ kJ \cdot kg^{-1}$

Question: Can the figures given and calculated in Example 4.2 also be found in the humidity chart?

The humidity chart shows that ambient air always contains several grams or tenfolds of grams of water per kilogram of dry air. For a specific drying operation, it is sometimes necessary to lower or raise this moisture content. Lowering the moisture content is discussed first. It will be illustrated by means of an example. Polycarbonate granules can be dried convectively down to 0.015% by weight of water with ambient air heated up to 120°C in approximately 4 h. On drying these granules convectively with air of 120°C having a dew point of –30°C, the drying time is 1.5 h [3]. To arrive at this dry air, cooling to 5°C and separating the condensate is a first step. However, this cooled air still contains 0.0054 kg·kg^{-1}. Further moisture removal can be achieved by passing the air through silica gel, activated alumina, or molecular sieves in, for example, absorption wheels. In absorption wheels working with silica gel or molecular sieves, the sorption agent is fixed on a carrier. The rotational speed of the wheel is, for example, seven rotations per hour. The part of the wheel loaded with moisture passes continuously through a regenerating section. Regenerating air, at a temperature of 140°C for silica gel, passes through this section countercurrently with the drying air. The Swedish company Munters was the first company to introduce this technique into the market.

An example of deliberately raising the moisture content of the drying air has been noted by Dolinsky [4]. The bulk density of spray-dried detergents is increased by 30% when the drying air moisture content is raised from 0.012 kg·kg^{-1} to 0.150 kg·kg^{-1}.

4.6 WATER–MATERIAL INTERACTIONS

Sorption Isotherms

When a solid is exposed to a humid gas of constant temperature and humidity, it will either gain or lose moisture until equilibrium is reached. The equilibrium moisture content of the solid X_e (dry basis) is a function of the relative humidity of the gas, the temperature, and the nature of the the solid and the liquid. The variation of X_e with relative humidity is called a *sorption isotherm*. Often, the sorption isotherm does not coincide with the desorption isotherm, and the system is said to exhibit *hysteresis*. The desorption isotherm then always shows a greater equilibrium moisture content at a given relative humidity than that of the absorption isotherm.

Knowledge of sorption isotherms is important for both drying and subsequent storage. In drying, X_e is the lower limit of moisture content that can be reached upon drying for the given conditions of gas temperature and humidity. Concerning storage, there is an X_e value corresponding to the conditions under which the product is stored. The dryer is performing unnecessary evaporation if its value is greater than the final product moisture content of the drying operation.

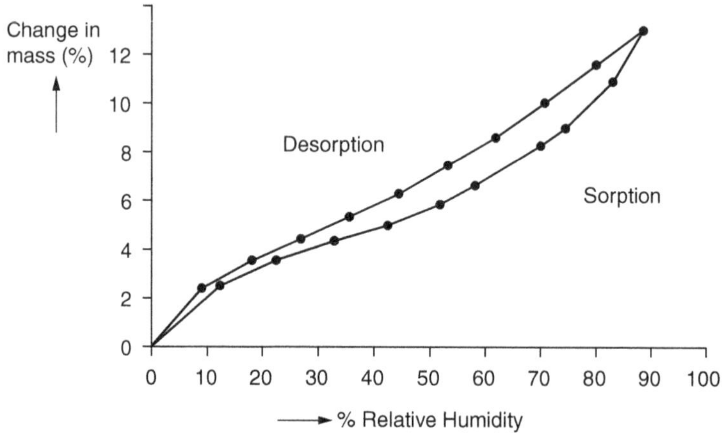

Figure 4.3 The sorption and desorption isotherms of photocopy paper 80 g·m^{-2} at 20°C. (Courtesy of Surface Measurement Systems Ltd, London, UK.)

Figure 4.3 contains the sorption and desorption isotherms of microcrystalline cellulose. The steep parts in the upper right-hand corner relate to the emptying or filling of capillaries. The moderately sloped lines relate to polymolecular layers of adsorbed water, while the steep parts in the lower left-hand corner represent a monomolecular layer of adsorbed water. Wood and starch also exhibit the types of sorption isotherms depicted in Figure 4.3. Wood typically contains 0.175 kg of water per kilogram of dry wood at 20°C and 80% relative humidity. Starch typically contains 0.125 kg of water per kilogram of dry starch at 20°C and 80% relative humidity. The capillaries of these materials are emptied on decreasing the relative humidity from 100% to 80% at 20°C. Typical capillary sizes in natural materials are in the range 10^{-7} to 10^{-8} m. Water cannot exert its saturated vapor pressure when present in these capillaries. The water vapor capillary depression is expressed by *Thomson's formula*:

$$\frac{p_c}{p_g^*} = \exp\left(\frac{-2\sigma}{r_c} \cdot \frac{v'}{R_D T}\right) \tag{4.5}$$

Example 4.4

$$\begin{aligned}
T &= 293\,\text{K} \\
\sigma &= 0.07275\,\text{N}\cdot\text{m}^{-1} \\
r_c &= 10^{-7}\,\text{m} \\
v' &= 10^{-3}\,\text{m}^3\cdot\text{kg}^{-1} \\
R_D &= 461.9\,\text{J}\cdot\text{kg}^{-1}\cdot\text{K}^{-1}(8314\,\text{J}\cdot\text{kmol}^{-1}\cdot\text{K}^{-1}) \\
\frac{p_c}{p_g^*} &= 0.99
\end{aligned}$$

Thus, the water vapor pressure in capillaries having a radius of 10^{-7} m is depressed by 1% at 20°C. The depression is 10% at 20°C for capillaries having a radius of 10^{-8} m.

Semiempirical or empirical correlations exist to relate X_e data to relative humidity and temperature for various materials. Papadakis et al. [5] give an overview of these correlations. Oakley [6] adopts one of these correlations and applies it to the desorption of milk powder:

$$X_e = A \cdot \exp\left(-BT \cdot \ln \frac{1}{\Psi}\right) \tag{4.6}$$

In this correlation, $A = 0.1499$ kg·kg^{-1} and $B = 2.306 \cdot 10^{-3}$ K^{-1} for milk powder. Ψ is the relative humidity expressed as a fraction.

Example 4.5 Calculate the equilibrium moisture content of milk powder at 20°C and 60% relative humidity.

$$X_e = 0.1499 \cdot \exp\left(-2.306 \cdot 10^{-3} \cdot 293 \cdot \ln \frac{1}{0.6}\right) = 0.106\,\text{kg}\cdot\text{kg}^{-1}$$

Milk powder typically contains approximately 0.04 kg·kg^{-1} on leaving a spray dryer. Thus, the actual moisture content of the spray-dried product is lower than the equilibrium moisture content under storage conditions. Automated instruments exist that measure the water sorption isotherms of materials in a relatively short time.

Note: One may wonder how desert plants stay green and alive even after long periods of drought. The explanation is that the water in these plants is bound to such a degree that its vapor pressure is lower than the ordinary water vapor partial pressure in the desert atmosphere.

Dissolution

The materials described above interact with water; however, they are insoluble. Many materials are water soluble, and they attract water when the relative humidity at a given temperature exceeds a certain critical value. The attraction results in a solution of the material in water, behavior often termed *hygroscopicity*.

Example 4.6 A saturated solution of sodium chloride in water has, at temperatures in the range 0 to 100°C, a saturated vapor pressure that is 75% of the saturated vapor pressure of water. The latter value is constant because the solubility of sodium chloride is only weakly temperature dependent in the range 0 to 100°C. Solid sodium chloride forms a solution of sodium chloride in water if it is exposed, in this temperature range, to relative humidities exceeding 75%. An unsaturated brine is formed, for example, if the relative humidity is 80% and the saturated vapor pressure of this brine is 80% of the saturated water vapor pressure. If, subsequently, this unsaturated brine is exposed to a relative humidity below 75% (e.g., 70%), the brine dries out completely and the original solid sodium chloride is obtained again.

Other water-soluble materials, such as potassium nitrate and caprolactam, behave similarly. The solubility of potassium nitrate is strongly temperature dependent: A saturated solution contains 9% by weight of potassium nitrate at 0°C, whereas a saturated solution contains 71% by weight of the material at 100°C. Thus, potassium nitrate attracts water at elevated temperatures beginning at low relative humidities. Caprolactam is hygroscopic and attracts water under ambient conditions.

Hydrates

Anhydrous materials or relatively low hydrates can absorb water vapor from air to become hydrated. The material remains solid during absorption. Conversely, hydrates can lose water vapor to the air to become dehydrated. White copper sulfate, an anhydrous material, is an example of the first category; it absorbs water vapor gradually when left in air but remains solid. The anhydrous material changes into the pentahydrate, also known as bluestone or blue vitriol. Its color changes from white to blue during the absorption. The decahydrate of sodium carbonate, washing soda, is an example of the second category. When these crystals are exposed to ambient conditions, efflorescence occurs. The clear glassy particles gradually change to the monohydrate, a white powder. The water vapor pressure in the ordinary atmosphere is greater than that of the monohydrate but less than that of the decahydrate.

4.7 DRYING WITH AN AUXILIARY MATERIAL

An auxiliary material, often air, is used for convective drying. Nitrogen is sometimes used, usually for process safety reasons. When air is used, it is taken from the atmosphere, heated up, and after having been used, is discharged to the atmosphere. The temperature of the discharged air is higher than the temperature of the ambient air, and this heating up of drying air represents a loss.

Example 4.7 What is the energy consumption in kilojoules per kilogram of evaporated water in the drying operation in Section 4.5?

One kilogram of dry air can absorb $0.023 - 0.006 = 0.017$ kg of water. Air warming requires $1{\cdot}1(70 - 20) + 0.006{\cdot}1.9(70 - 20) = 50.6\ \mathrm{kJ{\cdot}kg^{-1}}$ (i.e., per kilogram of dry air).

$$\frac{50.6}{0.017} = 2976.5\ \mathrm{kJ{\cdot}kg^{-1}}$$

(here per kilogram of evaporated water). The heat of evaporation at 28°C is 2440 kJ·kg^{-1} (see Fig. 4.2). So, in a way, the efficiency of this drying operation is

$$\frac{2440}{2976.5} \cdot 100 = 82.0\%$$

This efficiency includes neither effects such as having to evaporate bound moisture and to heat solids, nor heat losses. This efficiency concerns the price that nature asks us to pay for the use of an auxiliary material: that is, air.

This effect can also be described by noting that the drying air was warmed from 20°C to 70°C, whereas cooling down occurred from 70°C to 28°C only. A net effect is that air was heated at the drying operation from 20 to 28°C. The temperature at which the air leaves at actual convective drying operations is higher than the adiabatic saturation temperature because bound moisture must also be evaporated. Thus, in this example 82.0% is a relatively high value. This type of efficiency can be kept relatively high by working with high air-inlet temperatures.

Example 4.8

Case 1 $TA_{\text{in}} = 800°\text{C}$, $TA_{\text{out}} = 150°\text{C}$. Air enters the burner at 20°C.

$$\text{Efficiency} = \frac{800 - 150}{800 - 20} \cdot 100 = 83\%$$

Case 2 $TA_{\text{in}} = 200°\text{C}$, $TA_{\text{out}} = 75°\text{C}$. Air enters the burner at 20°C.

$$\text{Efficiency} = \frac{200 - 75}{200 - 20} \cdot 100 = 69\%$$

However, these ratios can be compared directly only if the heat absorbed by the solids can be neglected. This is because the solid will leave the dryer at a higher temperature in case 1 than in case 2.

It is thus economical to use high air-inlet temperatures. Furthermore, because the drying operation utilizes less air at high air-inlet temperatures than at lower air-inlet temperatures, the drying equipment becomes smaller and this leads to lower investments. It is thus understandable that drying equipment manufacturers try to work with high air-inlet temperatures. However, product quality and process safety may impose restrictions. It is often stated that as long as the material to be dried is wet, the material temperature will not exceed the adiabatic saturation temperature. Thus, it would be possible to work with high air-inlet temperatures. It is true that the product temperature cannot rise much above the adiabatic saturation temperature if the flow regime is cocurrent. However, any incrustations at the feed position will become much warmer. Furthermore, the product temperature can become relatively high if the flow regime is countercurrent.

The solid-outlet temperature will be lower than the air-outlet temperature for cocurrent flow. The difference will be greater for short-residence-time dryers (flash and spray dryers) than for long-residence-time dryers (direct-heat rotary dryers). The solid-outlet temperature may exceed the air-outlet temperature for countercurrent flow and for cross-flow in plug-flow fluid-bed dryers.

4.8 GAS VELOCITIES

Table 4.1 gives the terminal velocities of spherical particles in air at 15, 90, and 205°C [7]. Beek et al. [2] contains methods to calculate terminal velocities of spherical particles.

Rotary Dryers

The empty-tube drying-gas velocity is selected by considering the possible entrainment of solids. Table 4.1 helps the selection. Typical values are between 0.5 and 2.5 $m{\cdot}s^{-1}$.

Flash Dryers

The empty-tube drying-gas velocity is selected after also considering the necessity to entrain all the solids. Typical values are between 10 and 30 $m{\cdot}s^{-1}$. It is possible to

Table 4.1 Terminal Velocities of Spherical Particles in Air

	$\Delta\rho = (\rho_S - \rho_F)$ (kg·m^{-3})					
	1000			2000		
Diameter (μm)	15°C	90°C	205°C	15°C	90°C	205°C
50	0.074	0.062	0.052	0.14	0.12	0.10
100	0.25	0.22	0.19	0.46	0.41	0.37
250	0.91	0.87	0.83	1.51	1.48	1.43
500	1.98	2.01	1.99	3.17	3.20	3.26
600	2.39	2.43	2.46	3.78	3.84	3.94
700	2.79	2.87	2.91	4.33	4.51	4.70
800	3.14	3.23	3.35	4.85	5.07	5.27
900	3.48	3.66	3.79	5.42	5.64	5.91
1000	3.84	4.00	4.21	5.97	6.19	6.50
1250	4.73	4.91	5.12	7.10	7.53	7.86
1500	5.37	5.76	5.98	7.78	8.65	9.40
2000	6.71	7.18	7.71	9.72	10.6	11.7
2500	7.71	8.45	9.20	11.3	12.5	13.7
3000	8.75	9.56	10.5	14.8	14.0	15.5
5000	11.6	13.1	14.6	16.7	18.6	21.2

mount a cage mill to mill any lumps at the lowest point. The milled product can be reentrained subsequently.

Fluid-Bed Dryers

The spent drying gas expands into the freeboard after passing through the product. The gas velocity in the disengaging height is selected so as to avoid excessive dust entrainment. Typical values are between 0.25 and 1 $m \cdot s^{-1}$.

Spray Dryers

The empty-tube gas-outlet velocities for spray dryers with rotary atomization are typically between 0.1 and 0.5 $m \cdot s^{-1}$ in the cylindrical part. For spray dryers with single-fluid nozzle atomization, 0.25 to 1 $m \cdot s^{-1}$ is normal. The lower range for rotary atomization is due to the relatively large diameter of the chambers with atomizing wheels. The diameter must be relatively large in this case to avoid wet material hitting the wall.

Note well that the spent-gas outlet flow contains evaporated water, the water content being high for high-temperature flows and low for low-temperature flows. Table 4.1 should be used with care since it is valid for air only.

4.9 HEAT LOSSES

Two categories of heat losses can be distinguished: (1) losses during normal, steady-state operation, and (2) additional losses that can be observed over relatively long periods (e.g., a year).

Losses During Normal, Steady-State Operation

These losses can be calculated from performance data reported in the literature and as supplied by manufacturers of drying equipment [8,9]. The losses calculated are in agreement with values obtained by plant measurements.

Causes of the losses are (1) conduction to, for example, supporting structures, (2) convection of plant air, (3) radiation, and (4) ingress of air cooling the drying air or, alternatively, the loss of high-temperature drying air.

Factors influencing these losses are:

1. The insulation of the dryer.
2. The material being dried. If the material being dried is a product, there is a tendency to give more attention to the dryer system than if a by-product is being dried.

3. The size of the dryer. A large dryer is more important and tends to receive more attention than a smaller system. A small dryer offers, per cubic meter of dryer volume, more area (m^2) for convection and radiation losses.
4. The pressure in the dryer. Efficient convective dryers usually work with a small underpressure (e.g., 5 to 10 mm water gauge) in the drying chamber. This requires two fans, one for drying air and one for spent drying gas. Simpler dryers work with one fan only. Overpressure would lead to uncontrolled emissions and could lead to incrustations at the dryer inlet. However, underpressure leads to the ingress of cold air. The heat economy is affected if cold air is mixed with the drying air because of the aspects discussed in Section 4.7. However, the heat economy is not affected if cold air is mixed with the spent drying gas. For process safety reasons, to prevent ingress air, there is a small overpressure in the drying chamber when a combustible solvent is evaporated. Nitrogen is used as the drying gas in that case.
5. The construction of the dryer. An open design can lead to losses (flash dryers), whereas a closed construction tends to minimize heat losses (spray dryers).

Surprisingly, the steady-state losses are barely influenced by the temperature of the drying gas. This is because, irrespective of the value of the drying air-inlet temperature, the drying process always occurs at a temperature of $100 \pm 50°C$.

The steady-state heat losses (normal load) of rotary, flash, and fluid-bed dryers amount to about 20 to 30% of the enthalpy gain of the feed; in spray dryers the corresponding figure is about 10 to 15%, with the closed design probably responsible for the difference.

Additional Losses

Additional losses are caused by starting up the dryer (with the heat used to warm the metal), cleaning the dryer and evaporation of the flushing water, and dryer control. Experience has shown that often the desired final-moisture content can be kept constant by maintaining a constant gas-outlet temperature that is higher than the adiabatic saturation temperature. Many dryers reduce the drying air temperature automatically to maintain a prescribed outlet temperature if the feed flow decreases. The efficiency of the drying process expressed, for example, in normal cubic meters of natural gas per metric ton of water, then decreases. The effect occurs at all air-inlet temperatures.

Example 4.9

	Situation			
	1	2	3	4
TA_{in} (°C)	160	200	590	800
TA_{out} (°C)	80	80	150	150
Efficiency (%)	57	67	77	83

The efficiencies are calculated as $(TA_{\text{in}} - TA_{\text{out}})/(TA_{\text{in}} - 20)\cdot 100$. The example illustrates (1) the decrease of efficiency on lowering TA_{in}, and (2) the dependence of the effect on the temperature levels. In the example, the $(\Delta T = TA_{\text{in}} - TA_{\text{out}})$ values were reduced to about two-thirds of the original values to assure that the comparison is fair.

The remedy would be to throttle the drying air and keep TA_{in} constant. However, this is generally impossible for convective dryers since a certain airflow is essential for their proper operation. One might argue that a lower TA_{in} leads to lower convection and radiation losses, as the drying process is carried out at a lower temperature. However, the net effect is an efficiency loss.

Depending on the dryer type, the additional heat losses may range from 10 to 75% of the steady-state total-dryer heat requirement. The lower values are found for fully loaded systems requiring little cleaning, and the higher values are found for systems with low throughput or frequent outage, or which often require cleaning. The additional heat losses for the various dryer types are discussed.

4.10 ELECTRICAL ENERGY CONSUMPTION

Fan Power

Pressure losses are experienced by the air and evaporated water flow passing through the dryer system. Pressure losses are located primarily in (1) the air-preparation system, (2) the dryer, and (3) the spent-air cleaning system.

The air-preparation system contains an indirect or direct heater. Since the air flows past finned tubes in an indirect heater, the pressure loss is almost negligible; however, this is not the situation in direct-heater systems.

The pressure loss of the air on passing through a direct heater (usually, gas fired) is a maximum of 50 mm water gauge when the gas, admitted via a sparger, burns in the total airflow. This type of burner can be used to provide temperatures of up to 500°C. If higher temperatures are required, a combustion chamber is used in which the gas is burned with a 10 to 20% excess of air compared to the stoichiometry of the reaction. The Δp value for this airflow is between 250 and 500 mm water gauge. Large combustion chambers (e.g., 5000 kW) exhibit high pressure losses, and smaller chambers show lower pressure drops. Secondary air is admixed with the flue gas to obtain the desired drying air temperature, and the pressure drop for this flow is about 150 mm water gauge.

The pressure drop across a dryer is strongly dependent on dryer type:

1. It is negligible for a rotary dryer, which is, in effect, an empty tube with curtains of falling solids.
2. It is negligible for a flash dryer in which the very diluted air–solids mixture flows through an empty tube.
3. It is substantial for a fluid-bed dryer in which the air must overcome the resistance of the distribution plate and the product layer. For shallow, large-diameter

beds of height 0.1 or 0.2 m, for example, the distributor pressure drop should be between 50 and 100% of the bed pressure drop. This is necessary to obtain stable fluidization (i.e., there are no preferential air passages). Deeper beds use a distribution plate offering between 10 and 20% of the product layer pressure drop [10].

4. It is negligible for a spray dryer, which is, in fact, an empty vessel with a very diluted solids–air mixture flowing through it.

The typical spent-air cleaning system may contain one or more of these appliances: a cyclone, a filter, an electrofilter, and a scrubber. A cyclone collects particles larger than 5 to 10 μm, and the associated pressure drop is 100 to 200 mm water gauge. A dust filter collects finer particles (0.1 to 1 μm); the equipment is larger than that of a cyclone and needs normal air velocities with respect to the filtering area of between 0.5 and 1.5 $m{\cdot}min^{-1}$. The pressure drop can approach 200 mm water gauge. An electrofilter is smaller than a filter. Its efficiency is usually slightly worse than that of a filter. A wet scrubber can be very simple, as in the case of a scrubber for air containing sodium chloride, with a pressure drop that is, for all practical purpose, negligible. On the other hand, wet scrubbers exist that have pressure drops of up to 1000 mm water gauge.

The spent-air cleaning system processes all solids for flash and many spray dryers but only the dust for rotary and fluid-bed dryers. The fans for rotary, flash, and spray dryers can be located in the air-preparation system, in the air-cleaning system, or in both. With fluid-bed dryers it is customary to have fans for air admission and for exhaust.

When knowing the air pressure drop that a fan must overcome, it is relatively simple to calculate the fan power. The formula is

$$\text{Fanpower} = \frac{\phi_v \cdot \Delta p}{3600 \cdot \eta \cdot 1000} \qquad \text{kW}$$

where $\eta = 0.6{\cdot}0.85{\cdot}0.95 = 0.4845$ (usually taken as 0.5), with 0.6 the fan efficiency and 0.85 the motor efficiency, and 0.95 takes further losses into account (e.g., lines). Δp should be expressed in $N{\cdot}m^{-2}$ (1 mm water gauge = 10 $N{\cdot}m^{-2}$). The formula is basically for incompressible fluids; however, compared to atmospheric pressure, the pressure losses are relatively small, so the formula can be used for all practical purposes.

Motive Power

For rotary dryers rotating at normal speed the motive power is

$$0.3 \cdot \frac{\pi}{4} \cdot D^2 \cdot L \qquad \text{kW}$$

(The peripheral speed is 0.25 to 0.5 m·s^{-1}; see [8], and for further details, see Chapter 6.) In flash dryers, a sling or cage mill is required to destroy agglomerates, typically drawing 10 to 20 kW.

Spray dryers require substantial motive power for rotary wheels or nozzles (e.g., 8 kWh/t (metric tonne) of feed for rotary atomization). Nozzle power consumption can be calculated by using the formula for fans when the flow and the pressure drop are known. η is 0.8 in this case.

Sundry Power

Electrical power is required for screws, locks, vibrating conveyors, oil pumps, and so on. Locks normally require 0.6 kW, whereas vibrating conveyors and screws require between 1 and 10 kW.

4.11 MISCELLANEOUS ASPECTS

Method of Heating

Heating can be direct or indirect. Direct heating involves the combustion of oil or gas and the passage of the combustion gases into the dryer. The composition is usually affected only slightly by the products of combustion, because of the large mass of airflow. The stoichiometric, adiabatic, and isochoric combustion of methane results in a temperature of 2655°C if the initial temperature is 20°C. The drying-gas temperature can, on utilizing direct combustion, be as high as 1000°C. The material of construction imposes a restriction here as, for example, steel starts to glow at 800°C. Incomplete combustion of oil or gas can lead to contamination of the product with soot. It has, on heating the drying air by direct oil combustion, caused black specks on salt at drying in a fluid-bed dryer. The direct combustion of pulverized coal at drying operations is not recommended. It has caused a coal dust explosion in the dust separation filter of a fluid-bed dryer for a mineral. Unburned coal particles had accumulated over time in the dust filter. When, at the drying of food and feed, direct heating is used, it is essential to avoid the formation of NO_x [(harmful) nitrogen oxides]. These oxides are formed at high temperatures. In low-NO_x burners, natural gas burns completely with an air excess of, for example, a factor of 3. This lowers the flame temperature. In these burners, the combustion occurs in, for example, a ceramic body with fine channels.

Indirect heating of drying air is usually carried out by means of steam or thermal oil. It is possible to burn natural gas and to exchange heat indirectly with drying air, however, the heat exchangers become quite large because of the poor heat transfer coefficient. Steam heating is not normally used to provide temperatures above 200°C because of the high steam pressures required to produce high steam condensation temperatures; for example, 16-bar absolute steam condenses at about 200°C. Heat-exchange oil may be used to achieve temperatures up to 350°C, with the oil being heated indirectly by a flame or by electricity. The use of indirect heating is inherently safer than the use of direct heating.

Dew-Point Off-Gas

The spent drying gas contains water vapor, and condensation occurs when this gas cools down to the dew point. This unwanted phenomenon has to be avoided. Good insulation of the equipment and lines handling spent drying gas is a useful measure. A further possibility is the deliberate addition to this gaseous flow of a relatively small amount of warm or hot drying air. Last but not least, it is important to prevent ingress air.

Example 4.10 The spent drying air leaving a convective dryer has a temperature of 80°C and a relative humidity of 30%. The dew point of this flow is 52°C. This flow is mixed with ambient air having a temperature of 15°C and a relative humidity of 50%. The mass ratio spent drying air to ingress air is 5 : 1. The temperature of the mixed gas becomes 71°C. The dew point of the latter flow is 49°C. Thus, the margin necessary to avoid wall condensation has come down from 80 – 52 = 28 K to 71 – 49 = 22 K.

Residence Times

The residence time of the solids is relatively long in fluid-bed dryers and rotary dryers, whereas the residence time of the solids is relatively short in both flash dryers and spray dryers. The gas residence time is short in all dryer types.

Dryer Type	Gas	Solids
Fluid-bed	0.2 s	5–30 min
Rotary	[a]	15–60 min
Flash	1 s	1 s
Spray	25 s	25 s

[a] The gas residence time in seconds is approximately equal to the dryer length in meters.

Materials of Construction

Carbon and stainless steel are the usual materials of construction for convective dryers. Direct-heat rotary dryers are usually made of carbon steel, as they are often used for minerals and sand.

Investment

In comparing fixed and variable costs, the concepts of capital charge and maintenance rate should be used.

Example 4.11 The investment for a large industrial dryer is $1,000,000. The annual production is 120,000 t, the capital charge is 30%, and the maintenance rate is

5%. Fixed cost per metric tonne is

$$\frac{1{,}000{,}000 \cdot 0.35}{120{,}000} = \$2.92$$

This cost can be compared with the energy cost per metric tonne in optimization calculations.

To arrive at an investment figure, an installation cost must be known as well as the capital cost of the equipment. As a rule of thumb, the equipment cost should be multiplied by a factor of 2.75 if the equipment is constructed from carbon steel or by a factor of 2.25 if the equipment is manufactured from stainless steel to obtain the total installed cost (including instrumentation, auxiliaries, building space allocated, supporting structure, engineering, etc.). These figures apply to spray, flash, rotary, and fluid-bed dryers. In specific cases, however, different factors should be used; for example, an outdoor plant is cheaper than an indoor installation.

Variable Costs

Variable costs are costs due to the consumption of natural gas, steam, and electricity.

4.12 MATERIAL BALANCE (kg·h^{-1})

$$\begin{aligned} W_{\text{out}} &= 0.01 \cdot \text{Cap} \cdot A_2 \\ \text{Sol} &= \text{Cap} - W_{\text{out}} \\ W_{\text{in}} &= \frac{\text{Sol} \cdot A_1}{100 - A_1} \\ \text{Evap} &= W_{\text{in}} - W_{\text{out}} \end{aligned}$$

4.13 HEAT BALANCE (kJ·h^{-1})

The following formula is applicable for rotary and flash dryers:

$$TA_{\text{out}} = a \cdot TA_{\text{in}} + b \ (^\circ\text{C})$$

TA_{out} is the dependent variable in this formula; TA_{in} is the independent variable. The values of a and b depend on the type of dryer. These values can be found in the relevant chapters. The relationship between TA_{in} and TA_{out} is such that when TA_{in} varies from 100 to 1000°C, TA_{out} varies from 50 to 150°C only.

For spray drying, $TA_{\text{out}} = 88.39 \cdot {}^{10}\log TA_{\text{in}} - 112.35$ °C (see Chapter 8). When $TA_{\text{in}} = 1000$°C in this formula, $TA_{\text{out}} = 152.8$°C, whereas $TA_{\text{out}} = 64.4$°C when $TA_{\text{in}} = 100$°C. The reader is referred to Chapter 5 for TA_{out} values in fluid-bed drying.

The heat transferred in the dryer from the air to the process stream is the net heat. Heat is also transferred from the air in other directions (e.g., to the surroundings), but that heat is not taken into account at the moment. The enthalpy of the process stream changes due to this transfer. This enthalpy change consists of three parts:

1. The evaporation of the water and the heating of the water vapor
2. The heating of the solid
3. The heating of the water remaining in the product

Expressed quantitatively in the same order, we have

$$\begin{aligned} Q_1 &= \text{Evap}(2500 + 1.9{\cdot}TA_{\text{out}} - 4.2{\cdot}T_f) \\ Q_2 &= \text{Sol}{\cdot}c_s(TA_{\text{out}} - c - T_f) \\ Q_3 &= W_{\text{out}}{\cdot}4.2(TA_{\text{out}} - c - T_f) \end{aligned}$$

0°C is taken as the point of reference for the calculation of Q_1; 2500 is the heat of evaporation of water at 0°C in kJ·kg^{-1}, and 1.9 and 4.2 are the specific heats of steam and water in kJ·kg^{-1}·K^{-1}, respectively. It would also be possible to take a different temperature (e.g., 50°C) as the point of reference for the calculation of Q_1, Q_2, and Q_3. The expression for Q_1 would then become

$$Q_1 = \text{Evap}\,[2385 + 1.9(TA_{\text{out}} - 50) - 4.2(T_f - 50)]$$

where 2385 is the heat of evaporation at 50°C in kJ·kg^{-1} and 1.9 and 4.2 are again the specific heats of water and steam in kJ·kg^{-1}·K^{-1}, respectively.

Example 4.12

$$\begin{aligned} \text{Evap} &= 100\ \text{kg}{\cdot}\text{h}^{-1} \\ TA_{\text{out}} &= 80^\circ\text{C} \\ T_f &= 20^\circ\text{C} \end{aligned}$$

For 0°C as the point of reference:

$$Q_1 = 100(2500 + 1.9{\cdot}80 - 4.2{\cdot}20) = 256{,}800\,\text{kJ}{\cdot}\text{h}^{-1}$$

For 50°C as the point of reference:

$$Q_1 = 100[2385 + 1.9{\cdot}(80 - 50) - 4.2(20 - 50)] = 256{,}800\,\text{kJ}{\cdot}\text{h}^{-1}$$

The two results are identical. This can be explained by the fact that the heat of evaporation becomes smaller as the temperature rises.

The variable c in the expressions for Q_2 and Q_3 takes account of the fact that the leaving product is usually cooler than the leaving gas. Its value depends on the dryer type and the flow regime. For more information on the solid specific heat, see Section 4.14. The net heat, Q_{tot_1}, is obtained by adding Q_1, Q_2, and Q_3.

More heat than the net heat needs to be put into the airflow, for two reasons:

1. Only part of the temperature increase of the airflow is extracted from it in the dryer.
2. There are heat losses.

The heat to be supplied to the dryer externally is Q_{tot_2} and it can be calculated as follows:

$$Q_{tot_2} = Q_{tot_1} \frac{TA_{in} - 10}{TA_{in} - TA_{out}} \cdot d$$

In this expression, the air-intake temperature is assumed to be 10°C. The value of d depends on the dryer type and allows for steady-state heat losses.

4.14 SPECIFIC HEAT OF SOLIDS

Values for the specific heat of solids can be found in the literature, or alternatively, by using Kopp's law, which is a method for quick approximation. *Kopp's law* states that the specific heat of a chemical compound at room temperature is approximately equal to the sum of the atomic heat capacities divided by the molecular weight. Gases should be at atmospheric pressure.

kJ·(kiloatom)$^{-1}$·K^{-1}			
Carbon	7.5	Oxygen	16.8
Hydrogen	9.6	Phosphorus	22.6
Boron	11.3	Fluorine	21.0
Silicon	15.9	Other elements	26.0

Source: Alstom Power, Inc., Warrenville, IL.

For example, we use the following data to calculate the specific heat of sodium sulfate (Na_2SO_4):

Element	Atomic Weight	Atomic Heat Capacity
Na_2	46.00	52.0
S	32.06	26.0
O_4	+64.00	+67.2
	142.06	145.2

Thus, the specific heat is 145.2/142.06 = 1.022 kJ·kg^{-1}·K^{-1}.

Perry's Handbook [8] gives 32.8 $cal^{-1}\cdot mol^{-1}\cdot K^{-1}$ (0.967 $kJ\cdot kg^{-1}\cdot K^{-1}$), that is, a 6% error from the value derived from Kopp's law. The law is also applicable to hydrates. Values obtained by Kopp's rule are safe in dryer and kiln heat calculations if increased by 10%, up to 260°C, and by 20%, up to 540°C.

4.15 GAS FLOWS AND FAN POWER

The airflow to the dryer can be calculated as follows:

$$\text{Airflow}_1 = \frac{Q_{\text{tot}_2}}{1.05\cdot 1.25(TA_{\text{in}} - 10)} \qquad \text{m}^3\cdot\text{h}^{-1}$$

where 1.05 is the specific heat of air in $kJ\cdot kg^{-1}\cdot K^{-1}$ (mean 0 to 600°C) and 1.25 is the specific mass of air in $kg\cdot m^{-3}$ at 10°C and atmospheric pressure. The air specific heat is 1.0 $kJ\cdot kg^{-1}\cdot K^{-1}$ for air temperatures of 100 to 200°C. The air specific mass at atmospheric pressure can be calculated using the formula

$$R_A = \frac{355}{273 + TA_{\text{out}}} \qquad \text{kg}\cdot\text{m}^{-3}$$

The power consumption of the fan conveying the air to the dryer is

$$\text{Power}_{G1} = \frac{\text{Airflow}_1\cdot\Delta p}{3600\cdot 1000\cdot 0.5} \qquad \text{kW}$$

The airflow leaving the dryer is:

$$\text{Airflow}_2 = \text{Airflow}_1\cdot\frac{1.25}{R_A}\cdot f \qquad \text{m}^3\cdot\text{h}^{-1}$$

where f accounts for the attraction of ingress air and its value depends on the dryer type. The evaporated water gas flow should be added to this airflow to obtain the gas flow removed by the exhaust fan:

$$WA_{\text{flow}} = \frac{\text{Evap}}{R_W} \qquad \text{m}^3\cdot\text{h}^{-1}$$

where R_W is the specific mass of steam at TA_{out} and atmospheric pressure. Its value can be calculated using the formula

$$R_W = \frac{220}{TA_{\text{out}} + 273} \qquad \text{kg}\cdot\text{m}^{-3}$$

$$\text{Gasflow}_2 = \text{Airflow}_2 + WA_{\text{flow}} \qquad \text{m}^3\cdot\text{h}^{-1}$$

Finally, the power consumption of the fan removing the gas flow from the drying system can be expressed as

$$\text{Power}_{G2} = \frac{\text{Gasflow}_2 \cdot \Delta p}{3600 \cdot 1000 \cdot 0.5} \qquad \text{kW}$$

4.16 DIRECT HEATING OF DRYING AIR

The direct heating of drying air by natural gas combustion will be discussed in some detail because burning natural gas produces water with which the drying gas, on leaving the burner, is preloaded.

The heat of combustion of methane is 804,000 kJ·kmol^{-1}. The division of Q_{tot_2} by this number yields the required methane flow in kmol·h^{-1}. The reaction equation for the combustion of methane is

$$CH_4 + 2O_2 \rightarrow CO_2 + 2H_2O$$

As the molecular weights of methane and water are 16 and 18 kg·kmol^{-1}, respectively, the water flow produced by the combustion of methane in kg·h^{-1} is the methane flow in kg·h^{-1} multiplied by 2·18/16. This water flow is part of the gas flow leaving the burner or furnace.

The material of this section is illustrated by means of an example in Section 5.5.

REFERENCES

[1] Poling, B.E., Prausnitz, J.M., O'Connell, J.P. (2001). *The Properties of Gases and Liquids*, McGraw-Hill, New York, pp. 7.4, A.59.

[2] Beek, W.J., Muttzall, K.M.K., van Heuven, J.W. (1999). *Transport Phenomena*, Wiley, Chichester, UK, pp. 314–317, 107–110.

[3] Röben, K.W. (1994). Continuously dried air dries plastics. *Kunststoffe*, *84*, 858–861 (in German).

[4] Dolinsky, A.A. (2001). High-temperature spray drying. *Drying Technology*, *19*, 785–806.

[5] Papadakis, S.E., Bahu, R., McKenzie, K.A., Kemp, I.C. (1993). Correlations for the equilibrium moisture content of solids. *Drying Technology*, *11*, 543–553.

[6] Oakley, D.E. (2004). Spray dryer modeling in theory and practice. *Drying Technology*, *22*, 1371–1402.

[7] Nonhebel, G., Moss, A.A.H. (1971). *Drying of Solids in the Chemical Industry*, Butterworth, London, p. 244.

[8] Perry, R.H., Green, D.W. (2008). *Perry's Chemical Engineers' Handbook*, 8th ed., McGraw-Hill, New York, pp. 12–77, 12–61, 2–162.

[9] Williams-Gardner, A. (1971). *Industrial Drying*, Leonard Hill, London, pp. 83, 113, 149, 164, 166, 168, 193, 194.

[10] Geldart, D., Baeyens, J. (1985). The design of distributors for gas-fluidized beds. *Powder Technology*, *42*, 67–78.

5

CONTINUOUS FLUID-BED DRYING

Section 5.1 provides a general introduction to continuous fluid-bed drying. In Section 5.2, theoretical aspects of fluidity are covered; for example, the concept of minimum fluidization velocity is discussed. Theoretical aspects of drying in a rectangular dryer are dealt with in Section 5.3. Drying theory is used to size a rectangular dryer with a shallow bed for a drying operation in which the bulk of the moisture is free. The removal of bound moisture in this dryer type is treated in Section 5.4. In Section 5.5 we discuss the design of a circular fluid-bed dryer with an example to illustrate the procedure.

5.1 GENERAL DESCRIPTION

A continuous fluid-bed dryer is essentially equipment in which a continuous feed of wet particulate material is dried by contact with warm or hot air that is blown through to maintain the material in a fluidized state. Because the drying action depends on fluidization, the moisture content of the feed is normally lower than that used in flash dryers and in rotary dryers.

Two principal types of fluid-bed dryers are in use: a circular type with a deep bed (0.5 to 2.0 m) and a rectangular form that usually has a bed up to 0.2 m deep. Circular fluid-bed dryers were introduced in the United States and in Canada in 1948. Operating with a high air-inlet temperature (e.g., 800°C), they were used to dry coal, limestone, blast furnace slag, and similar materials [1]. The equipment is not vibrated and the fluidized bed has the characteristics of a continuous-stirred tank reactor. The

---

Drying in the Process Industry, First Edition. C.M. van 't Land.
© 2012 John Wiley & Sons, Inc. Published 2012 by John Wiley & Sons, Inc.

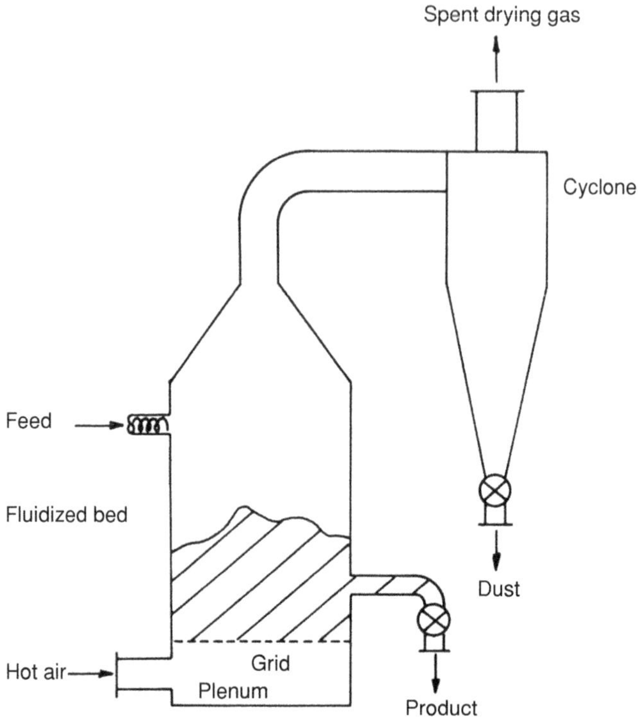

Figure 5.1 Round fluid-bed dryer.

composition of the product stream is the same as the composition of the bed. A consequence is that incrustation is not likely to occur since the dryer contents are dry. Neither are there any restrictions in scaling up the apparatus. Although the electrical consumption is high, since the fan drive is used to fluidize the bed, one of the principal attractions of this system is its simplicity (see Fig. 5.1).

The mode of operation of the rectangular fluid-bed dryer is shown in Figure 5.2. Usually, the dryer has a small bed height, leading to a plug-flow characteristic. It is not easy to accomplish fluidization of the wet material near the feed inlet, and this can lead to improper transportation and incrustation. Various modifications can be made to remedy this fault: (1) more air in the feed section, (2) directing the blown air so as to create a transport effect, and (3) vibration of the dryer. The latter option exists because this equipment was developed from vibrating conveyors that can be used for many types of particulate materials and hence is applicable to a wide range of particle sizes. Vibration assists fluidization, safeguards the transport independent of the air supply, prevents the occurrence of incrustations, and may lead to improved heat and mass transfer. However, it limits scaling up to a maximum size of 8×1.6 m^2, for example. The maximum temperature for vibrating equipment is approximately 300°C, whereas the air temperature can be as high as 800°C for stationary equipment. Rectangular dryers with deep beds (e.g., 0.5 to 1 m) exhibit considerable backmixing,

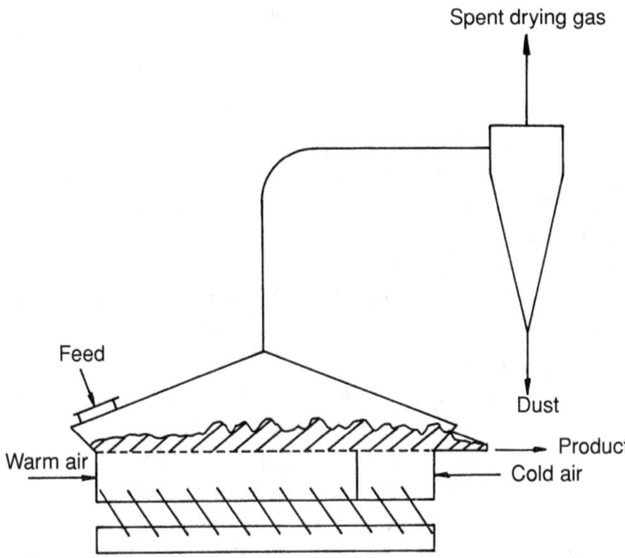

Figure 5.2 Rectangular fluid-bed dryer (vibrated).

so relatively dry material can already be found in the feed section, which decreases the tendency for incrustation and makes vibration superfluous. Tailor [2] gives an example of such a dryer for potassium chloride in which the air-inlet temperature in the first section is 427°C.

To some extent, fluid-bed dryers are intermediate between direct-heat rotary and flash dryers:

1. Flash dryers are sensitive to load variations, whereas direct-heat rotary dryers are relatively insensitive.
2. Flash dryers have residence times of a few seconds, direct-heat rotary dryers approximately 30 min, and fluid-bed dryers with a small bed height have a residence time of several minutes.
3. The moderate feed disintegration is intermediate to that caused in flash dryers (maximum feed disintegration) and rotary dryers (mild disintegration).

The size of fluid-bed dryers becomes disproportionally large for materials having a weight-average particle size of $\leq 100\ \mu$m.

An interesting aspect is the installation of heat-exchanging surfaces in tall fluid beds; the heat transfer is good and the heat transferred by conduction decreases the amount of heated air required. Moreover, the efficiency of the heat transfer by conduction exceeds the efficiency of the heat transfer by convection. The reason is that, at contact drying, an auxiliary material is not required (see Section 4.6). However, the heat exchange surfaces must not suffer from incrustation. This means, generally speaking, that this option is less suitable when the solid material is soluble in the liquid to be evaporated. A typical application is the use of rectangular fluid beds with heating panels in the polymer industry. Typical values for the wall-to-bed heat transfer coefficients are in the range 200 to 600 $W \cdot m^{-2} \cdot K^{-1}$. It is recommended that actual heat transfer coefficients be established by means of drying tests rather than by reliance on published correlations.

It is possible to combine a fluid-bed dryer and a fluid-bed cooler in the same equipment, which cannot be done with a flash dryer and which requires special internals in a rotary dryer.

Also, fluid-bed dryers can combine drying and particle-size enlargement (i.e., granulation). *Fluid-bed granulation* can be defined as a particle-forming process that is achieved by spraying a liquid into a fluidized bed already containing particles. The liquid may contain dissolved material. There are two possible nucleation mechanisms: (1) the sprayed droplets may evaporate and dissolved matter may be left behind, or (2) attrition may occur between fluidized granules. There are also two possible growth mechanisms: (1) a wetted particle may attach itself to a second particle by liquid bridging which, with subsequent drying, creates a solid bridge (*agglomeration*); or (2) the wetted particle may dry and the residual matter may attach to the original particle (*layering*). Fluid-bed granulation is capable of producing particulate solids of between 0.5 and 3 mm. However, adding all the solid material to be solidified in a solution is an energy-intensive process. Another option for fluid-bed processing is coating.

5.2 FLUIDIZATION THEORY

The term *fluidization* was coined to describe a certain mode of contacting granular solids with fluids. To illustrate, let the solid be a well-rounded silica sand contained in a cylindrical vessel with a porous bottom. The fluid is air. As the air passes upward, through the porous bottom and bed, measurements can be taken of the bed height and of the pressure drop across the bed as a function of the airflow rate. A profile such as that shown in Figure 5.3 will be obtained.

As long as the bed is stationary, and for dry particles, the pressure loss as a function of the airflow rate is given by Ergun's expression [3]. When the pressure reaches its maximum value, the bed starts to expand. In this condition, the individual sand granules are disengaged from each other and may be moved readily around by the expenditure of much less energy than would be required if the bed was not suspended by the airstream. The mobility of the aerated sand column resembles that of a high-viscosity liquid. This condition is known as the *fluidization point*, and the corresponding airflow rate is termed the *minimum fluidization velocity*. With a further increase in the airflow, the point at which the first bubbles appear can be observed. This point locates the bubbling point and the minimum bubbling velocity. A further increase in the flow rate leads to the maximum bed-expansion ratio.

The first large-scale fluidization application was in the United States in 1940 for cracking oil vapor. Fluid cracking catalysts have a weight-average particle size of about 80 μm and a bulk density of about 1100 $kg \cdot m^{-3}$. The description and

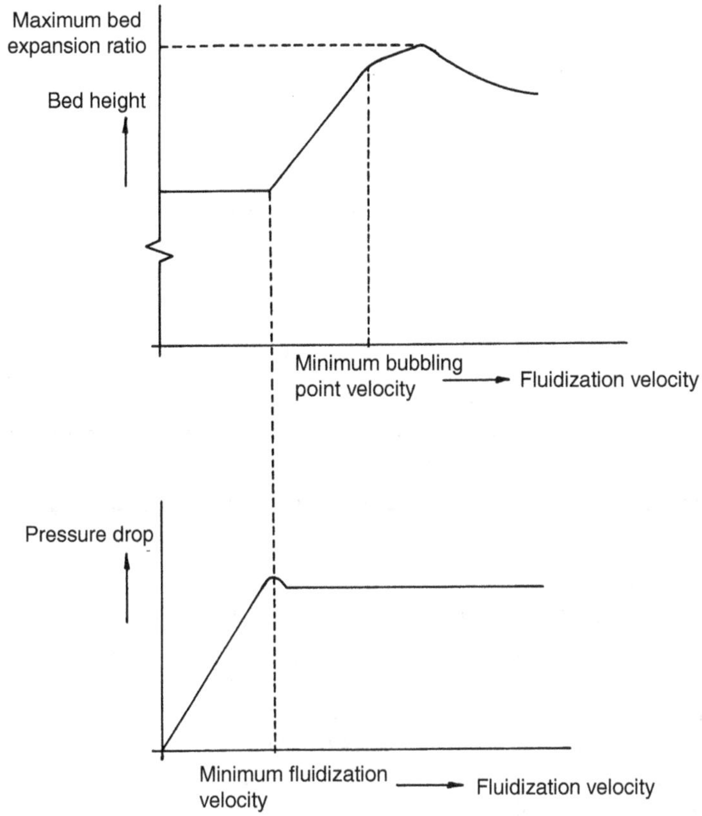

Figure 5.3 Diagrams for ideally fluidizing solids.

successful operation of the fluid cracking units precipitated a number of fundamental and applied studies in fluidization fields.

Heat and mass transfer in a fluidized bed are efficient. Figure 5.3 can be used to show how granular materials display ideal fluidization characteristics; however, not all materials exhibit this behavior. For example, channeling (preferential airflow) is a frequently occurring unwanted phenomenon, and Figure 5.4 shows typical pressure drop and flow data for moderately channeling solids (with flour, particle size $\leq$ 20 μm, exhibiting the channeling). The design of the gas-inlet device has a profound effect on channeling; thus, with porous plates, by means of which gas distribution into a bed tends toward uniformity, channeling tendencies are much reduced when compared with a multiorifice distributor in which the gas is introduced through a relatively small number of geometrically spaced holes. For shallow, large-diameter beds, the distributor pressure drop should be between 50 and 100% of the bed pressure drop. Higher beds (e.g., between 0.5 and 1 m) require lower pressure drops across the bottom plate. Here it can be between 10 and 20% of the bed pressure drop. An article by Geldart and Baeyens [4] gives more details. It is well known that

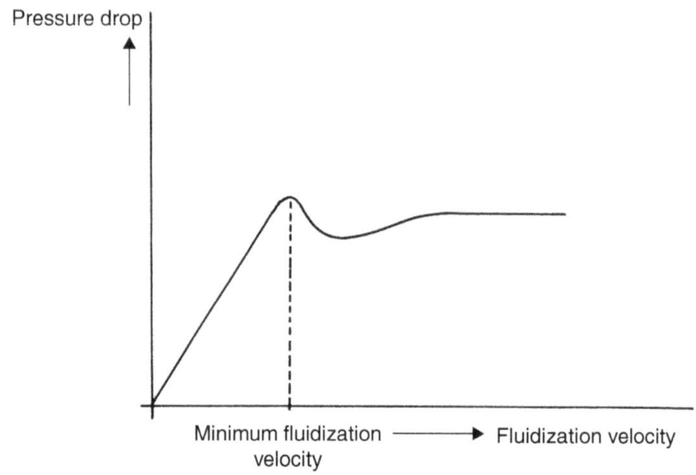

Figure 5.4 Typical pressure drop/flow diagram for moderately channeling solids.

if a gas distributor gives a pressure drop that is too low, the result is poor fluidization; that is, some parts of the bed will receive much less gas than others and may be temporarily or permanently defluidized, while in other parts, the gas forms permanent spouts or channels.

Conidur is a suitable commercial distributor for fluid-bed dryers and coolers. A typical distributor with fine triangular holes has perforations of 0.2 mm (side length of the equilateral triangles), the sheet thickness is 0.5 mm, and the percentage of open area is 5.5. A typical approach velocity for this sheet is 1 $m{\cdot}s^{-1}$. Sheets with precision-manufactured slotted holes are also available. They are usually more suitable for high gas velocities because of the high free area, which is typically 10%.

Example 5.1 With a bed height of 0.1 m, containing a material of bulk density of 1000 $kg{\cdot}m^{-3}$, the pressure drop across the bed is equal to $1000{\cdot}g{\cdot}0.1 \approx 1000\ N{\cdot}m^{-2}$ (100 mm water gauge). The pressure drop across the distributing device must be about 700 $N{\cdot}m^{-2}$ to avoid preferential gas passage through the bed.

A shallow, wet bed of particulate material is difficult to fluidize. Such a wet bed is present, for example, in the first few meters of a rectangular plug-flow fluid-bed dryer for vacuum-pan salt. As the average moisture content decreases, fluidization becomes feasible. However, particle sizes of 1 mm or larger exhibit spouting beds rather than fluid beds. It is not customary to distinguish between these phenomena; an upward flow through a particulate mass to be dried means that we are dealing with a fluid-bed dryer if the material moves. The reason undoubtedly is that the terms *fluid bed* and *fluidization* are generally associated with efficient heat and mass transfer.

Concerning fluid-bed drying, two gas mass velocities are needed: (1) the minimum value for uniform fluidization, which is likely to be approximately twice the minimum fluidization velocity for a perforated plate distributor; and (2) the minimum

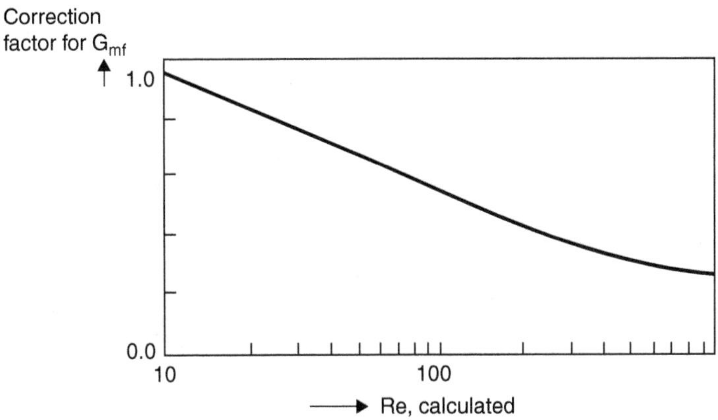

Figure 5.5 Correction factor for results obtained with Leva's equation when Re > 10. (Courtesy of Informations Chimie, Paris, France.)

value for substantial elutriation, which is approximately equal to the terminal velocity of the fines in the product. The latter value fixes the practical maximum gas velocity. First, the minimum fluidization velocity is discussed.

The ability to predict reliably the point of incipient fluidization is of basic importance in virtually all fluidized-process studies and design. For the estimation of the minimum fluidization mass velocity Leva [5] recommends

$$G_{\mathrm{mf}} = 0.0093 \cdot \overline{d}_p^{1.82} \cdot \frac{[\rho_F(\rho_S - \rho_F)]^{0.94}}{\mu^{0.88}} \qquad \mathrm{kg \cdot m^{-2} \cdot s^{-1}} \tag{5.1}$$

This equation has been supported by many experimental studies. The direct application of the correlation is limited to applications where

$$\mathrm{Re} = \frac{G_{\mathrm{mf}} \cdot \overline{d}_p}{\mu} < 10$$

If Re > 10, G_{mf} must be corrected using Figure 5.5. G_{mf} must be multiplied by the value found. If the material is vesicular, the solid specific mass must be corrected. The composite diameter is the Sauter diameter, also called the surface-to-volume diameter.

Example 5.2 A perforated plate is charged with dry vacuum-pan salt (NaCl). The bed height is 0.2 m and the bulk density is 1200 kg·m^{-3}. The specific mass is 2160 kg·m^{-3}. The particles are assumed to be cubes. Sieve analysis of a sample of 100 g was carried out and the results are given in Table E5.2A. The number of particles in a size range, n, has been calculated from the mass in the size range (Table E5.2A).

The minimum fluidization mass velocity is calculated. First, the Sauter diameter is calculated. This diameter is also called the surface-to-volume diameter.

Table E5.2A

Size Range (μm)	d_p	g	n	$n \cdot d_p^2$	$n \cdot d_p^3$
>600	—	0	0	0	0
500–600	550	9	$2.50 \cdot 10^4$	$7.58 \cdot 10^9$	$4.17 \cdot 10^{12}$
420–500	460	31	$1.47 \cdot 10^5$	$3.12 \cdot 10^{10}$	$1.44 \cdot 10^{13}$
250–420	335	35	$4.31 \cdot 10^5$	$4.84 \cdot 10^{10}$	$1.62 \cdot 10^{13}$
125–250	187.5	20	$1.40 \cdot 10^6$	$4.94 \cdot 10^{10}$	$9.26 \cdot 10^{12}$
63–125	94	5	$2.79 \cdot 10^6$	$2.46 \cdot 10^{10}$	$2.31 \cdot 10^{12}$
<63	—	+0	0	+0	+0
		100		$1.61 \cdot 10^{11}$	$4.64 \cdot 10^{13}$

The Sauter diameter is

$$\overline{d}_p = \frac{4.64 \cdot 10^{13}}{1.61 \cdot 10^{11}} = 288\ \mu\text{m} \quad \text{and} \quad \frac{1}{\overline{d}_p} = \frac{1.61 \cdot 10^{11}}{4.64 \cdot 10^{13}} = 0.00347\ \mu\text{m}^{-1}$$

The following calculation is a practical alternative calculation method. X is the mass fraction.

Table E5.2B

Size Range (μm)	d_p	g	X	X/d_p
>600	—	0	0	0
500–600	550	9	0.09	0.000164
420–500	460	31	0.31	0.000674
250–420	335	35	0.35	0.001045
125–250	187.5	20	0.20	0.001067
63–125	94	5	0.05	0.000532
<63	—	+0	+0	+0
		100	1.0	0.003482

The Sauter diameter is

$$\overline{d}_p = \frac{1}{0.003482} = 287\ \mu\text{m}$$

The proof proceeds as follows. Take, for example, the fraction 250 to 420 μm. The mass fraction in this size range is X_3.

$$X_3 = \frac{1.62 \cdot 10^{13}}{4.64 \cdot 10^{13}} = \frac{n \cdot d_p^3}{\sum n \cdot d_p^3}$$

$$\frac{X_3}{335} = \frac{4.84 \cdot 10^{10}}{4.64 \cdot 10^{13}} = \frac{n \cdot d_p^2}{\sum n \cdot d_p^3}$$

$$\sum \left(\frac{X}{d_p}\right) = \frac{\sum n \cdot d_p^2}{\sum n \cdot d_p^3}$$

$$\bar{d}_p = \frac{\sum n \cdot d_p^3}{\sum n \cdot d_p^2} = \frac{1}{\sum \left(\frac{X}{d_p} \right)}$$

The implicit assumption is that the *form factor* of the particles is constant. The form factor of cubical particles is 1, whereas the form factor of spherical particles is $\pi/6$.

The salt is fluidized with air at 20°C, and the air discharges from the bed under atmospheric pressure. Fluidization pressure drop $\Delta p = \rho_b \cdot g \cdot h = 1200 \cdot 9.81 \cdot 0.2 = 2354\ \text{N} \cdot \text{m}^{-2}$. The air pressure just above the distributor is $101{,}300 + 2534 = 103{,}834\ \text{N} \cdot \text{m}^{-2}$.

$$\rho_F = \frac{355}{273 + 20} \cdot \frac{103{,}834}{101{,}300} = 1.24\ \text{kg} \cdot \text{m}^{-3}\ \text{(see Section 4.15)}$$

The atmospheric pressure is 101,300 $\text{N} \cdot \text{m}^{-2}$.

$$G_{\text{mf}} = 0.0093(2.87 \cdot 10^{-4})^{1.82} \cdot \frac{[1.24(2160 - 1.24)]^{0.94}}{\left(1.8 \cdot 10^{-5}\right)^{0.88}} = 0.083\ \text{kg} \cdot \text{m}^{-2} \cdot \text{s}^{-1}$$

The viscosity of the fluidizing air at 20°C is $1.8 \cdot 10^{-5}\ \text{N} \cdot \text{s} \cdot \text{m}^{-2}$.

$$v_{\text{mf}} = \frac{0.083}{1.24} = 0.067\ \text{m} \cdot \text{s}^{-1}$$

$$\text{Re} = \frac{G_{\text{mf}} \cdot \bar{d}_p}{\mu} = \frac{0.083 \cdot 2.87 \cdot 10^{-4}}{1.8 \cdot 10^{-5}} = 1.32$$

No correction is required.

A more recent correlation concerning the minimum fluidization velocity [6] is

$$v_{\text{mf}} = \frac{0.0009(\rho_S - \rho_F)^{0.934} \cdot g^{0.934} \cdot \bar{d}_p^{1.8}}{\mu^{0.87} \cdot \rho_F^{0.066}} \quad \text{m} \cdot \text{s}^{-1}$$

For Example 5.2, $v_{\text{mf}} = 0.055\ \text{m} \cdot \text{s}^{-1}$ can be calculated with this correlation. $\bar{d}_p$ is, also in this correlation, the Sauter diameter. However, for the purposes of this book, it is not necessary to investigate the concept of minimum fluidization velocity in further detail. Similar emperical relationships exist for the minimum bubbling velocity. The fluidization velocity actually chosen is usually many times greater than the minimum fluidization velocity. Table 5.1 [7] provides an estimate of the maximum superficial air velocities through vibrating conveyor screens.

Table 5.1 Estimate of Maximum Superficial Air Velocities Through Vibrating Conveyor Screens ($m{\cdot}s^{-1}$)

		Specific Gravity	
Mesh Size	Micrometers	2.0	1.0
200	74	0.22	0.13
100	149	0.69	0.38
50	297	1.4	0.89
30	590	2.6	1.8
20	840	3.2	2.5
10	2000	6.9	4.6
5	4000	11.4	7.9

Source: Columns 1, 3, and 4: Courtesy Carrier Vibrating Equipment, Inc. Column 2 gives the corresponding hole sizes in micrometers. The mesh size reflects the U.S. number, National Bureau of Standards LC-584 and ASTM E-11.

Using this table we find that the maximum superficial air velocity for the material of the example is about 20 times that of the minimum fluidization velocity. Table 5.1 corresponds roughly with Table 4.1. A value of 1.4 $m{\cdot}s^{-1}$ would indicate elutriation of particles (having a specific gravity of 2.0) smaller than 297 μm. Typically, the air velocity below the distributor would be 1 $m{\cdot}s^{-1}$ for the product mentioned in Example 5.2. If the air temperature is 170°C, 5% by weight of the product will be elutriated (i.e., all particles smaller than 125 μm).

5.3 DRYING THEORY FOR RECTANGULAR DRYERS

The drying process in a rectangular dryer with plug flow and a shallow product layer will now be considered. Figure 5.2 depicts a typical drying plant, with the wet product entering at the left and moving to the right; drying, postdrying, and cooling occur subsequently. Surface moisture (free moisture) is evaporated on the left of the bed. Both the gas flow and the product attain the adiabatic saturation temperature (i.e., the temperature the gas acquires on cooling down and accepting the water vapor). The gas sensible heat is converted into the latent heat of evaporation.

This part of the bed ends when the product has reached the critical moisture content. The next section of the dryer is used to bring down the bound moisture to the desired moisture content, as along the section the material temperature increases.

Cooling occurs along the right-hand section of the bed. Cooling is often required to prevent caking of the product in the bunkers or silos. Caking may be caused by evaporation of water from the material and subsequent condensation of this moisture on cold walls. Figure 5.6 shows a normal temperature and moisture profile for the dryer.

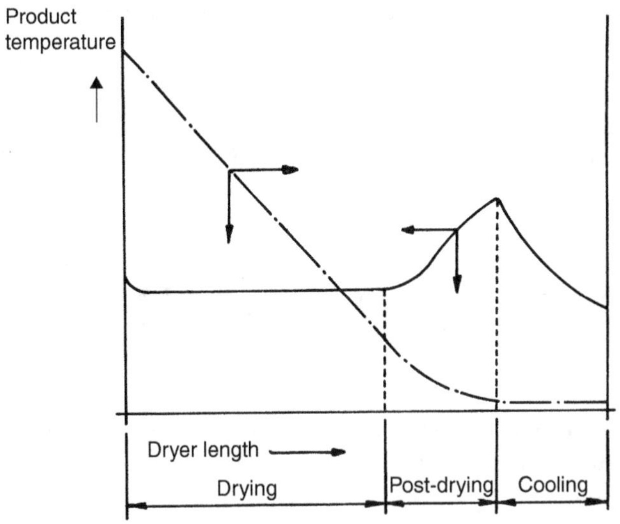

Figure 5.6 Temperature and moisture profile for a continuous rectangular fluid-bed dryer.

Sizing of the First Part (Drying)

To size the first part (drying), the humidity chart is used (see Fig. 4.2). The design equation is

$$\phi_{H_2O} = B \cdot L_1 \cdot v_F \cdot \rho_F (x_o - x_i) \qquad \text{kg}\cdot\text{s}^{-1} \tag{5.2}$$

Typically, x_o and x_i are found as follows. For example, atmospheric air enters at 10°C with a relative humidity of 30% and is warmed to 50°C in an indirect heater. In the humidity chart the drying process is represented by a sloped line. It is assumed that the evaporation of water occurs from a solution of the solid in water, with the solution exerting 90% of the saturated vapor pressure of water. Thus, the adiabatic saturation temperature is 22°C.

$$x_o = 1.4 \cdot 10^{-2} \text{ kg of water per kg of dry air}$$
$$x_i = 2 \cdot 10^{-3} \text{ kg of water per kg of dry air}$$

The velocity of the warm gas v_F, can be selected from either Table 5.1 or Table 4.1. Cooling down to 22°C causes a lower velocity. A further reduction may occur by widening the freeboard.

The design equation leads to the value of L_1, the length of the first part of the bed. An implicit assumption is that the feed enters at the adiabatic saturation temperature. L_1 is multiplied by 1.15 to allow for the steady-state heat losses. An additional load factor is needed to allow for the extra heat losses in the long run; 1.25 is a usual value. Table 5.2 summarizes cases where the simple model was found to be applicable,

Table 5.2 Applicability of the Model for the Evaporation of Free Water on Fluid-Bed Drying

Product	Scale of Test[a]	h_{exp} (mm)	v_F ($m \cdot s^{-1}$)	TA_{IN} (°C)	A_1 (% by weight)
Cubic vacuum-pan salt	p	150	0.9	85	2.0
Dendritic salt	p	100	1	180	15
Open-pan salt	i	150[b]	0.4	110	15
Solid peroxydicarbonate (batchwise drying)	i	700[b]	0.7	40	20–25

[a] p, pilot-plant scale; i, industrial scale.
[b] Approximation.

using equation (5.2) and multiplying L_1 by 1.15. Vaněček [8] states that an adiabatic saturation temperature is usually obtained. Table 5.3 lists the physical properties of materials mentioned in Table 5.2.

Sizing of the Second Part (Postdrying)

To simplify the situation, assume that only heating occurs and that no additional moisture is evaporated. Furthermore, ideally, mixed cross-flow is assumed to exist without backmixing. Figure 5.7 illustrates this starting point. At any position in the bed, the temperature of the product leaving the mixing cell equals the temperature of the leaving air. The average residence time of the product in the cell is longer than the average residence time of the air. The temperature of the product increases, while the product flows from the product inlet to the product outlet. Thus, the driving force for heat transfer decreases with the length traveled as the heating air temperature

Table 5.3 Physical Properties of Fluid-Bed-Processed Materials

(Product)	d_p (mm)	ρ_S ($kg \cdot m^{-3}$)	ρ_b ($kg \cdot m^{-3}$)	c_s ($kJ \cdot kg^{-1} \cdot K^{-1}$)	λ_s ($W \cdot m^{-1} \cdot K^{-1}$)
Cubic vacuum-pan salt	0.4	2160	1250	0.870	5.8
Dendritic salt	0.2	2160	800	0.870	5.8
Granular salt	1.5	2160	1250	0.870	5.8
Open-pan salt	0.7	2160	800	0.870	5.8
Solid peroxydicarbonate	0.1	1100	450		
Anhydrous citric acid	0.45	1665	900	1.210	
Citric acid monohydrate	0.65	1540	800	1.470	
CAN (a fertilizer)	2[a]	1800[b]	1070	1.60[c]	2.0[d]
Sand	—	2600	—	0.840	

[a] Datum from [9].
[b] At 20°C.
[c] Temperature range 32.3–84.2°C.
[d] Datum from [10].

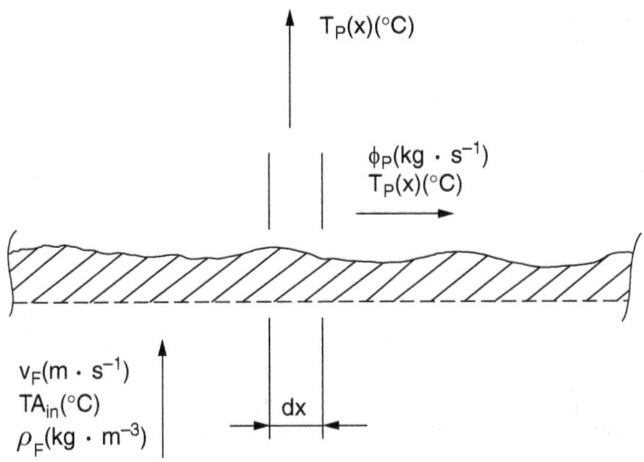

Figure 5.7 Cross-flow air heating of particulate material.

is constant. In such a case, $-\ d[TA_{\text{in}} - T_p(x)]/dx$ is proportional to $TA_{\text{in}} - T_p(\text{x})$. On integrating the corresponding differential equation, an expression containing a natural logarithm is obtained. A heat balance for a small length dx is

$$\rho_F \cdot v_F \cdot B \cdot dx \cdot c_p[TA_{in} - T_{\text{p}}(\text{x})] = \phi_p \cdot c_s \cdot dT_p(x)$$

Rearranging and integrating between $x = 0$, $T_p(x = 0)$ and $x = L_2$, $T_p(x = L_2)$ yields

$$L_2 = \frac{c_s \cdot \phi_p}{c_p \cdot v_F \cdot \rho_F \cdot B} \ln \frac{TA_{\text{in}} - T_p(x = 0)}{TA_{\text{in}} - T_p(x = L_2)} \qquad \text{m} \tag{5.3}$$

Equation (5.3) is analogous to Newton's law of cooling (1701). In the literature, Newton's law of cooling is synonymous with the following statement: The rate of cooling of a warm body at any moment is proportional to the temperature difference between the body and its surrounding medium (air). Newton measured the temperature of a hot block of iron, when placed in a uniformly blowing wind, as a function of time. There are two differences: Equation (5.3) relates to heating instead of cooling, and it relates to a continuous process instead of a batch process. However, heating is the inverse of cooling and it obeys the same laws. Furthermore, Newton's time coordinate is, in the second part of the plug-flow fluid-bed dryer, replaced by a length coordinate. A certain length traveled in a plug-flow fluid-bed dryer is unambiguously coupled to a certain dwelling time.

L_2 is multiplied by 1.15 to allow for the steady-state heat losses. An additional load factor is needed to allow for the extra heat losses in the long run; 1.25 is a usual value. Table 5.4 relates to the applicability of the combination of the model for the evaporation of free water and the model for the evaporation of bound moisture. Table 5.3 lists the physical properties of the materials mentioned in Table 5.4.

Table 5.4 Applicability of the Combination of the Models for the Evaporation of Free Water and Heating-Up in Cross-Flow on Fluid-Bed Drying

Product	Scale of Test[a]	h_{exp} (mm)	v_F (m·s^{-1})	TA_{in} (°C)	A_1 (% by weight)
Cubic vacuum-pan salt	i	60–200	1.0–1.1	165–210	2.0
	i	150	0.85	110–170	1.9
	i	200	1.25	160	2.4
	p	150	0.8	95	0.9
	p	150	1.4	62	2.7
	i	150	0.4	110	3.5
Dendritic salt	p	100	1	165	19
Granular salt	p	150	1.7	108	1.25
Anhydrous citric acid	p	100	1/1.2	94/60	2.0
Citric acid monohydrate	p	70/200	1.25/0.95	60	2.5

[a]p, pilot-plant scale; i, industrial scale.

Sizing of the Third Part (Cooling)

Equation (5.3) is applicable. Table 5.5 contains data concerning the applicability of a cross-flow cooling model; using equation (5.2) and multiplying L_3 by 1.15. Most of the data were collected within Akzo Nobel. Table 5.3 lists physical properties of the materials included in Table 5.5.

The Biot number provides the possibility of an assessment of the relative importance of the heat transfer to the particle and the heat dissipation in the particle itself.

$$\mathrm{Bi} = \frac{\alpha_0}{\lambda_S/(d_{50}/2)}$$

The Biot number for the fluid-bed cooling of CAN (calcium ammonium nitrate) is estimated in the appendix to this chapter. The estimation shows that the resistance to heat transfer is located mainly in the boundary layer surrounding the particle. This means that there will be hardly a temperature gradient within the particle. Furthermore, the estimation shows that the coefficient for the heat transfer from the particles

Table 5.5 Applicability of the Model for Cross-Flow Cooling in a Plug-Flow Fluid-Bed Cooler

Product	Scale of Test[a]	h_{exp} (mm)	v_F (m·s^{-1})	TA_{in} (°C)
Anhydrous citric acid	p	100	0.7/1	15
Citric acid monohydrate	p	70/200	1	15
CAN (a fertilizer)[b]	i	—	1.2	34
Cubic vacuum-pan salt	i	200	0.9	26

[a]p, pilot-plant scale; i, industrial scale.
[b]Data from [9].

to the air is high: 294 $W{\cdot}m^{-2}{\cdot}K^{-1}$. This figure can be compared to the range for the overall coefficients for the heat transfer from central heating radiators to air: 5 to 15 $W{\cdot}m^{-2}{\cdot}K^{-1}$. The area offered by 100 kg of 3.3-mm spherical CAN particles is 101 m^2. The large bed area and the high heat transfer coefficient render the observations of Table 5.5 understandable.

VDI [11] describes a more sophisticated model. In this source the latter model is applied to the industrial batchwise fluid-bed cooling of roasted coffee beans from 300°C to 35°C. There are 60 kg of beans per square meter. The specific airmass flow is 3 $kg{\cdot}m^{-2}{\cdot}s^{-1}$ and the air is at 20°C.

Further data:

$$d_p = 6 \text{ mm}$$
$$\rho_s = 630 \text{ kg}{\cdot}\text{m}^{-3}$$
$$c_s = 1700 \text{ J}{\cdot}\text{kg}^{-1}{\cdot}\text{K}^{-1}$$
$$\lambda_S = 0.10 \text{ W}{\cdot}\text{m}^{-1}{\cdot}\text{K}^{-1}$$

VDI [11] estimates that cooling for 150 s will reduce the temperature of the beans to a temperature in the range 25.3 to 34.3°C. This result is not, however, compared to experimental data. We now calculate what the simple model would predict. Equation (5.3) cannot be used directly because it was derived for continuous cooling. At batchwise cooling the driving force for heat transfer decreases with the time elapsed. In such a case, $-\,d[T_p(t) - TA_{\text{in}}]/dt$ is proportional to $T_p(\text{t}) - TA_{\text{in}}$. A heat balance for 1 m^2 of bed area in a small time dt is

$$\rho_F{\cdot}v_F{\cdot}1{\cdot}c_p{\cdot}dt[T_p(t) - TA_{\text{in}}] = -M{\cdot}c_s{\cdot}dT_p(t)$$

Rearranging and integrating between $t = 0$, $T_p(t = 0)$ and $t = t$, $T_p(t = t)$ yields

$$t = \frac{M{\cdot}c_s}{\rho_F{\cdot}v_F{\cdot}1{\cdot}c_p}{\cdot}\ln\frac{TA_{\text{in}} - T_p(t = 0)}{TA_{\text{in}} - T_p(t = t)} \quad \text{s} \tag{5.4}$$

This equation predicts cooling to 23.5°C for a cooling time of 150 s:

$$150 = \frac{60{\cdot}1700}{3{\cdot}1000}{\cdot}\ln\frac{20 - 300}{20 - 23.5}$$

Sixty kilograms of roasted coffee beans per square meter corresponds to an expanded bed height of 30 to 40 cm. Note that the diameter of the beans is relatively large (6 mm) and the thermal conductivity is low (0.10 $W{\cdot}m^{-1}{\cdot}K^{-1}$). Still, the simple model is approximately correct. The fluid-bed dryers and coolers are operated with relatively shallow layers, such as a 200-mm expanded bed height.

Figure 5.8 depicts a typical vibrating fluid-bed-dryer system.

Figure 5.8 Vibrating fluid-bed-dryer system in the manufacturer's shop. (Courtesy of Carrier Vibrating Equipment, Inc., Louisville, KY.)

Example 5.3 A vibrating fluid-bed dryer for vacuum-pan salt is to be designed.

Product

General: 20 $t \cdot h^{-1}$ of vacuum-pan salt from the dryer
TP_{in}: 50°C
d_{50}: 400 μm
A_1: 2.5% water by weight (as fed)
A_2: $< 0.1\%$ water by weight
c_s: 0.87 $kJ \cdot kg^{-1} \cdot K^{-1}$
Solubility solid: 0.33 kg of NaCl per kg of water
ρ_b: 1250 $kg \cdot m^{-3}$

Process

Ambient temperature: 10°C
Ambient relative humidity: 25%
TA_{in}: 170°C by indirect heat exchange with steam
Unfluidized bed height: 0.2 m
Fluid-bed dryer width: 1.25 m

Steam data

P: 14 bar absolute
ΔH: 1964.4 kJ·kg^{-1}

See Sections 4.13 and 4.15 for additional data on water, steam, and air.

The water vapor pressure of saturated brine is 75% of the water vapor pressure between 0 and 100°C. This is related to the fact that the solubility of NaCl in water is almost constant between 0 and 100°C; a saturated brine contains 25% by weight of NaCl in this temperature range.

To dry, the salt must be heated to 65°C.

Mass balance (kg·h^{-1})

	In	Out
Salt	19,980	19,980
Water	+ 512	+ 20
	20,492	20,000

$$\text{Evap} = 512 - 20 = 492\ \text{kg·h}^{-1} = 0.1367\ \text{kg·s}^{-1}$$

Sizing of the first part In equation (5.2), $v_F = 1.0$ m·s^{-1} at atmospheric pressure (see Table 5.1).

$$R_A = \frac{355}{273 + 170} = 0.801\ \text{kg·m}^{-3}$$
$$B = 1.25\ \text{m}$$
$$x_i = 0.002\ \text{kg·kg}^{-1}\ \text{(humidity chart)}$$

The adiabatic saturation temperature is 46°C and $x_o = 0.051$ kg·kg^{-1} (humidity chart; the relative humidity of the leaving gas is 75%).

$$L_1 = \frac{0.1367}{1.25 \cdot 1.0 \cdot 0.801(0.051 - 0.002)} = 2.79\ \text{m}$$

Take a length of 1.15·2.79 = 3.21 m. Note that the sensible heat made available by cooling the feed from 50°C to 46°C is neglected.

Sizing of the second part The salt is heated to 65°C. $v_F = 1.0$ m·s^{-1} at atmospheric pressure. By equation (5.3),

$$L_2 = \frac{0.870 \cdot 5.55}{1.0 \cdot 1.0 \cdot 0.801 \cdot 1.25} \ln \frac{170 - 46}{170 - 65} = 0.80\ \text{m}$$

Take a length of 1.15·0.80 = 0.92 m.

Sizing of the third part The salt is cooled to 40°C with air of 25°C.

$$R_A = \frac{355}{273 + 25} = 1.19\ \text{kg}\cdot\text{m}^{-3}$$

$v_F = 1.0\ \text{m}\cdot\text{s}^{-1}$ at atmospheric pressure. By equation (5.3),

$$L_3 = \frac{0.870\cdot 5.55}{1.0\cdot 1.0\cdot 1.19\cdot 1.25}\ln\frac{25 - 65}{25 - 40} = 3.18\ \text{m}$$

Take 1.15·3.18 = 3.66 m.

Area review The total dryer length is 3.21 + 0.92 + 3.66 = 7.79 m and the width is 1.25 m. The drying area is 5.2 m^2 and the cooling area is 4.6 m^2.

Steam consumption Steam consumption is based on an ambient temperature of 10°C. The drying airflow is (3.21 + 0.92)1.25·1.0·0.801·3600 = 14,887 $\text{kg}\cdot\text{h}^{-1}$. The heat flow is 14,887·1.0(170 – 10) = 2,381,920 $\text{kJ}\cdot\text{h}^{-1}$, which is exchanged in the air heater. Heat losses from the air heater are neglected.

$$\frac{2{,}381{,}920}{1964.4} = 1213 \text{ kg of steam per hour}$$

$$\frac{2{,}381{,}920}{492} = 4841 \text{ kJ per kg of evaporated water}$$

$$\frac{1213}{492} = 2.47 \text{ kg of steam per kg of evaporated water}$$

The long-term consumption (including startup, shutdown, cleaning, and low load) is obtained by multiplying by 1.1:

$$1.1\cdot 4{,}841 = 5325 \text{ kJ per kg of evaporated water}$$

$$1.1\cdot 2.47 = 2.72 \text{ kg of steam per kg of evaporated water}$$

$$2.72\cdot\frac{492}{20} = 66.9 \text{ kg of steam per metric ton of product}$$

Electrical energy consumption

Drying airflow

$$R_A = \frac{355}{273 + 10} = 1.25\ \text{kg}\cdot\text{m}^{-3}$$

$$\frac{14{,}887}{1.25} = 11{,}910\ \text{m}^3\cdot\text{h}^{-1}\text{(drying airflow) at atmospheric pressure}$$

Bed height: 0.2 m

$\rho_b = 1250\ \text{kg}\cdot\text{m}^{-3}$

Pressure loss due to the bed: $1250\cdot 0.2\cdot 9.81 = 2452.5\ \text{N}\cdot\text{m}^{-2}$

Pressure loss due to the distribution plate: $0.5\cdot 2452.5 = 1226\ \text{N}\cdot\text{m}^{-2}$

Total pressure loss: $2452.5 + 1226 = 3678.5\ \text{N}\cdot\text{m}^{-2}$

$$P = \frac{11,910\cdot 3,678.5}{3600\cdot 1000\cdot 0.5} = 24.3\ \text{kW}$$

Choose a fan with a 30-kW motor.

Cooling airflow

$$3.66\cdot 1.25\cdot 1.0\cdot 1.19\cdot 3,600 = 19,599\ \text{kg}\cdot\text{h}^{-1}$$

$$\frac{19,599}{1.19} = 16,470\ \text{m}^3\cdot\text{h}^{-1}$$

$$P = \frac{16,470\cdot 3678.5}{3600\cdot 1000\cdot 0.5} = 33.7\ \text{kW}$$

Install a fan with a 40-kW motor.

Exhaust gas flow The mass flow of drying and cooling air is 14,887 + 19,599 = 34,486 $\text{kg}\cdot\text{h}^{-1}$. The total airmass flow that is handled is $1.1\cdot 34,486 = 37,935\ \text{kg}\cdot\text{h}^{-1}$ (the factor 1.1 is due to the attraction of ingress air). The exhausted gases will be at about 55°C.

$$R_A = \frac{355}{273 + 55} = 1.08\ \text{kg}\cdot\text{m}^{-3}$$

Exhausted air flow: $\frac{37,935}{1.08} = 35,125\ \text{m}^3\cdot\text{h}^{-1}$

$$R_W = \frac{220}{273 + 55} = 0.67\ \text{kg}\cdot\text{m}^{-3}$$

Exhausted water vapor flow: $\frac{492}{0.67} = 734\ \text{m}^3\cdot\text{h}^{-1}$

Total exhausted gas flow: $35,125 + 734 = 35,859\ \text{m}^3\cdot\text{h}^{-1}$

Cyclone pressure drop: $1500\ \text{N}\cdot\text{m}^{-2}$

$$P = \frac{35,859\cdot 1500}{3600\cdot 1000\cdot 0.5} = 29.9\ \text{kW}$$

Select a fan with a 40-kW motor.

Miscellaneous

Vibration: 10 kW consumption

Rotary locks: $2\cdot 1.5 = 3$ kW consumption

Total electric energy consumption: $24.3 + 33.7 + 29.9 + 10 + 3 = 100.9$ kW

$$\frac{100.9}{20} = 5.0 \text{ kWh per t of salt}$$

Long-term consumption figure: $1.1 \cdot 5.0 = 5.5$ kWh per t of salt

Final remark The figures calculated are representative of actual continuous rectangular fluid-bed dryers for vacuum-pan salt.

Drying Curves

A drying curve can be drawn after a batch fluid-bed drying experiment has been carried out. The scale could be 1 L or 10 L, for example. The dryer is charged with a certain amount of wet material, and warm air is passed through the distributor and the bed from the bottom to the top. The temperature, the water content in $kg \cdot kg^{-1}$, and the velocity of this air are kept constant. The exhaust gas temperature as a function of time is recorded and samples from the material processed are taken intermittently and analyzed for their moisture content in $kg \cdot kg^{-1}$ (see Fig. 5.9). A design of an industrial dryer can be based on a drying curve that has been selected for scaling up. A drying curve presents basically the moisture content of the material as a function of time. If, for example, it is determined that the material must be processed batchwise for 15 min to reach the moisture specification, the material must also reside for 15 min in the plug-flow fluid-bed dryer. Traveling a certain length in a plug-flow fluid-bed dryer is coupled unambiguously with having resided for a certain amount of time in such a dryer. The implicit assumption when making this statement is that the temperature, the water content, and the velocity of the drying air

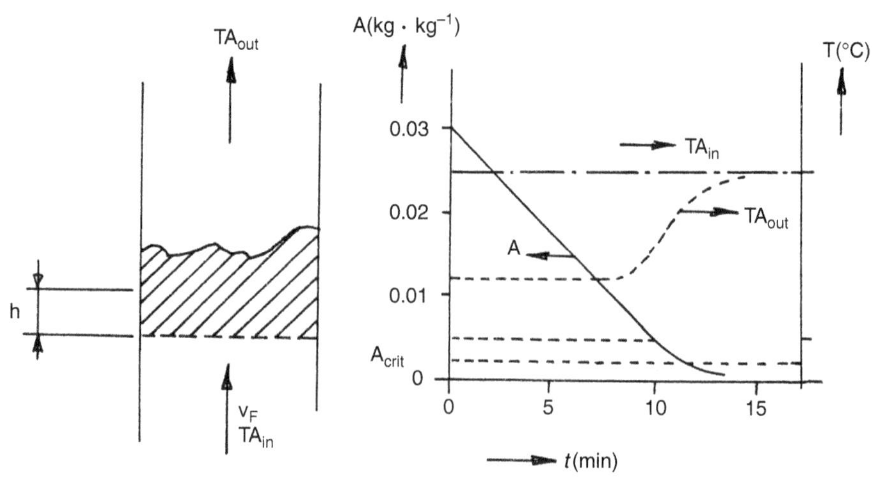

Figure 5.9 Drying curve.

are kept constant. A further assumption is that the product layer thickness is kept constant.

It is possible to check the results of the first drying step by means of the following equation:

$$G_{H_2O} = G_{air}(x_o - x_i) \qquad \text{kg} \tag{5.5}$$

G_{H_2O} is the amount of water evaporated in the first drying step; G_{air} is the amount of dry air passed through the batch in the first drying step; and x_o and x_i are, respectively, the leaving and entering air moisture content in kg·kg^{-1}. x_o is found by means of the humidity chart.

The result of the second drying step can be checked directly by means of equation (5.4). The latter equation was derived for batchwise cooling and is also applicable to batchwise warming. It is also possible to cool the material batchwise in a small dryer. The result of the cooling step can be checked by using equation (5.4).

Transport of the material through a plug-flow fluid-bed dryer should be checked in a pilot-plant plug-flow fluid-bed dryer. These tests also provide an opportunity to check any caking tendencies. Finally, it is advisable to work on at least two different laboratory scales to get a perspective on wall effects.

Example 5.4 The vacuum-pan salt of Example 5.3, could be tested in a 1-L laboratory fluid-bed dryer. The unfluidized bed height is also 0.2 m. The feed enters with a water content of 2.5% by weight at a temperature of 50°C. The bulk density of dry vacuum-pan salt is 1250 kg·m^{-3}. It is assumed that the same bulk density applies for the centrifuge cake. Ambient air at a temperature of 10°C is heated to 170°C by means of an electric heater. The ambient air contains 0.002 kg of water vapor per kilogram of dry air. The air velocity below the distributor is 1 m·s^{-1} at atmospheric pressure. The characteristics of the distributor of the laboratory fluid-bed dryer (e.g., percentage free area; shape, size, and pattern of the holes) are the same as the characteristics of the distributor of the large plug-flow fluid-bed dryer. The amount of material in the first part of the plug-flow fluid-bed dryer is 1.25·3.21·0.2·1250 = 1003 kg. This corresponds to a residence time of 3 min. Thus, a processing time of 3 min can be expected in the laboratory fluid-bed dryer as well. The amount of material in the second part is 1.25·0.92·0.2·1250 = 288 kg. This corresponds with a residence time of almost 1 min. By the same token, a processing time of 1 min to raise the product temperature from the adiabatic saturation temperature (i.e., 46°C) to 65°C, with air having a temperature of 170°C can be expected in the laboratory fluid-bed dryer. This prediction can be checked by means of equation (5.4). Finally, the amount of material in the third part (i.e., the cooling section) is 1.25·3.66·0.2·1250 = 1144 kg. This corresponds with a residence time of 3.4 min. The time to decrease the product temperature from 65°C to 40°C with air of

25°C would thus also be 3.4 min in the laboratory batch fluid-bed dryer. Again, this prediction can be checked by means of equation (5.4).

5.4 REMOVAL OF BOUND MOISTURE FROM A PRODUCT IN A RECTANGULAR DRYER

In Section 5.3 we dealt with drying duties in which the bulk of the moisture is free moisture, which is not always the case. It may be that the bulk of the water is bound. In this section we show how to proceed under such circumstances.

Example 5.5 We wish to obtain 500 $kg \cdot h^{-1}$ of a granular organic formula from a continuously functioning rectangular fluid-bed dryer. The feed enters the dryer with a water content of 25% by weight at 15°C. The product moisture content must be less than 0.5% by weight. Fifty percent by weight of the particles is larger than 1750 μm, with no particles smaller than 1000 μm. The maximum allowable product temperature is 100°C. The material's specific heat is 0.95 $kJ \cdot kg^{-1} \cdot K^{-1}$ and the material is insoluble in water. The wet bulk density is 595 $kg \cdot m^{-3}$ and the dry bulk density is 523 $kg \cdot m^{-3}$. The specific heat of the air is 1.0 $kJ \cdot kg^{-1} \cdot K^{-1}$.

Screening experiments are carried out in a small, circular batch fluid-bed dryer.

Fluid-bed diameter: 0.2 m

Feed weight: 1.85 kg

Initial stationary product-height calculation:

$$\text{Feed volume: } \frac{1.85}{0.595} = 3.1 \text{ L}$$

$$\text{Product height: } \frac{3.1}{\frac{\pi}{4} \cdot 2^2} = 1 \text{ dm } (0.1 \text{ m})$$

An optimum set of conditions is selected. The air velocity under the distributor is 2.0 $m \cdot s^{-1}$ and the air temperature is 90°C. By maintaining these conditions for 10 min, an in-spec product is obtained. In this 10-min period the following observations are made.

1. The bed's temperature gradually rises from 15°C to 90°C.
2. Simultaneously, the product's water content diminishes in weight from 25% to 0.2%.
3. Some dust carry-over occurs.
4. Particle breakage does not occur.

The optimum set of conditions is used for scaling up.

Drying

Experiment mass balance (kg)

	In	Out
Product	1.39	1.39
Water	+ 0.46	+ 0.00
	1.85	1.39

The water evaporation is 0.46 kg. The specific water evaporation capacity is

$$\frac{0.46 \cdot 60}{10 \cdot \frac{\pi}{4} \cdot 0.2^2} = 87.9 \text{ kg} \cdot \text{h}^{-1} \cdot \text{m}^{-2}$$

Industrial drying duty mass balance (kg·h^{-1})

	In	Out
Product	500	500
Water	+ 167	+ 0
	667	500

Water evaporation: 167 kg·h^{-1}

Area required: $\frac{167}{87.9} = 1.90 \text{ m}^2$

Take $1.15 \cdot 1.90 \approx 2.2 \text{ m}^2$.

Cooling Subsequent cooling from 90°C to 40°C is desired. The maximum ambient temperature is 30°C. Equation (5.3) is referred to. The air velocity is 1.5 m·s^{-1} at atmospheric pressure. The air specific mass at 30°C is

$$\frac{355}{273 + 30} = 1.17 \text{ kg} \cdot \text{m}^{-3} \quad \text{(see Section 4.15)}$$

and the area is

$$\frac{0.95 \cdot 500}{3600 \cdot 1.0 \cdot 1.5 \cdot 1.17} \cdot \ln \frac{90 - 30}{40 - 30} = 0.13 \text{ m}^2$$

Take 0.2 m^2

Total dryer and cooler sizes

$$2.2 + 0.2 = 2.4\ \text{m}^2.\ \text{Take}\ 4{\cdot}0.6\ \text{m}^2$$

Dryer length: 3.65 m
Cooler length: 0.35 m

Calculation of steam and of electrical-energy consumption follows the steps in Example 5.3. The transport and drying functions should be checked in a pilot dryer (size, e.g., $3{\cdot}0.2\ \text{m}^2$).

Final remark The test drying was actually carried out, however, the industrial dryer was not installed.

5.5 CIRCULAR FLUID-BED DRYERS

Introductory remarks on the circular fluid-bed dryer and an illustration depicting such a dryer were included in Section 5.1 (see Fig. 5.1). The composition of the material in the bed equals the composition of the material in the exit flow. Because the contents are almost ideally mixed, a substantial amount of the material has a residence time of, for example, 10% of the average residence time. Thirteen combinations of air-inlet and gas-outlet temperatures were reviewed [1,12,13]. The air-inlet temperatures were between 200 and 800°C, and the gas-outlet temperatures varied between 55 and 120°C. A high air-inlet temperature did not necessarily cause a relatively high gas-outlet temperature. By the same token, a low air-inlet temperature did not necessarily lead to a relatively low gas-outlet temperature.

The temperature of the fluidized bed is very uniform, thus permitting the performance of subtle drying operations [13, 14]. In both of these references, the replacement of a direct-heat rotary dryer by a circular fluid-bed dryer is described. Both undercuring and overcuring are impossible because of the accurate temperature control (large heat capacity).

It is not possible to dry and cool with the same equipment. An important consideration is the substantial power consumption of the fan. Figure 5.10 depicts a circular fluid-bed dryer and a circular fluid-bed cooler. Note that a flash dryer is being used here as a predryer. Continuous fluid-bed dryers/granulators also have a thick product layer and are often circular (see Section 3.2).

Example 5.6 A stationary circular fluid-bed dryer for sand is to be designed.

Product

General: 40 $\text{t}{\cdot}\text{h}^{-1}$ of sand from the dryer
TP_{in}: 10°C

Figure 5.10 High-capacity fluid bed (center) with flash dryer (left) as predrying stage and product cooling in fluid bed (center right). Product: polypropylene. (Courtesy of GEA Process Engineering A/S, Søborg, Denmark.)

d_{50}: 400 μm
A_1: 5% water by weight (as fed)
A_2: 0.1% water by weight
c_s: 0.840 kJ·kg^{-1}·K^{-1}
Solubility solid: nil
ρ_b: 1500 kg·m^{-3}

Process

Ambient temperature: 10°C
Ambient relative humidity: 50%
TA_{in}: 750°C by burning natural gas

TA_{out}: 120°C
Unfluidized bed height: 0.5 m
c_p: 1.05 kJ·kg^{-1}·K^{-1}

Natural gas data

Composition: 85% by volume of methane, 15% by volume of nitrogen
Methane heat of combustion: 804 kJ·mol^{-1}
ρ_g: 0.80 kg·nm^{-3}

See Sections 4.13 and 4.15 for data on water, steam, and air.

Pilot-plant drying tests were not carried out. It is assumed that the dried-product temperature equals the gas exit temperature. The carry-over is assumed to be negligible.

Mass balance (kg·h^{-1})

	In	Out
Sand	39,960	39,960
Water	+ 2,103	+ 40
	42,063	40,000

$$\text{Evap} = 2103 - 40 = 2063$$

Net heat (kJ·h^{-1})

$$Q_1 = 2063(2500 + 1.9{\cdot}120 - 4.2{\cdot}20) = 5{,}454{,}572$$
$$Q_2 = 39{,}960{\cdot}0.840(120 - 10) = 3{,}692{,}304$$
$$Q_3 = 40{\cdot}4.2(120 - 10) = +\ 18{,}480$$
$$Q_{tot_1} = 9{,}165{,}356$$
$$Q_{tot_2} = 1.1{\cdot}\frac{750 - 10}{750 - 120}{\cdot}9{,}165{,}356 = 11{,}842{,}222$$

Q_{tot_2} is the heat to be supplied in the combustion chamber by the combustion of natural gas. A heat loss factor of 1.1 appears to be satisfactory for steady-state operation.

$$\text{kJ per kg of evaporated water: } \frac{11{,}842{,}222}{2063} = 5740$$

Due to startup, shutdown, and cleaning, for example, the long-term consumption figure is probably a factor of 1.1 higher: 1.1·5740 = 6314 kJ per kilogram of evaporated water.

Table E5.6

	Stoichiometric Natural-Gas Combustion ($kg \cdot h^{-1}$)	
	In	Out
Natural gas		
Methane	235.7	0.0
Nitrogen	72.8	72.8
Dry air		
Oxygen	942.7	0.0
Nitrogen	3299.3	3299.3
Water vapor		
In air	17.0	17.0
By combustion	0.0	530.2
Carbon dioxide	+ 0.0	+ 648.2
	4567.5	4567.5

Dew-point exhaust gas The amount of heat transferred in the combustion chamber is 11,842,222 $kJ \cdot h^{-1}$. The methane flow required is 11,842,222/804 = 14,729 $mol \cdot h^{-1}$. On using the data of Section 4.16 and on realizing that the natural gas composition has been simplified to 85% by volume of methane and 15% by volume of nitrogen Table E5.6 can be calculated.

The natural gas consumption is

$$\frac{14{,}729}{0.85} \cdot 0.0224 = 388.3 \text{ nm}^3 \cdot \text{h}^{-1} \text{ (molar volume 22.4 nL)}$$

The total gas mass flow from the combustion chamber is

$$\frac{11{,}842{,}222}{1.05(750 - 10)} = 15{,}241.0 \text{ kg} \cdot \text{h}^{-1}$$

The specific heat of the hot gas has been taken as 1.05 $kJ \cdot kg^{-1} \cdot K^{-1}$. The secondary airmass flow is 15,241.0 – 4567.5 = 10,673.5 $kg \cdot h^{-1}$ (composed of 10,631.0 $kg \cdot h^{-1}$ of dry air and 42.5 $kg \cdot h^{-1}$ of water vapor).

Air does not leak into the hot gas flow as there is, up to and including the fluidized layer, overpressure. It is now possible to make the following table ($kg \cdot h^{-1}$):

Process Flow	Dry	Water	Total
Combustion	4,020.3	547.2	4,567.5
Secondary air	10,631.0	42.5	10,673.5
Evaporation	+ 0.0	+ 2,063.0	+ 2,063.0
	14,651.3	2,652.7	17,304.0

$$\text{Water content: } \frac{2652.7}{14{,}651.3} = 0.181 \text{ kg per kg of dry air}$$

The exhaust gases must be cooled from 120°C to 63°C to get condensation. This provides an ample margin.

Sizing drying-gas preparation unit

$$Q_{\text{tot}_2} = 11{,}842{,}222 \text{ kJ}\cdot\text{h}^{-1}$$

Buy a combustion chamber having a capacity of 14,400 MJ·h^{-1} or 4 MW (21.5% spare capacity). The gas mass flow from the combustion chamber is 15,241.0 kg·h^{-1}. The gas mass flow to the combustion chamber (excepting the natural gas flow) is approximately equal to this gas mass flow. The pressure drop due to the fluidized bed is 1500·9.81·0.5 = 7357.5 N·m^{-2}. The pressure drop due to the air distributor is approximately 0.25·7357.5 = 1839.4 N·m^{-2}.

$$7357.5 + 1839.4 + 2500 = 11{,}696.9 \text{ N}\cdot\text{m}^{-2}$$
$$2500 \text{ N}\cdot\text{m}^{-2}: \text{ combustion chamber's} \Delta p$$
$$R_A = \frac{355}{273 + 10} = 1.25 \text{ kg}\cdot\text{m}^{-3}$$
$$\text{Airflow}_1 = \frac{15{,}241.0}{1.25} = 12{,}192.8 \text{ m}^3\cdot\text{h}^{-1}$$

This is an approximation of the volumetric airflow to the combustion chamber, as the pressure increase generated by the fan is approximately 10% of the atmospheric pressure.

$$\text{Power}_{\text{G1}} = \frac{12{,}192.8 \cdot 11{,}696.9}{3600 \cdot 1000 \cdot 0.5} = 79.2 \text{ kW}$$

Install a fan with a 100-kW motor.

Dryer sizing To avoid excessive dust formation, a superficial velocity just above the fluidized bed of 0.8 m·s^{-1}(at 120°C and atmospheric pressure) is taken. This velocity is reduced to 0.4 m·s^{-1} in a disengaging section. The gas mass flow leaving the fluidized layer, excluding water vapor, is 17,304.0 – 2652.7 = 14,651.3 kg·h^{-1}. The composition is approximately equal to the composition of dry air.

$$R_A = \frac{355}{273 + 120} = 0.90 \text{ kg}\cdot\text{m}^{-3}$$
$$\text{Airflow}_2 = \frac{14{,}651.3}{0.90} = 16{,}279.2 \text{ m}^3\cdot\text{h}^{-1}$$
$$R_W = \frac{220}{273 + 120} = 0.56 \text{ kg}\cdot\text{m}^{-3}$$

$$WA_{\text{flow}} = \frac{2652.7}{0.56} = 4737.0\ \text{m}^3{\cdot}\text{h}^{-1}$$
$$\text{Gasflow}_2 = 16{,}279.2 + 4737.0 = 21{,}016.2\ \text{m}^3{\cdot}\text{h}^{-1}$$
$$\frac{\pi}{4}{\cdot}D^2{\cdot}0.8 = \frac{21{,}016.2}{3600}$$
$$D = 3.05\ \text{m; use } D = 3.00\ \text{m.}$$

The diameter of the disengaging section becomes $\sqrt{2}{\cdot}3.00 = 4.24$ m; take 4.25 m.

Sizing exhaust-gas unit

Cyclone pressure drop: 1500 $\text{N}{\cdot}\text{m}^{-2}$

$$\text{Power}_{\text{G2}} = \frac{21{,}016.2{\cdot}1500}{3600{\cdot}1000{\cdot}0.5} = 17.5\ \text{kW}$$

Install a fan with a 25-kW motor.

Survey

Fluid-bed diameter: 3 m

Diameter of the disengaging section: 4.25 m

Capacity of the combustion chamber: 4 MW

Natural gas consumption (load factor 1.1): $\frac{1.1{\cdot}388.2}{2.063} = 207.0\ \text{nm}^3$ per t of evaporated water

On using the load factor, the long-term losses are also taken into account. The electricity consumption (load factor 1.1) is

$$\frac{1.1(79.2 + 17.5 + 20.0)}{2.063} = 62.2\ \text{kWh per t of evaporated water}$$

(20.0: allowance)

5.6 APPENDIX: CALCULATION OF THE BIOT NUMBER OF A CAN PARTICLE IN A FLUID-BED COOLER

$$d_p = 3.3{\cdot}10^{-3}\text{m}$$

Calcium ammonium nitrate physical properties

$$\rho_S = 1800\ \text{kg}{\cdot}\text{m}^{-3}$$
$$\lambda_S = 2.0\ \text{W}{\cdot}\text{m}^{-1}{\cdot}\text{K}^{-1}[10]$$

Air physical properties (70°C)

$$\rho_F = 1.03\ \text{kg}\cdot\text{m}^{-3}\ \text{at atmospheric pressure}$$
$$c_p = 1010\ \text{J}\cdot\text{kg}^{-1}\cdot\text{K}^{-1}$$
$$\mu = 2.05\cdot 10^{-5}\ \text{N}\cdot\text{s}\cdot\text{m}^{-2}$$
$$\lambda_G = 0.029\ \text{W}\cdot\text{m}^{-1}\cdot\text{K}^{-1}$$

The product is at 100°C while the air heats up from 40°C to 100°C.

The VDI approach [11] is followed to calculate α_0, the heat transfer coefficient. The coefficient for the heat transfer from the particles in a fluidized bed to the gas is approximately equal to the heat transfer coefficient of a particle falling at terminal velocity [11].

The heat transfer through the laminary boundary layer is distinguished from the heat transfer in the turbulent wake. By using Ar, the *Archimedes number*, calculation of the terminal velocity can be avoided. The Archimedes number is also known as the *Grashof number*.

$$\text{Ar} = \frac{g\cdot d_p^3\cdot \rho_F(\rho_S - \rho_F)}{\mu^2} = 1{,}554{,}405$$

$$\text{Re} = 18\left(\sqrt{1 + \frac{1}{9}\sqrt{\text{Ar}}} - 1\right)^2 = 2104$$

$$\text{Pr} = \frac{\nu}{a} = \frac{\mu\cdot c_p}{\lambda_G} = 0.71$$

$$\text{Nu}_{\text{lam}} = 0.664\cdot\text{Re}^{1/2}\cdot\text{Pr}^{1/3} = 27.2$$

$$\text{Nu}_{\text{turb}} = \frac{0.037\cdot\text{Re}^{0.8}\cdot\text{Pr}}{1 + 2.443\cdot\text{Re}^{-0.1}(\text{Pr}^{2/3} - 1)} = 15.6$$

$$\text{Nu} = 2 + \sqrt{27.2^2 + 15.6^2} = 33.4$$

$$\alpha_0 = \frac{\lambda_G}{d_p}\cdot\text{Nu} = 294\ \text{W}\cdot\text{m}^{-2}\cdot\text{K}^{-1}$$

$$\frac{2\cdot\lambda_S}{d_{50}} = 1212\ \text{W}\cdot\text{m}^{-2}\cdot\text{K}^{-1}$$

$$\text{Bi} = \frac{\alpha_0\cdot d_{50}}{2\cdot\lambda_S} = \frac{294}{1212} = 0.24$$

Conclusion: The resistance to heat transfer is located mainly in the boundary layer surrounding the particle.

Note: A lower value of λ_S would have shown that the heat conduction in the particle cannot be neglected.

REFERENCES

[1] Beeken, D.W. (1960). Thermodrying in fluidised beds. *British Chemical Engineering*, *5*, 484–487.

[2] Tailor, J.P. (1969). Drying and bulk handling of potassium chloride. *Transactions of the Third Symposium on Salt*, Northern Ohio Geological Society, Cleveland, OH.

[3] Beek, W.J., Muttzall, K.M.K., van Heuven, J.W. (1999). *Transport Phenomena*, Wiley, Chichester, UK, pp. 119–122.

[4] Geldart, D., Baeyens, J. (1985). The design of distributors for gas-fluidized beds. *Powder Technology*, *42*, 67–78.

[5] Leva, M. (1959). *Fluidization*, McGraw-Hill, New York, pp. 62–74

[6] Baeyens, J., Geldart, D. (1974). Predictive calculations of flow parameters in gas fluidized beds and fluidization behaviour of various powders. In *Fluidisation et ses applications, Compte-Rendu du Congrès International*, Cepadues-Editions, Toulouse, France, pp. 263–273.

[7] Perry, R.H., Green, D.W. (2008). *Perry's Chemical Engineers' Handbook*, 8th ed., McGraw-Hill, New York, p. 12–83.

[8] Vaněček, V., Markvart, M., Drbohlav, R. (1966). *Fluidized Bed Drying*, Leonard Hill, London, p. 118.

[9] Winterstein, G., Rose, K., Viehweg, H., Schreyer, L. (1964). Cooling of particulate materials in a rectangular fluid bed. *Chemische Technik*, *16*, 106–107 (in German).

[10] Osman, M.B.S., Dakroury, A.Z., Dessouky, M.T., Kenawy, M.A., El-Sharkawy, A.A. (1996). Measurement of thermophysical properties of ammonium salts in the solid and molten states. *Journal of Thermal Analysis*, *46*, 1697–1703.

[11] VDI (1994). *VDI Wärmeatlas*, VDI Verlag, Düsseldorf, Germany, pp. Mf1–Mf4 (in German).

[12] Williams-Gardner, A. (1971). *Industrial Drying*, Leonard Hill, London, pp. 193–194.

[13] Vreeland, R., Bacchetti, J.A. (1982). Equipment plugging eliminated with fluid bed dryer. *Chemical Processing, 45*, 24–25.

[14] Viebrock, W.A., Hodel, A.E. (1983). Fluid-bed dryer eliminates calcining, reduces heat requirements by 7.5%. *Chemical Processing*, *46*, 20–21.

6

CONTINUOUS DIRECT-HEAT ROTARY DRYING

A general description of direct-heat rotary drying is provided in Section 6.1, followed by a discussion of the screening design methods in Section 6.2. An example concludes the chapter.

6.1 GENERAL DESCRIPTION

In a direct-heat rotary dryer, a continuous feed of wet particulate material is dried by contact with heated air while being transported along the interior of a rotating cylinder (see Fig. 6.1), with the rotating shell acting as a conveying device and as a stirrer. In addition, at least two other types of continuous rotary dryers exist: (1) an indirect-heat rotary dryer in which heat is transferred indirectly (e.g., a steam-tube dryer, treated in greater detail in Chapter 10), and (2) an indirect–direct rotary dryer, often considered a hybrid in which heat is transferred indirectly (by conduction and radiation) and directly, as illustrated in Figure 6.2. The construction is more complicated than that of a direct-heat rotary dryer, but the heat losses to the surrounding area are minimized because the outer wall is not in contact with the hottest gas.

Generally, refractory-lined rotary kilns are used to create a heat treatment for a process stream and are not straightforward dryers. The hot gases may enter at temperatures of up to 1650°C. This temperature range is also used in refractory-lined rotary incinerators.

Air-swept rotary coolers may be used to cool a product to an appropriate temperature for bagging and storage. Some rotary dryers also have a cooling element: Drying

Drying in the Process Industry, First Edition. C.M. van 't Land.
© 2012 John Wiley & Sons, Inc. Published 2012 by John Wiley & Sons, Inc.

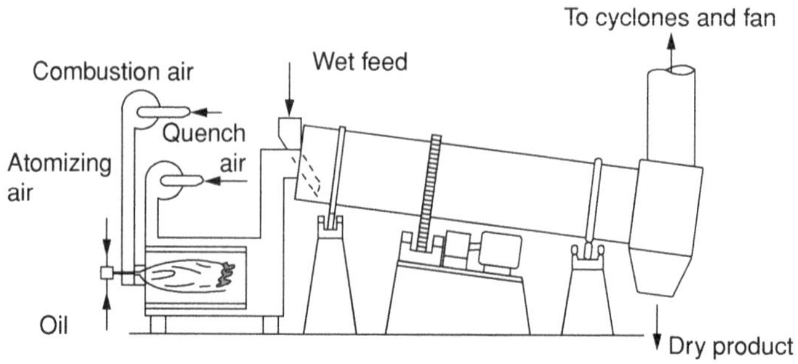

Figure 6.1 Typical cascading direct-heat rotary dryer arranged for concurrent operation. (From [1].)

occurs concurrently in the first section and cooling is performed countercurrently in the second. The spent gases can be combined and exhausted at two-thirds of the shell length, for example. Rotary dryers with coolers are used to process wet sugar from a centrifuge.

Direct-heat rotary dryers can be classified according to their relationship between product and gas flow: (1) concurrent (parallel), (2) countercurrent, and (3) cross-flow (Roto-Louvre dryers).

The Roto-Louvre dryer is a distinctive dryer in which the material to be processed rests on the bottom of a rotating cylinder through which drying gas is passed that also passes through the bed (see Fig. 6.3). This dryer is suitable for friable material since the movement is gentle.

Direct-heat rotary dryers are universally applicable for particulate materials, but normally they cannot process solutions, slurries, or pastes. Transport and drying functions are independent of load over wide variations. The dryer needs little attention, with incrustations being prevented by the use of shell knockers (which are optional). Consequently, direct-heat rotary dryers are suitable for applications where supervision is minimal. However, rotary dryers are mechanically relatively complicated, with the running gear, seals, feed breechings, and discharge breechings requiring attention. Maintenance costs per annum can reach 10% of the investment cost. Compare this to the 5% figure usually found for other convective dryers.

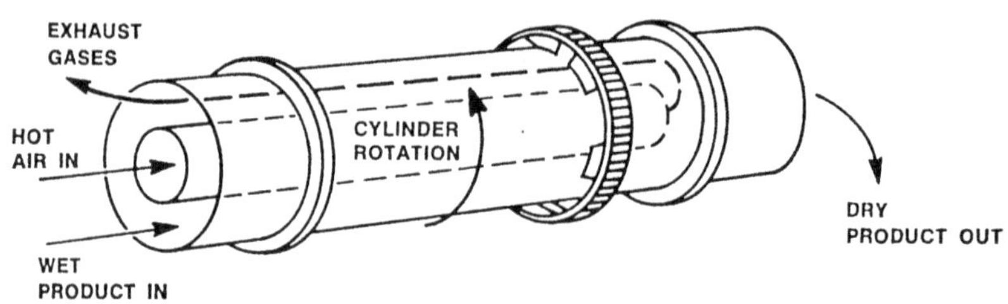

Figure 6.2 Indirect–direct rotary dryer. (Courtesy of Alstom Power Inc., Warrenville, IL.)

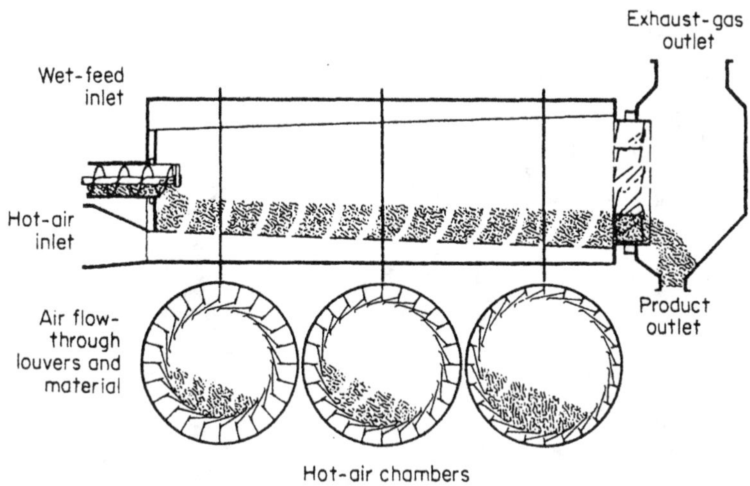

Figure 6.3 Roto-Louvre dryer. (From [2].)

Air-velocity choice is important; a standard guide is to choose an air velocity that provides a maximum carry-over of 5 to 10% by weight of the dry solid. Probably 90% of all rotary dryers operate at gas velocities below 2 $m \cdot s^{-1}$. The minimum particle size of materials processed in direct-heat rotary dryers is approximately 100 μm.

Substantial holdup is found in rotary dryers, with residence times between several minutes and 1 h being common. Typical volumetric loadings vary between 10 and 15%. This can affect process safety. Decomposition or reaction of the processed material can have serious consequences because of the large mass. At some locations, sugar beet pulp is dried in direct-fired rotary dryers with gases entering at 800 to 900°C. If the operation is interrupted, the pulp is in stationary contact with the hot shell wall and may catch fire. The direct-heat rotary dryer should not be used for the evaporation of solvents. If there is leakage of air into the dryer, the creation of explosive mixtures may result; reverse leakage may lead to a fire.

Both concurrently, and countercurrent direct-heat rotary dryers are equipped with flights on the interior of the shell for lifting and showering the solids through the gas stream during passage. These lifting flights are offset every 0.6 to 2 m to ensure continuous and uniform curtains of solids in the gas chamber. The shape of the flights is chosen on the basis of the handling characteristics of the solids (see Fig. 6.4). The cruciform flights reduce the falling height of the material and hence the entrainment; however, cleaning them can be a problem. In addition, to prevent entrainment, flights can be left out at the product end on cocurrent drying.

Diameter Rotary dryers are generally between one and several meters in diameter. Any equipment exceeding this size range can create difficulties in rail or road transport. On-site manufacturing can be an effective alternative.

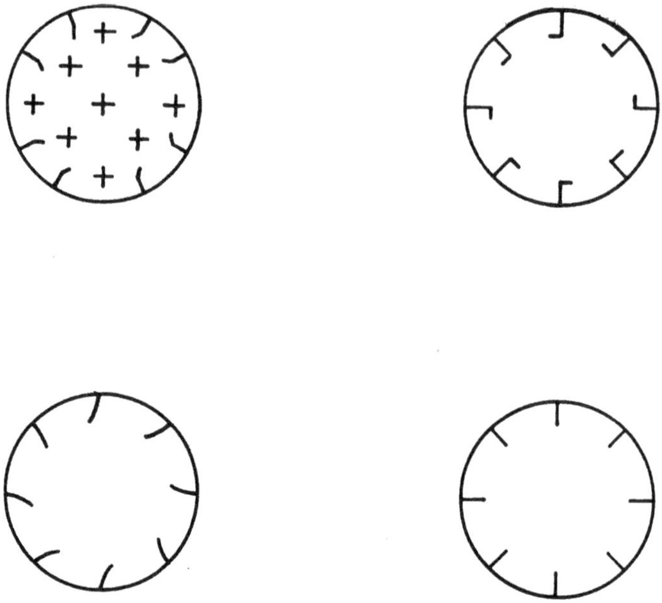

Figure 6.4 Possible direct-heat rotary dryer lifting flight arrangements.

Length The length of a direct-heat rotary dryer is usually five to eight times its diameter. The factor is two to four for Roto-Louvre dryers.

Slope The slope of the cylinder from feed end to product end is usually between 0 and -5°. However, positive slopes occur as well, these being used for cocurrent drying of relatively light materials.

Peripheral Speed Speed values between 0.1 and 0.5 $m{\cdot}s^{-1}$ are used, with speeds of 0.35 to 0.4 $m{\cdot}s^{-1}$ being quite common. Some larger dryers rotate with speeds between 0.5 and 1 $m{\cdot}s^{-1}$.

Number of Flights The flight count per circle is $2.4{\cdot}D$ to $3.0{\cdot}D$ (D in feet).

Flight Depth $D/12 - D/8$.

In rotary dryers, dust separation is often accomplished by cyclones. However, baghouses, wet scrubbers, and electrostatic separators are also used. Figures 6.5 to and 6.7 illustrate several aspects of rotary dryers and rotary dryer operation.

Testing on a small-scale direct-heat rotary dryer will, for a certain throughput, determine (1) the gas inlet and outlet temperatures, (2) the feed- and product-moisture contents, (3) the permissible superficial gas velocity, and (4) the residence time required. The diameter of the small-scale dryer should not be less than 0.5 m.

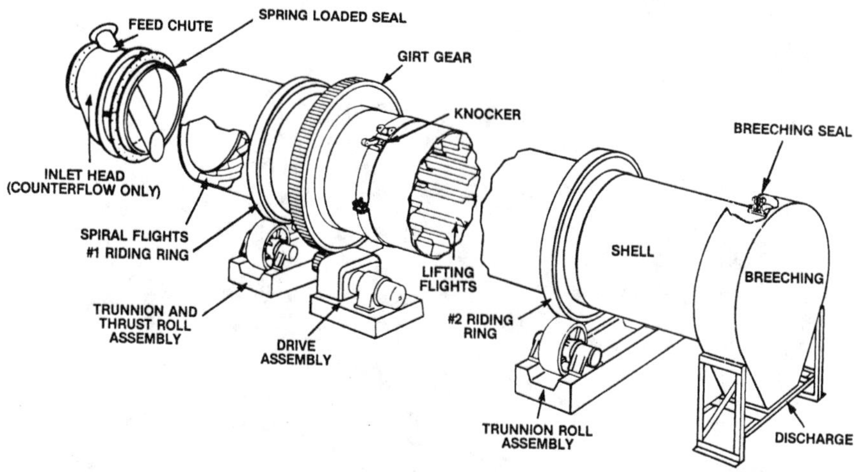

Figure 6.5 Component arrangements of a countercurrent direct-heat rotary dryer. (Courtesy of Alstom Power Inc., Warrenville, IL.)

Scaling up occurs on the basis of dryer cross-sectional area with the *L/D* ratio being kept constant. The background is explained in Section 6.2.

6.2 DESIGN METHODS

In this section we deal with approximate design methods not specifically requiring test data. However, the availability of test data will produce more reliable designs. The methods outlined in this section can also be used for the interpretation of test data.

The *mass balance* regarding the product is made first (see Chapter 4).

The *drying gas-exit temperature* must be known to calculate the heat transferred to the product flow, Q_{tot_1}. The proposed relationship between TA_{in} and TA_{out} is

$$TA_{\text{out}} = 0.05 \cdot TA_{\text{in}} + 64.5\ ^{\circ}\text{C} \tag{6.1}$$

Figure 6.8 contains a graphical presentation of plant measurements; the data were obtained from the literature [2, 3] and observed for industrial dryers.

The data for counter and parallel airflow (11 points; one point was measured for four different dryers) were analyzed statistically. The relationship $TA_{\text{out}} = 0.0495 \cdot TA_{\text{in}} + 63.76\ ^{\circ}\text{C}$ holds (linear regression), with the correlation coefficient being 0.916. Both constants are chosen slightly higher. Note that accepting a TA_{out} value by employing equation (6.1) can, in low-temperature drying, lead to discrepancies. *Low-temperature drying* is defined as drying with an air-inlet temperature approximately between 40 and 100°C. For example, the relation predicts

Figure 6.6 Direct-heat rotary dryer during manufacture. (Courtesy of Swenson Technology, Inc., Monee, IL.)

Figure 6.7 Direct-heat rotary dryer installed in the user's plant. (Courtesy of Swenson Technology, Inc., Monee, IL.)

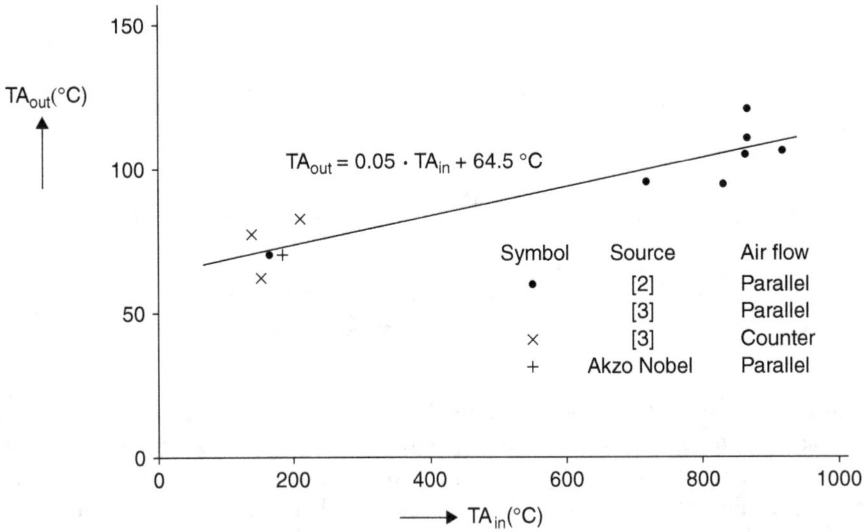

Figure 6.8 Relationship between air-inlet and air-outlet temperatures for direct-heat rotary dryers.

$TA_{out} = 69.5°C$ when $TA_{in} = 100°C$. The useful temperature drop is $100 - 69.5 = 30.5$ K. If, in actual fact, $TA_{out} = 45°C$, when $TA_{in} = 100°C$, the useful temperature drop is considerably larger: 55 K. The useful temperature drop is important for the dryer design.

A different approach is noted in *Perry's Handbook* [2]. The direct-heat rotary dryer is treated as a heat exchanger having a constant wall temperature T_W. The latter temperature equals the adiabatic saturation temperature. The general design equation is

$$Q_{tot_1} = U{\cdot}A(\Delta T)_m \qquad \text{kJ·h}^{-1}$$

Q_{tot_1}, the net heat, is supplied by the air that cools down:

$$Q_{tot_1} = \frac{\pi \cdot D^2{\cdot}G{\cdot}c_p(TA_{in} - TA_{out})}{4{\cdot}1.25} \qquad \text{kJ·h}^{-1}$$

G is the specific dry airmass flow in kg·m^{-2}·h^{-1}. Airmass velocities in rotary dryers usually range from 0.5 to 5.0 kg·m^{-2}·s^{-1} [2]. Experiments do show what should be the air rate with regard to dusting. An air rate of 1.4 kg·m^{-2}·s^{-1} (5040 kg·m^{-2}·h^{-1}) can usually be used safely with 420-μm solids [2]. That value corresponds to approximately 1.5 m·s^{-1} at 100°C and atmospheric pressure.

The factor 1.25 takes account of the heat losses due to convection, conduction, and radiation (see Chapter 4). This factor can be calculated as an average value from

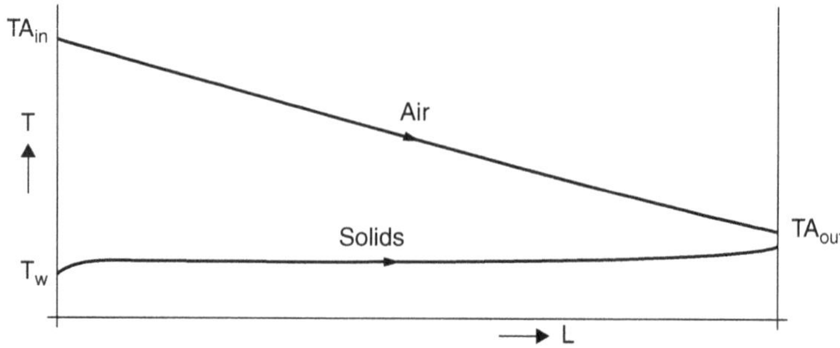

Figure 6.9 Temperature profiles of a cocurrent direct-heat rotary dryer.

measurements reported in the literature [2, 3]. These sources describe the steady-state heat consumption for the drying operation [4, 5]. However, literature data and plant measurements concerning very large industrial dryers (3 to 5 m in diameter) indicate a value of 1.1 instead of 1.25, which is probably due to the attention paid to important industrial drying operations. Furthermore, the m^2/m^3 ratio is relatively small for large dryers. On the other hand, performance data for small dryers (e.g., 1 m in diameter) in the literature point to values of up to 1.4.

An additional load factor is needed to allow for the extra heat losses in the long run; 1.25 is recommended. $(\Delta T)_m$ is the logarithmic mean temperature difference:

$$(\Delta T)_m = \frac{(TA_{\text{in}} - T_W) - (TA_{\text{out}} - T_W)}{\ln[(TA_{\text{in}} - T_W)/(TA_{\text{out}} - T_W)]} = \frac{TA_{\text{in}} - TA_{\text{out}}}{\ln[(TA_{\text{in}} - T_W)/(TA_{\text{out}} - T_W)]} \quad \text{K.}$$

(See Fig. 6.9.) Note that the logarithmic mean temperature difference is obtained by taking the differences between the drying gas temperatures and the adiabatic saturation temperature at the inlet and exit ends of the dryer shell. Use of material temperatures will yield conservative results, particularly when the dry solids are superheated after drying. The earlier equations lead readily to

$$\ln \frac{TA_{\text{in}} - T_W}{TA_{\text{out}} - T_W} = \frac{4 \cdot 1.25 \cdot U \cdot A}{\pi \cdot D^2 \cdot G \cdot c_p} = \frac{4 \cdot 1.25 \cdot U \cdot S \cdot L}{\pi \cdot D^2 \cdot G \cdot c_p} = \frac{L}{L_e} = N_t \tag{6.2}$$

A is the area offered by the product in the dryer and S is the area offered by the product in the dryer per meter of dryer length. Equation (6.2) expresses the fact that the temperature difference between the gas and the solid to be dried decreases with a factor e (= 2.718) over a length

$$L_e = \frac{\pi \cdot D^2 \cdot G \cdot c_p}{4 \cdot 1.25 \cdot U \cdot S} \quad \text{m}$$

This length, which is characteristic of the heat transfer process, is known as 1 HTU, the *height of a transfer unit*. In this case (constant wall temperature), the height of a transfer unit is the height (or the length) that it takes to reduce the driving force by a factor e. The ratio L/L_e is termed the *number of transfer units* (NTU).

For direct-heat rotary dryers, N_t is in general between 1.5 and 2.5 [2]. Assuming that $N_t = 2$, it is possible, using equation (6.2), to calculate TA_{out}. The two methods are compared in two examples.

Example 6.1

$TA_{\text{in}} = 145°\text{C}$
Ambient air temperature: 10°C
Relative humidity of the atmosphere: 60%

Heating is indirect by means of steam.

Method 1

$$TA_{\text{out}} = 0.05 \cdot 145 + 64.5 = 71.8°\text{C}$$

Method 2. Sixty percent relative humidity at 10°C indicates 0.005 kg of water vapor per kilogram of dry air. The adiabatic saturation temperature of air at 145°C with 0.005 kg of water vapor per kilogram of dry air is 39°C (see the humidity chart).

$$N_t = \ln \frac{TA_{\text{in}} - T_W}{TA_{\text{out}} - T_W}$$

N_t	1.5	2.0	2.5
TA_{out} (°C)	62.7	53.3	47.7

Example 6.2

$TA_{\text{in}} = 300°\text{C}$
Ambient air temperature: 10°C
Relative humidity of the atmosphere: 60%

Heating is direct by means of gas combustion.

Method 1

$$TA_{\text{out}} = 0.05 \cdot 300 + 64.5 = 79.5°\text{C}$$

Method 2 The combustion of natural gas and the mixing of the hot gases with atmospheric air until 300°C is obtained results in drying gas with approximately 0.02 kg of water vapor per kilogram of dry air. The corresponding adiabatic saturation temperature is 56°C.

N_t	1.5	2.0	2.5
TA_{out} (°C)	110.4	89.0	76.0

There is a tendency for method 1 to predict a relatively high value for low drying-air temperatures. The predictions of both methods are more or less in line for higher drying-air temperatures.

It is assumed that the product-exit temperature equals TA_{out}, which is about right for cocurrent operation but may not be entirely correct for countercurrent operation. However, the effect on Q_{tot_1} is limited.

Assessment of the *dryer diameter* is the next step. The basic equation is

$$1.25 \cdot Q_{tot_1} = \frac{\pi}{4} \cdot D^2 \cdot G \cdot c_p (TA_{in} - TA_{out}) \qquad \text{kJ} \cdot \text{h}^{-1}$$

Q_{tot_1} is the net heat (see Chapter 4), G is chosen as a function of the particle size, c_p is approximately 1.05 kJ·kg^{-1}·K^{-1}, and calculation of the temperature difference has been discussed. It is thus possible to arrive at D, the dryer diameter.

Calculation of the *dryer length* is now attempted. The basic equation for the calculation of heat exchangers is

$$Q_{tot_1} = U \cdot A(\Delta T)_m \qquad \text{kJ} \cdot \text{h}^{-1}$$

The basic equation for the calculation of heat exchangers is replaced by

$$Q_{tot_1} = U \cdot a \cdot V(\Delta T)_m \qquad \text{kJ} \cdot \text{h}^{-1} \tag{6.3}$$

where a is the specific area offered by the product in m^2·m^{-3} (m^3 here is m^3 of dryer volume), and $U \cdot a$ is termed the volumetric heat-transfer coefficient in kJ·h^{-1}·m^{-3}·K^{-1}. The background is that it is difficult to assess U and A separately. Measurements enable the calculation of $U \cdot a$ values and correlation of the data with operational conditions. This approach can also be found with gas absorption, for example, where $k_l \cdot a$ values are calculated. Calculating $k_l \cdot a$ values is as far as we can get, because it is neither possible to arrive at k_l values (mass transfer coefficient values) nor to arrive at a values. Data for evaluating $U \cdot a$ have been developed [6–8]. The relationships can be reduced to

$$U \cdot a = K \cdot \frac{G^n}{D} \qquad \text{kJ} \cdot \text{h}^{-1} \cdot \text{m}^{-3} \cdot \text{K}^{-1} \tag{6.4}$$

G is expressed in kg·m^{-2}·h^{-1}.

The data were compared [9] and it was found that all experimental evidence can be described by $n = 0.67$. Note that the power of G equals the power of the Reynolds number in the usual Nusselt correlations for heat transfer. The volumetric heat transfer coefficient is inversely proportional to the dryer diameter. A possible explanation is as follows. On scaling up convective rotary dryers, the circumferential velocity is kept more or less constant. A value of 0.35 $m{\cdot}s^{-1}$ is normal. The consequence is that, on scaling up, the dryer rotational speed in min^{-1} is reduced. For a circumferential velocity of 0.35 $m{\cdot}s^{-1}$, the rotational speed is 6.6 min^{-1} for a 1-m-diameter dryer and 2.2 min^{-1} for a 3-m dryer. A higher rotational speed will result in more efficient heat transfer. Each particle processed in a rotary dryer is exposed to the action of the drying gas, for example, 1 s, 6.6 min^{-1} in a 1-m dryer and 2.2 min^{-1} in a 3-m dryer. The comparison for equation (6.4) [9] comprised (1) experimental and commercial dryers, (2) countercurrent and cocurrent operation, (3) the drying and heating of different materials, (4) drying air temperatures up to at least 300°C, and (5) a maximum drying airflow of 4.4 $kg{\cdot}m^{-2}{\cdot}s^{-1}$; however, 1 to 1.5 were more typical values.

K is a variable and depends on:

1. The physical properties of the material dried
2. The flight count
3. The flight depth
4. The flight load
5. The rotational speed

Perry's Handbook [2] recommends $K = 3.5$ when using SI units. The AICE [10] recommends $3.75 \le K \le 5.25$, with high $K = 5.25$ for dryers operating at a high fillage and having a full set of lifting flights.

A few additional observations should be made concerning dryer length. Q_{tot_1} is proportional to the dryer capacity, which is proportional to the dryer cross-sectional area and hence to D^2. The implicit assumption is then that the drying gas velocity and temperature are kept constant. On substituting equation (6.4) into equation (6.3), it follows that Q_{tot_1} is proportional to $D{\cdot}L$. Both observations can be true only if D is proportional to L. The consequence is that dryer capacity scales with $V^{2/3}$. This is also found in practice, where the dryer length is usually between five and eight times the diameter. The fact that, on scaling up, the circumferential velocity of the dryer is kept approximately constant provides an explanation. Further data concerning heat transfer in direct-heat rotary dryers have been reviewed by Friedman and Marshall [12]. Finally, equations (6.3) and (6.4) enable calculation of the dryer volume. The dryer length follows from the dryer volume and the dryer diameter.

The *residence time* in a semitechnical rotary dryer was investigated [12]. A relationship based on this work is provided [2]:

$$\tau = \frac{0.23{\cdot}L}{S{\cdot}N^{0.9}{\cdot}D} \pm \frac{2{\cdot}B{\cdot}L{\cdot}G}{F} \qquad \text{min} \tag{6.5}$$

$$B = 5{\cdot}\bar{d}_p^{-0.5}$$

AICE [10] recommends the same relationship. F is the specific dry product mass flow in $kg{\cdot}m^{-2}{\cdot}h^{-1}$. Further data concerning the residence times of both solid and gas in direct-heat rotary dryers are reviewed in *Perry's Handbook* [11].

The first part of the right-hand side of equation (6.5) is the more important contribution; the second part concerns a correction for the air conveyance. Thus, basically, the residence time is proportional to the dryer length. τ is approximately inversely proportional to the peripheral speed, because $N^{0.9}{\cdot}D$ is about equal to $N{\cdot}D$. Furthermore, the residence time is inversely proportional to the slope. The second part of the right-hand side of equation (6.5) indicates that it is more important for fine particles than for coarse particles. According to equation (6.5), τ is not dependent on the throughput.

If, on scaling up, the circumferential velocity, the slope, and the specific air and product mass flows (in $kg{\cdot}m^{-2}{\cdot}h^{-1}$) are kept constant, the residence time is approximately proportional to the dryer length. On scaling a dryer diameter up from 1 m to 3 m, the residence time increases from 10 min to approximately 30 min. However, the number of times that an individual particle is lifted stays constant. The explanation is that, on keeping the circumferential velocity constant, the rotational speed of a 3-m dryer in min^{-1} is three times smaller than the rotational speed of a 1-m dryer. Also see the text to elucidate equation (6.4).

The plus sign in equation (6.5) refers to countercurrent flow and the negative sign to cocurrent operation. Equation (6.5) can be used for air velocities not appreciably exceeding 1 $m{\cdot}s^{-1}$.

Based on a review of the literature data [2, 3], private communications [4], and measurements at the plant site, an empirical formula for the *motive power consumption* of a dryer with lifting flights is suggested:

$$\text{Power}_{\text{rot}} = 0.3{\cdot}\frac{\pi}{4}{\cdot}D^2{\cdot}L \qquad \text{kW} \tag{6.6}$$

However, lower values are found for slowly rotating dryers (e.g., sugarbeet pulp dryers).

AIChE [10] also recommends a relationship for the motive power consumption of a dryer with lifting flights:

$$\text{Power}_{\text{rot}} = \frac{N[34.3{\cdot}D{\cdot}w + 1.39(D + 0.6)W + 0.73{\cdot}W]}{134{,}040} \qquad \text{kW} \tag{6.7}$$

The expression $D + 0.6$ is meant to estimate the diameter of the cylinder riding ring; w and W are, respectively, the holdup and the holdup plus dryer rotating weight, both in kilograms. Furthermore, it was assumed that the flights are sufficient to shower all the material in the cylinder.

Example 6.3 A direct-heat rotary dryer for a mineral is to be designed.

Product

General: 15 $t \cdot h^{-1}$ of mineral from the dryer
TP_{in}: 15°C
d_{50}: 600 μm
A_1: 12% water by weight (as fed)
A_2: 3% water by weight (as discharged)
c_s: 0.8 $kJ \cdot kg^{-1} \cdot K^{-1}$
Solubility solid: nil
ρ_b: 1400 $kg \cdot m^{-3}$

Process

Ambient temperature: 10°C
Ambient relative humidity: 50%
TA_{in}: 700°C by burning natural gas
Airflow: cocurrent
G: 4000 $kg \cdot m^{-2} \cdot h^{-1}$
c_p: 1.05 $kJ \cdot kg^{-1} \cdot K^{-1}$

Natural gas data

Composition: 85% by volume of methane, 15% by volume of nitrogen
Methane heat of combustion: 804 $kJ \cdot mol^{-1}$
ρ_g: 0.80 $kg \cdot nm^{-3}$

See Sections 4.13 and 4.15 for data on water, steam, and air.

Pilot-plant drying tests were not carried out. It is assumed that the dried-product temperature equals the gas exit temperature. The carry-over is assumed to be negligible.

Mass balance (kg·h^{-1})

	In	Out
Solids	14,550	14,550
Water	+ 1,984	+ 450
	16,534	15,000

$$\text{Evap} = 1984 - 450 = 1534$$

Net heat $(kJ{\cdot}h^{-1})$

$$TA_{\text{out}} = 0.05{\cdot}700 + 64.5 = 99.5°C$$

$$\begin{aligned} Q_1 &= 1534(2500 + 1.9{\cdot}99.5 - 4.2{\cdot}15) = && 4{,}028{,}361 \\ Q_2 &= 14{,}550{\cdot}0.8(99.5 - 15) = && 983{,}580 \\ Q_3 &= 450{\cdot}4.2(99.5 - 15) = && +\ \underline{159{,}705} \\ Q_{\text{tot}_1} &= && 5{,}171{,}646 \end{aligned}$$

$$Q_{\text{tot}_2} = 1.25 \cdot \frac{700 - 10}{700 - 99.5} \cdot 5{,}171{,}646 = 7{,}428{,}051$$

Q_{tot_2} is the heat to be supplied in the combustion chamber by the combustion of natural gas. See Section 6.2 for a discussion of the heat loss factor, 1.25.

$$\text{kJ per kg of evaporated water: } \frac{7{,}428{,}051}{1534} = 4842$$

Due to startup, shutdown, and cleaning, for example, the long-term consumption figure is probably a factor of 1.25 higher: 1.25·4842 = 6053 kJ per kilogram of evaporated water.

Dew-point exhaust gas The amount of heat transferred in the combustion chamber is 7,428,051 $kJ{\cdot}h^{-1}$. The methane flow required is 7,428,051/804 = 9239 $mol{\cdot}h^{-1}$. On using the data of Section 4.16, on realizing that the natural gas composition has been simplified to 85% by volume of methane and 15% by volume of nitrogen, and using a molar volume of 22.4 nL, Table E6.3 can be calculated.

Table E6.3

	Stoichiometric Natural-Gas Combustion ($kg{\cdot}h^{-1}$)	
	In	Out
Natural gas		
Methane	147.8	0.0
Nitrogen	45.7	45.7
Dry air		
Oxygen	591.3	0.0
Nitrogen	2069.5	2069.5
Water vapor		
In air	10.6	10.6
By combustion	0.0	332.6
Carbon dioxide	+ 0.0	+ 406.5
	2864.9	2864.9

The natural gas consumption is

$$\frac{9239}{0.85} \cdot 0.0224 = 243.5 \text{ nm}^3 \cdot \text{h}^{-1}$$

The total gas mass flow from the combustion chamber is

$$\frac{7{,}428{,}051}{1.05(700 - 10)} = 10{,}252.7 \text{ kg} \cdot \text{h}^{-1}$$

The specific heat of the hot gas has been taken as 1.05 $\text{kJ} \cdot \text{kg}^{-1} \cdot \text{K}^{-1}$. The secondary airmass flow is 10,252.7 – 2864.9 = 7387.8 $\text{kg} \cdot \text{h}^{-1}$ (composed of 7358.4 $\text{kg} \cdot \text{h}^{-1}$ of dry air and 29.4 $\text{kg} \cdot \text{h}^{-1}$ of water vapor). Ingress air into the dryer amounts to 10% of the total amount of gases from the combustion chamber: $0.1 \cdot 10{,}252.7 = 1025.3$ $\text{kg} \cdot \text{h}^{-1}$ (composed of 1021.2 $\text{kg} \cdot \text{h}^{-1}$ of dry air and 4.1 $\text{kg} \cdot \text{h}^{-1}$ of water vapor). It is now possible to make the following table ($\text{kg} \cdot \text{h}^{-1}$):

Process Flow	Dry	Water	Total
Combustion	2,521.7	343.2	2,864.9
Secondary air	7,358.4	29.4	7,387.8
Ingress air	1,021.2	4.1	1,025.3
Evaporation	+ 0.0	+ 1,534.0	+ 1,534.0
	10,901.3	1,910.7	12,812.0

$$\text{Water content: } \frac{1910.7}{10{,}901.3} = 0.175 \text{ kg per kg of dry air}$$

The exhaust gases must be cooled from 99.5°C to 62°C in order to get condensation. This provides a satisfactory margin.

Sizing drying-gas preparation unit

$$Q_{\text{tot}_2} = 7{,}428{,}051 \text{ kJ} \cdot \text{h}^{-1}$$

Buy a combustion chamber having a capacity of 2.5 MW (21% spare capacity). The gas mass flow from the combustion chamber is 10,252.7 $\text{kg} \cdot \text{h}^{-1}$. The gas mass flow to the combustion chamber is approximately equal to this gas mass flow.

$$R_A = \frac{355}{273 + 10} = 1.25 \text{ kg} \cdot \text{m}^{-3}$$

$$\text{Airflow}_1 = \frac{10{,}252.7}{1.25} = 8202.2 \text{ m}^3 \cdot \text{h}^{-1}$$

$$\text{Power}_{G1} = \frac{8202.2 \cdot 2500}{3600 \cdot 1000 \cdot 0.5} = 11.4 \text{ kW}$$

2500: pressure loss of the combustion chamber in $\text{N} \cdot \text{m}^{-2}$

Use a fan with a motor of 15 kW.

Dryer diameter

$$\frac{\pi}{4}\cdot D^2\cdot 4000 = 10{,}252.7 \rightarrow D = 1.81\ \text{m}$$

Choose a dryer diameter of 2 m. G then becomes 3276 $kg\cdot m^{-2}\cdot h^{-1}$.

Dryer length Take $K = 5.3$.

$$U\cdot a = 5.3\cdot\frac{3276^{0.67}}{2.00} = 600.5\ \text{kJ}\cdot\text{m}^{-3}\cdot\text{h}^{-1}\cdot\text{K}^{-1}$$

The drying gas is obtained by the combustion of natural gas. The adiabatic saturation temperature of this drying gas is 73°C. This can be calculated by using the method discussed in Section 4.1. In this case, the gas cooling down also contains water vapor, providing sensible heat.

$$(\Delta T)_m = \frac{700 - 99.5}{\ln (700 - 73)/(99.5 - 73)} = 189.8\ \text{K}$$

$$V = \frac{Q_{tot_1}}{U\cdot a(\Delta T)_m} = \frac{5{,}171{,}646}{600.5\cdot 189.8} = 45.4\ \text{m}^3$$

$$L = 14.5\ \text{m}$$

The L/D ratio is 7.3, which is a normal value.

$$N_t = \ln\frac{700 - 73}{99.5 - 73} = 3.2$$

Motive power consumption

$$\text{Power}_{\text{rot}} = 0.3\cdot\frac{\pi}{4}\cdot D^2\cdot L = 13.7\ \text{kW}$$

Take a 25-kW motor for the rotation.

Residence time

$$B = 5\cdot\bar{d}_p^{-0.5} = 5\cdot 600^{-0.5} = 0.204$$

$$\text{N} = 4\ \text{min}^{-1}\ (\text{peripheral speed } 0.42\ \text{m}\cdot\text{s}^{-1})$$

$$\text{S} = 0.01\ (\text{slope})$$

$$\tau = \frac{0.23\cdot 14.5}{0.01\cdot 4^{0.9}\cdot 2.00} - \frac{2\cdot 0.204\cdot 14.5\cdot 10{,}252.7}{15{,}000} = 47.9 - 4.0 = 43.9\ \text{min}$$

In the second term, the total gas mass flow from the combustion chamber and the product flow have been taken into account.

Volumetric loading

$$\text{Solids volume in dryer: } \frac{43.9}{60} \cdot \frac{15{,}000}{1400} = 7.84\,\text{m}^3$$

$$\text{Volume \%: } \frac{7.84 \cdot 4}{\pi \cdot 2.00^2 \cdot 14.5} \cdot 100 = 17.2$$

Sizing exhaust gas unit The gas mass flow leaving the dryer, excluding water vapor, is 12,812.0 – 1910.7 = 10,901.3 kg·h^{-1}. The composition is approximately equal to the composition of dry air.

$$R_A = \frac{355}{273 + 99.5} = 0.95\,\text{kg}\cdot\text{m}^{-3}$$

$$\text{Airflow}_2 = \frac{10{,}901.3}{0.95} = 11{,}475.1\,\text{m}^3\cdot\text{h}^{-1}$$

Water vapor flow leaving the dryer: 1910.7 kg·h^{-1}

$$R_W = \frac{220}{273 + 99.5} = 0.59\,\text{kg}\cdot\text{m}^{-3}$$

$$WA_{\text{flow}} = \frac{1910.7}{0.59} = 3238.5\,\text{m}^3\cdot\text{h}^{-1}$$

$$\text{Gasflow}_2 = 11{,}475.1 + 3238.5 = 14{,}713.6\,\text{m}^3\cdot\text{h}^{-1}$$

$$\text{Power}_{\text{G2}} = \frac{14{,}713.6 \cdot 1500}{3600 \cdot 1000 \cdot 0.5} = 12.3\,\text{kW}$$

1500: pressure loss of the cyclone in N·m^{-2}

Use a fan with a motor of 15 kW.

Survey

Dryer diameter: 2.00 m

Dryer length: 14.5 m

Capacity of the combustion chamber: 2.5 MW

Natural gas consumption (load factor 1.25):

$$\frac{1.25 \cdot 243.5}{1.534} = 198.4\ \text{nm}^3 \text{per t of evaporated water}$$

The load factor was mentioned in Section 6.2. The long-term losses are, on using the load factor, also taken into account.

Electricity consumption (load factor 1.25):

$$\frac{1.25(11.4 + 12.3 + 13.7 + 10.0)}{1.534} = 38.6\,\text{kWh per t of evaporated water}$$

REFERENCES

[1] Nonhebel, G., Moss, A.H.H. (1971). *Drying of Solids in the Chemical Industry*, Butterworth, London, p. 181.

[2] Perry, R.H., Green, D.W., Maloney, J.O. (1997). *Perry's Chemical Engineers' Handbook*, 7th ed., McGraw-Hill, New York, pp. 12-52–12-64.

[3] Williams-Gardner, A. (1971). *Industrial Drying*, Leonard Hill, London, p. 149.

[4] Swenson (1985, 1986). Private communications.

[5] Buell (1985). Private communication.

[6] Miller, C.O., Smith, B.A., Schuette, W.H. (1924). Factors influencing the operation of rotary dryers. *Transactions of the American Institute of Chemical Engineers*, *38*, 841–864.

[7] Friedman, S.J., Marshall, W.R., Jr. (1949). Studies in rotary drying: II, Heat and mass transfer. *Chemical Engineering Progress, 45*, 573–588.

[8] Seaman, W.C., Mitchell, T.R., Jr. (1954). Analysis of rotary dryer and cooler performance. *Chemical Engineering Progress*, *50*, 467–475.

[9] McCormick, P.Y. (1962). Gas velocity effects on heat transfer in direct heat rotary dryers. *Chemical Engineering Progress*, *58*, 57–61.

[10] American Institute of Chemical Engineers (2006). *Equipment Testing Procedure: Continuous Direct-Heat Rotary Dryers*, AIChE, New York.

[11] Perry, R.H., Green, D.W. (2008). *Perry's Chemical Engineers' Handbook*, 8th ed., McGraw-Hill, New York, pp. 12-72–12-77.

[12] Friedman, S.J., Marshall, W.R., Jr. (1949). Studies in rotary drying: I. Holdup and dusting. *Chemical Engineering Progress*, *45*, 482–493.

7

FLASH DRYING

A general description of flash drying is provided in Section 7.1, followed by a discussion of the screening design methods in Section 7.2. Section 7.3 contains information on short-time convection drying (several seconds). The chapter concludes with an application of the design methods in Section 7.4.

7.1 GENERAL DESCRIPTION

Basically, a flash dryer consists of a long tube or duct carrying a gas at high velocity, a fan to propel the gas, a suitable feeder for addition and dispersion of particulate solids in the gas stream, and a cyclone collector or other separation equipment for final recovery of solids from the gas (see Fig. 7.1). Drying must be completed instantaneously in all applications; internal diffusion of moisture must not be limiting, and particle sizes must be small enough so that the thermal conductivity of the solids does not control during product heating and cooling. Flash dryers are normally fed by a flooded screw and a disperser that propels the wet feed into the drying tube. However, the solids feeder may be of any type; venturi sections, high-speed grinders, and dispersion mills are employed. As to the gas–solid separation, cyclones are preferred for low investment. The cyclone may be followed by a bag collector or wet scrubber if maximum dust recovery is required.

Generally, cooling and drying cannot be combined in one dryer. Consequently, the dried product is often cooled in a separate pneumatic transport system; however, other types of coolers are available. Flash-drying equipment is simple, occupying

Drying in the Process Industry, First Edition. C.M. van 't Land.
© 2012 John Wiley & Sons, Inc. Published 2012 by John Wiley & Sons, Inc.

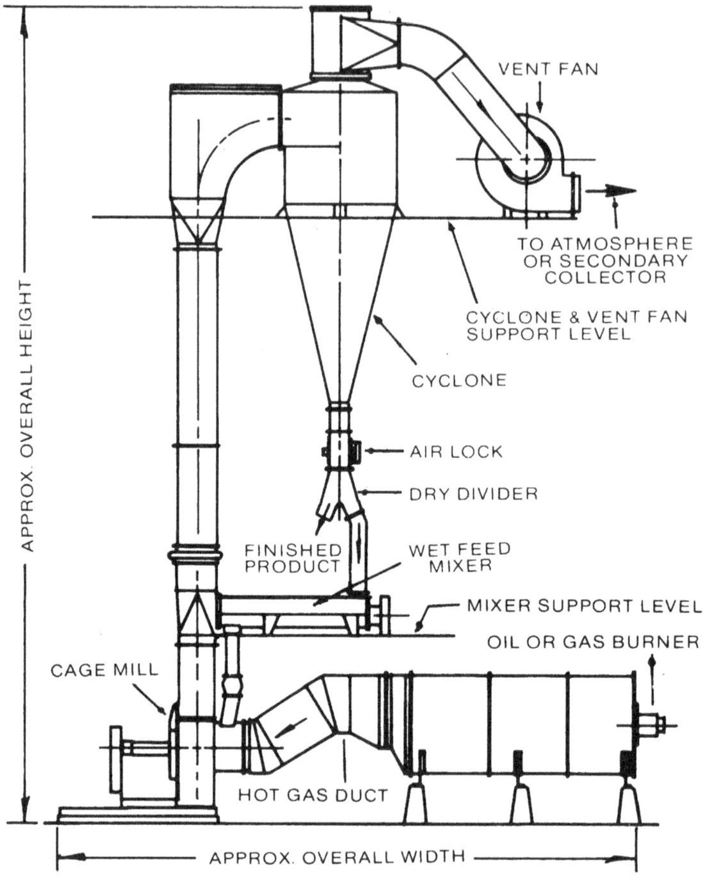

Figure 7.1 Single-stage pneumatic-conveyor dryer. (Courtesy of Alstom Power, Inc., Warrenville, IL.)

little space and having few moving parts. It is often feasible to combine drying and vertical transport. Typical gas exit velocities are in the range 10 to 30 $m{\cdot}s^{-1}$. Hence, the flash-drying gas velocities exceed those in rotary and fluid-bed dryers by a factor of approximately 10.

The maximum particle size that can be dried is 1 to 2 mm because larger particles are not entrained by air. Unlike rotary dryers, the flash dryer is susceptible to overloading since the material cannot be transported at high feed rates. Consequently, flash dryers require more attention than rotary dryers.

The high gas velocity can result in abrasion or dust formation, which, in turn, leads to increased maintenance costs. This factor must be considered since it is often stated that maintenance costs are lower for flash dryers than for rotary dryers. Although this finding is, on the surface, correct (because rotary dryers do have an important rotating element), it is not always borne out in practice. Generally, flash dryers may require up to 5% of the investment annually for maintenance purposes.

Dust formation is promoted by high gas velocity when the smaller particles are eroded from the mother particles. The short residence time is another source of dust formation. This applies when the moisture is a solution of the solid in the liquid and the evaporation proceeds at such a rate that the solid crystallizing from the liquid does not attach itself to the mother crystals but instead forms nuclei. Not only is dust formation a nuisance, it can also be a hazard. However, an attractive aspect concerning process safety is the minimal holdup of a flash dryer.

The composition of the feed must not cause sticking in the feed section. If the feed is not entirely suitable, it can be modified either by backmixing it with the product or by using a cage mill to disintegrate it while the drying gas is passed through the mill (approximate maximum air temperature 750°C). Furthermore, these two options can be combined. However, it must be realized that recycling material entails its additional thermal treatment. Another option is the installation of a slinger below the feed point so that falling particles are disintegrated.

It is possible to arrange several flash dryers in series to provide additional residence time or to dry at different conditions. In addition, it is also possible to arrange in series a flash dryer and a dryer providing a longer residence time (e.g., a fluid-bed dryer; this system is used in some PVC installations).

Usually, the air used for drying is sufficient for conveying; however, it is good practice not to exceed a solids/airmass ratio of 1. There are some exceptions to this generalization: for example, in the drying of relatively coarse material (500-μm particles) containing little moisture (e.g., 2 to 3% water by weight) at elevated temperature (ca. 350°C).

The thermal degradation of organic materials is dependent on three factors: (1) time, (2) temperature, and (3) concentration. A high air-inlet temperature can sometimes be chosen because most of the evaporation occurs while the particles are at the wet-bulb temperature. Another reason, specific to flash dryers, is the short drying time. The air-outlet temperature usually exceeds that of the product by 10 to 30 K.

It is often necessary to raise the temperature of the solid to a specific value to be certain of a final moisture content. Again, the particles must be small enough so that the thermal conductivity of the solids does not control during the heating operation. Mass transfer is often the rate-determining step in the final stage of the drying process.

There is usually a slight negative pressure at the feed point for two reasons: (1) the underpressure prevents incrustations in the feed section; and (2) the underpressure prevents dust emissions at the feed point. The drying system may contain only one fan, typically located between the cyclones and the wet scrubber. Alternatively, the introduction of both primary and secondary air into the combustion chamber may require additional fans.

The diameter of the drying tube can reach 1 m, and the length can vary between 10 and 30 m (see Fig. 7.2, which illustrates a conventional flash-dryer

Figure 7.2 Flash-dryer installation for the drying of potato starch. (Courtesy of Bepex International LLC, Minneapolis, MN.)

installation). An effective way to control the drying operation is via the fuel or the steam flow on the basis of the air-outlet temperature. The dryer must be able to cope with varying feed flows because there is usually no buffer between the liquid–solid separating system and the dryer; holdup of wet solids cannot be controlled easily. It is a good practice not to vary the airflow because this affects the transport function. Furthermore, it is important that the dryer be internally smooth to avoid incrustations.

The ring dryer is a development of flash drying technology to increase its applicability. Basically, the gas and the solids flow through a ring with a makeup and a purge. The ring dryer is equipped with an internal classifier which allows fine particles, which dry quickly, to leave, while larger particles, which dry relatively slowly, have an extended residence time within the system. Furthermore, the combination of the classifier with an internal mill can allow simultaneous grinding and drying with control of product particle size and moisture. The ring can have either a vertical or a horizontal configuration.

7.2 DESIGN METHODS

The equation recommended for calculation of the air-outlet temperature is

$$TA_{\text{out}} = 0.1875 \cdot TA_{\text{in}} + 35\ ^{\circ}\text{C} \tag{7.1}$$

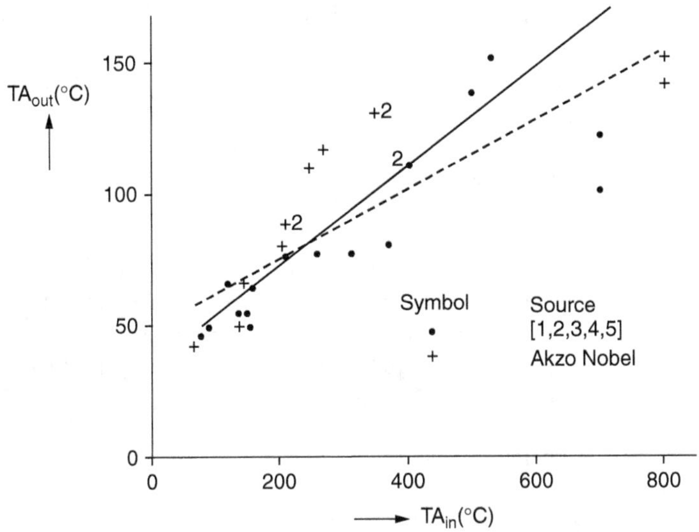

Figure 7.3 Relationship between air-inlet and air-outlet temperatures for flash dryers. Solid line: $TA_{out} = 0.1875 \cdot TA_{in} + 35\,^{\circ}C$; dashed line: $TA_{out} = 0.13 \cdot TA_{in} + 49.4\,^{\circ}C$.

Figure 7.3 plots this relationship, with 17 measured points taken from references 1 to 5 (dots) and 12 measured points observed for industrial dryers at Akzo Nobel (+signs). As can be seen in Figure 7.3, there is a considerable spread in the points measured. The corresponding data for rotary dryers are more homogeneous. The data were analyzed statistically. The relationship $TA_{out} = 0.13 \cdot TA_{in} + 49.4°C$ holds (linear regression), and the correlation coefficient is 0.82. Instead of this relationship, equation (7.1) is employed. This approach is adopted in order to be relatively safe. TA_{out} for flash dryers is often chosen to guarantee a desired product-moisture content; a large exit driving force is often mandatory to obtain the desired process result in a short time. A comparison of the two relationships:

	TA_{in}(°C)	
	200	600
$TA_{out} = 0.1875 \cdot TA_{in} + 35\,^{\circ}C$	72.5	147.5
$TA_{out} = 0.13 \cdot TA_{in} + 49.4\,^{\circ}C$	75.4	127.4

If a specific flash dryer has air- and product-outlet temperatures that deviate considerably from the predictions based on equation (7.1) (e.g., 20 to 30 K), and if the feed moisture content is low (2 to 4% by weight), the impact on the net heat (process-flow heat requirement) is considerable. This situation arises because, in these instances, the heat effect that is associated with the heating up of solids is important in relation to the evaporation heat requirement. The flash drying of sodium sulfate is an example.

The restricted applicability (see Fig. 7.3) of equation (7.1) at low-temperature drying as, for equation (6.1), pointed out in Section 6.2, must be mentioned here as well. Again, low-temperature drying is defined as drying with an air-inlet temperature approximately between 40 and 100°C. For example, the relation predicts that $TA_{\text{out}} = 53.75$°C when $TA_{\text{in}} = 100$°C. The useful temperature drop is $100 - 53.75 = 46.25$ K. If, in actual fact, $TA_{\text{out}} = 65$°C when $TA_{\text{in}} = 100$°C, the useful temperature drop is considerably smaller (i.e., 35 K). The useful temperature drop is important for dryer design.

A factor of 1.2 (coefficient d in Section 4.12) allows for the steady-state heat losses, which can be calculated as an average value from measurements reported in the literature [1,3,4]. The implicit assumption is that the primary energy consumption reported for the 17 dryers was measured over a relatively short period and does not reflect the long-term consumption. Assumptions:

1. The ambient temperature is 10°C.
2. $T_f = 20$°C.
3. $TP_{\text{out}} = TA_{\text{out}} - 30$ K.
4. Reported TA_{in} and TA_{out} values were used.

The arithmetic average of the ratio calculated consumption to consumption indicated is 1.0, with a standard deviation of 0.10. However, the aforementioned factor of 1.2 is an average. Higher values apply for relatively small evaporation loads (e.g., 1.4 for 100 to 200 kg of water evaporation per hour). For large evaporation loads (e.g., 4 to 6 t of water evaporation per hour), a value of 1.1 is often found. Addditional losses were mentioned in Section 4.9. They can be observed over relatively long periods (e.g., a year). It is possible to use a load factor. An analysis of the long-term primary-energy consumption of seven flash dryers in the process industry taught that the load factor varies between 1.25 and 1.75.

Tests in a flash dryer having a diameter of, for example, 5 cm should precede a design.

7.3 DRYING IN SECONDS

In this section we consider, by means of an example, the flash drying of an organic material consisting of spherical particles with a diameter of 250 μm and having an inlet moisture content of 10% by weight. It will be exhibited qualitatively that the drying of small particles in a flash dryer proceeds at a high rate.

Example 7.1 The basic equation for the design of heat exchangers, $Q_{\text{tot}_1} = U \cdot A(\Delta T)_m$, will be used. The calculation of U, the heat transfer coefficient, proceeds via an appropriate Nusselt correlation. The equation is used to calculate A, the area for heat transfer. The next step is the calculation of τ, the residence time. The air-inlet temperature is 225°C; the heat transfer is rate determining.

Basic data

Solid

Insoluble in water
250-μm spheres
$\rho_S = 1200\ \text{kg}\cdot\text{m}^{-3}$
$\lambda_S = 0.2\ \text{W}\cdot\text{m}^{-1}\cdot\text{K}^{-1}$
$c_s = 1.2\ \text{kJ}\cdot\text{kg}^{-1}\cdot\text{K}^{-1}$

Air

See Section 4.15.

$$\mu = \begin{cases} 2.18\cdot 10^{-5}\ \text{N}\cdot\text{s}\cdot\text{m}^{-2} & \text{at } 100^\circ\text{C} \\ 2.59\cdot 10^{-5}\ \text{N}\cdot\text{s}\cdot\text{m}^{-2} & \text{at } 200^\circ\text{C} \end{cases}$$

Temperature (°C)	$\lambda_S(\text{W}\cdot\text{m}^{-1}\cdot\text{K}^{-1})$
77	0.0300
127	0.0338
177	0.0373
227	0.0407

Water See Sections 4.13 and 4.15.

Operational data

Feed

20°C
$A_1 = 10\%$ by weight

Product

$2000\ \text{kg}\cdot\text{h}^{-1}$
60°C
$A_2 = 0.5\%$ by weight

Ambient air

10°C
60% relative humidity

Warm air

Indirect steam heating

225°C

$TA_{out} = 0.1875 \cdot 225 + 35 = 77.2°C$

Mass balance $(kg \cdot h^{-1})$

	In	Out
Solids	1990	1990
Water	+ 221	+ 10
	2211	2000

$$\text{Evap} = 221 - 10 = 211$$

Net heat $(kJ \cdot h^{-1})$

$$Q_1 = 211(2500 + 1.9 \cdot 77.2 - 4.2 \cdot 20) = 540{,}725$$
$$Q_2 = 1990 \cdot 1.2(60 - 20) = 95{,}520$$
$$Q_3 = 10 \cdot 4.2(60 - 20) = +\ 1{,}680$$
$$Q_{tot_1} = 637{,}925$$

Temperature profiles See Figure 7.4. The adiabatic saturation temperature is 46°C. See the humidity chart.

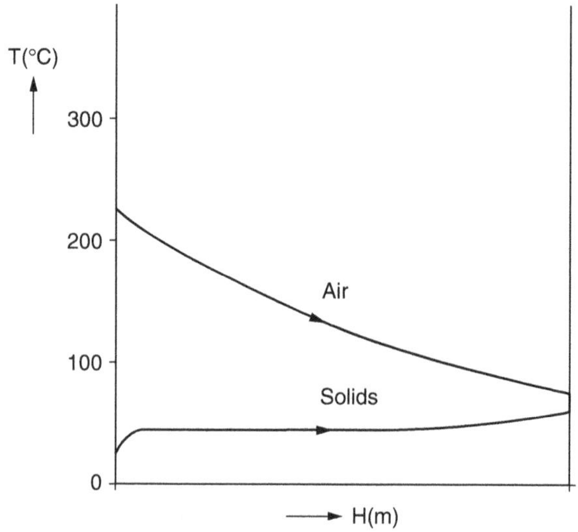

Figure 7.4 Temperature profiles plotted according to the sample calculation in Section 7.3.

Heat transfer The evaporation rate of droplets of nitrobenzene, aniline, and water in flowing air was investigated by Froessling [6]. Some data from these experiments:

Droplet sizes 0.1 to 0.9 mm,
Air velocities 0.2 to 7 $m \cdot s^{-1}$
Re 2 to 800

The evaporation rate of naphthalene spheres was also investigated, with the temperature of the experiments being 20°C. Heat transfer is analogous to mass transfer. The results obtained were used to obtain a Nusselt correlation:

$$\mathrm{Nu} = 2 + 0.552 \cdot \mathrm{Re}^{1/2} \cdot \mathrm{Pr}^{1/3} \tag{7.2}$$

If the medium around the sphere is stagnant, Re = 0, and the correlation degrades to Nu = 2. This is the Nusselt number for the heat transfer due to conduction. The heat transfer coefficient in the flash dryer will be calculated by means of equation (7.2).

It is assumed that the gas surrounding the particles is air at a temperature of 151°C (the arithmetic average of 225 and 77.2°C). It is also assumed that the spherical particles fall through stagnant air with their terminal velocity. The latter velocity is approximately 1 $m \cdot s^{-1}$ (see Table 4.1).

$$\mathrm{Re} = \frac{R_A \cdot v \cdot d_p}{\mu} = \frac{0.84 \cdot 1 \cdot 250 \cdot 10^{-6}}{2.39 \cdot 10^{-5}} = 8.79$$

$$\mathrm{Pr} = \frac{\mu \cdot c_p}{\lambda_G} = \frac{2.39 \cdot 10^{-5} \cdot 1000}{0.0355} = 0.673$$

$$\mathrm{Nu} = 2 + 0.552 \cdot 8.79^{1/2} \cdot 0.673^{1/3} = 3.43$$

$$U = \frac{\mathrm{Nu} \cdot \lambda_G}{d_p} = \frac{3.43 \cdot 0.0355}{250 \cdot 10^{-6}} = 487\ \mathrm{W \cdot m^{-2} \cdot K^{-1}}$$

$$(\Delta T)_m = \frac{(225 - 46) - (77.1 - 46)}{\ln (225 - 46)/(77.1 - 46)} = 84.5\ \mathrm{K}$$

Calculating $(\Delta T)_m$ in this way is a fair approximation [7].

$$A = \frac{Q_{\mathrm{tot}_1}}{U(\Delta T)_m} = \frac{637{,}925 \cdot 1{,}000}{3{,}600 \cdot 487 \cdot 84.5} = 4.31\ \mathrm{m}^2$$

Area passed through the dryer per hour:

$$\frac{1990 \cdot \pi (250 \cdot 10^{-6})^2 \cdot 6}{\pi (250 \cdot 10^{-6})^3 \cdot 1200} = 39{,}800\ \mathrm{m^2 \cdot h^{-1}}$$

Required residence time:

$$\tau = \frac{4.31 \cdot 3600}{39{,}800} = 0.4\ \mathrm{s}$$

This example exhibits qualitatively that the drying of small particles in a flash dryer proceeds at a high rate.

Additional remarks It was explained in Section 5.3 that the Biot number provides the possibility of an assessment of the relative importance of the heat transfer to the particle and of the heat dissipation in the particle itself.

$$\mathrm{Bi} = \frac{U \cdot d_p}{2 \cdot \lambda_S} = \frac{487 \cdot 250 \cdot 10^{-6}}{2 \cdot 0.2} = 0.30$$

If Bi < 1, the external resistance dominates.

Conclusion: The resistance to heat transfer is located mainly in the boundary layer surrounding the particle. A smaller λ_S or a larger particle size would, however, have shown that the heat conduction in the particle cannot be neglected.

Typical conductivities of inorganic materials are (1) salt (sodium chloride) 5.8 $W \cdot m^{-1} \cdot K^{-1}$ at 30°C, and (2) sand 1.4 $W \cdot m^{-1} \cdot K^{-1}$ at 30°C. Biot numbers for inorganic materials tend to be much smaller than those for organic materials.

The high rate of heat transfer in a flash dryer was discussed in this section. Flash drying was simulated by a particle falling at terminal velocity through a stagnant medium. This situation can also be considered a model for other types of convective drying, such as fluid-bed drying, continuous direct-heat rotary drying, spray drying, and conveyor drying. The heat transfer coefficient calculated amounts to 487 $W \cdot m^{-2} \cdot K^{-1}$. Probably, in an actual flash dryer, the heat transfer coefficient is greater because of additional turbulence. If the coefficient calculated is considered typical, it explains part of the superiority of convective drying over conduction drying. A typical heat transfer coefficient at conduction drying is 100 $W \cdot m^{-2} \cdot K^{-1}$. A further aspect, on comparing convective drying and conduction drying, is the area. A cylindrical conduction dryer having a diameter of 1 m and a length of 2 m has an area of 6.3 m^2. The same area is offered by 1 kg of normal vacuum-pan salt (NaCl).

7.4 APPLICATION OF THE DESIGN METHODS

Product

General: 5 $t \cdot h^{-1}$ of an inorganic material from the dryer

TP_{in}: 20°C

d_{50}: 600 μm

A_1: 15% water by weight (as fed)

A_2: 0.1% water by weight

c_s: 0.8 $kJ \cdot kg^{-1} \cdot K^{-1}$

Solubility solid: nil

Process

Ambient temperature: 10°C
Ambient relative humidity: 50%
TA_{in}: 600°C by burning natural gas
c_p: 1.05 kJ·kg^{-1}·K^{-1}

Natural gas data

Composition: 85% by volume of methane, 15% by volume of nitrogen
Methane heat of combustion: 804 kJ·mol^{-1}
ρ_g: 0.80 kg·nm^{-3}

See Sections 4.13 and 4.15 for data on water, steam, and air.

Pilot-plant drying tests were not carried out. It is assumed that the dried-product temperature is 30 K lower than the gas-exit temperature. The product is not recycled.

Mass balance (kg·h^{-1})

	In	Out
Solids	4995	4995
Water	+ 881	+ 5
	5876	5000

$$\text{Evap} = 881 - 5 = 876$$

Net heat (kJ·h^{-1})

$$TA_{\text{out}} = 0.1875 \cdot 600 + 35 = 147.5°C$$

$$\begin{aligned}
Q_1 &= 876(2500 + 1.9 \cdot 147.5 - 4.2 \cdot 20) = 2{,}361{,}915 \\
Q_2 &= 4995 \cdot 0.8(117.5 - 20) = 389{,}610 \\
Q_3 &= 5 \cdot 4.2(117.5 - 20) = + 2{,}048 \\
Q_{\text{tot}_1} &= 2{,}753{,}573 \\
Q_{\text{tot}_2} &= 1.2 \cdot \frac{600 - 10}{600 - 147.5} \cdot 2{,}753{,}573 = 4{,}308{,}353
\end{aligned}$$

Q_{tot_2} is the heat to be supplied in the combustion chamber by the combustion of natural gas. See Section 7.2 for a discussion of the heat loss factor 1.2.

$$\text{kJ per kg of evaporated water: } \frac{4{,}308{,}353}{876} = 4918$$

Table E7.2

	Stoichiometric Natural-Gas Combustion ($kg{\cdot}h^{-1}$)	
	In	Out
Natural gas		
Methane	85.7	0.0
Nitrogen	26.5	26.5
Dry air		
Oxygen	343.0	0.0
Nitrogen	1200.4	1200.4
Water vapor		
In air	6.2	6.2
By combustion	0.0	192.9
Carbon dioxide	+ 0.0	+ 235.8
	1661.8	1661.8

Due to startup, shutdown, and cleaning, for example, the long-term consumption figure is probably a factor of 1.5 higher: 1.5·4918 = 7377 kJ per kilogram of evaporated water.

Dew-point exhaust gas (°C) The amount of heat transferred in the combustion chamber is 4,308,353 $kJ{\cdot}h^{-1}$. The required methane flow is 4,308,353/804 = 5359 $mol{\cdot}h^{-1}$. On using the data of Section 4.16 and on realizing that the natural gas composition has been simplified to 85% by volume of methane and 15% by volume of nitrogen, Table E7.2 can be calculated.

The natural gas consumption is

$$\frac{5359}{0.85}\cdot 0.0224 = 141.2\ \mathrm{nm^3{\cdot}h^{-1}}\ \text{(molar volume 22.4 nl)}$$

The total gas mass flow from the combustion chamber is

$$\frac{4{,}308{,}353}{1.05(600-10)} = 6954.6\,\mathrm{kg{\cdot}h^{-1}}$$

The specific heat of the hot gas has been taken as 1.05 $kJ{\cdot}kg^{-1}{\cdot}K^{-1}$. The secondary air mass flow is: 6954.6 − 1661.8 = 5292.8 $kg{\cdot}h^{-1}$ (composed of 5271.7 $kg{\cdot}h^{-1}$ of dry air and 21.1 $kg{\cdot}h^{-1}$ of water vapor).

Ingress air into the dryer amounts to 20% of the total amount of gases from the combustion chamber: 0.2·6954.6 = 1390.9 $kg{\cdot}h^{-1}$ (composed of 1385.3 $kg{\cdot}h^{-1}$ of dry air and 5.6 $kg{\cdot}h^{-1}$ of water vapor). It is now possible to set up the following table ($kg{\cdot}h^{-1}$):

Process Flow	Dry	Water	Total
Combustion	1462.7	199.1	1661.8
Secondary air	5271.7	21.1	5292.8
Ingress air	1385.3	5.6	1390.9
Evaporation	+ 0.0	+ 876.0	+ 876.0
	8119.7	1101.8	9221.5

Water content: $\dfrac{1101.8}{8119.7} = 0.136$ kg per kg of dry air

The exhaust gases must be cooled from 147.5°C to 58°C to get condensation. This provides an ample margin.

Sizing drying-gas preparation unit

$$Q_{\mathrm{tot}_2} = 4{,}308{,}353\ \mathrm{kJ{\cdot}h^{-1}}$$

Buy a combustion chamber having a capacity of 5040 $\mathrm{MJ{\cdot}h^{-1}}$ or 1400 kW (17% spare capacity). The gas mass flow from the combustion chamber is 6954.6 $\mathrm{kg{\cdot}h^{-1}}$. The gas mass flow to the combustion chamber is approximately equal to this gas mass flow.

$$R_A = \frac{355}{273 + 10} = 1.25\ \mathrm{kg{\cdot}m^{-3}}$$

$$\mathrm{Airflow}_1 = \frac{6954.6}{1.25} = 5563.7\ \mathrm{m^3{\cdot}h^{-1}}$$

$$\mathrm{Power}_{\mathrm{G1}} = \frac{5563.7{\cdot}2500}{3600{\cdot}1000{\cdot}0.5} = 7.7\ \mathrm{kW}$$

2,500: pressure loss of the combustion chamber in $\mathrm{N{\cdot}m^{-2}}$

Select a fan with a motor of 10 kW.

Dryer sizing The gas flow leaving the dryer is considered. The upward gas flow, excluding water vapor is 8119.7 $\mathrm{kg{\cdot}h^{-1}}$. The composition is approximately equal to the composition of dry air.

$$R_A = \frac{355}{273 + 147.5} = 0.84\ \mathrm{kg{\cdot}m^{-3}}$$

$$\mathrm{Airflow}_2 = \frac{8119.7}{0.84} = 9666.3\ \mathrm{m^3{\cdot}h^{-1}}$$

Upward water vapor flow: 1101.8 $\mathrm{kg{\cdot}h^{-1}}$

$$R_W = \frac{220}{273 + 147.5} = 0.52\ \mathrm{kg{\cdot}m^{-3}}$$

$$WA_{\text{flow}} = \frac{1101.8}{0.52} = 2118.8\,\text{m}^3{\cdot}\text{h}^{-1}$$

$$\text{Gasflow}_2\text{: } 9666.3 + 2118.8 = 11{,}785.1\,\text{m}^3{\cdot}\text{h}^{-1}$$

Use a gas velocity of 20 m·s^{-1}.

$$\frac{\pi}{4}{\cdot}D^2{\cdot}20{\cdot}3600 = 11{,}785.1\,\text{m}^3{\cdot}\text{h}^{-1} \rightarrow D = 0.46\,\text{m}$$

Take $D = 0.45$ m. The height of the dryer is 12 m.

Sizing exhaust gas unit

$$\text{Power}_{\text{G2}} = \frac{11{,}785.1{\cdot}3000}{3600{\cdot}1000{\cdot}0.5} = 19.6\,\text{kW}$$

3000: pressure loss of the cyclone in N·m^{-2}

Select a fan with a motor of 25 kW.

Survey

Dryer diameter: 0.45 m
Drying-tube length: 12 m
Capacity of the combustion chamber: 1400 kW
Natural gas consumption (load factor 1.5):

$$\frac{1.5{\cdot}141.2}{0.876} = 241.8\,\text{nm}^3 \text{ per t of evaporated water}$$

See Section 7.2 for a discussion of the load factor. The long-term losses are, on using the load factor, also taken into account.

Electricity consumption (load factor 1.5):

$$\frac{1.5(7.7 + 19.6 + 10.0)}{0.876} = 63.9\,\text{kWh per t of evaporated water}$$

(10.0: allowance)

REFERENCES

[1] Williams-Gardner, A. (1971). *Industrial Drying*, Leonard Hill, London, pp. 164, 166.

[2] Nonhebel, G., Moss, A.H.H. (1971). *Drying of Solids in the Chemical Industry*, Butterworth, London, p. 223.

[3] Noden, D. (1972). Trend towards use of dispersion dryers. *Chemical Process Engineering*, *53*, 48–52.

[4] Noden, D. (1974). Efficient energy utilization in drying. *Processing*, 25–27.

[5] Meedom, H. (1977). A compact system for the combustion of filter cakes. *Zement–Kalk–Gips* (*Cement–Limestone–Gypsum*), *30*, 369–371 (in German).

[6] Froessling, N. (1938). On the evaporation of falling droplets. *Gerlands Beiträge zur Geophysik* (*Leipzig*), *52*, 170–216 (in German).

[7] Perry, R.H., Green, D.W. (2008). *Perry's Chemical Engineers' Handbook,* 8th ed., McGraw-Hill, New York, pp. 12-98–12-104.

8

SPRAY DRYING

In Section 8.1, a general description of spray drying is given. Sections 8.2 through 8.4 deal with feed atomization. The single-fluid nozzle is discussed in Section 8.2, and the rotary atomizer and the pneumatic nozzle are discussed in Sections 8.3 and 8.4 respectively. The quality of spray-dried products is treated in Section 8.5, and the heat of crystallization in Section 8.6. Section 8.7 covers the recovery of the powder and Section 8.8 the product transport by pneumatic conveying. Design methods are dealt with in Section 8.9. The chapter concludes with an example.

8.1 GENERAL DESCRIPTION

Spray drying equipment accepts a feed in the fluid state (solution, suspension, or paste) and converts it into a dried particulate form by spraying the fluid into a warm or hot drying medium (usually, air).

There are four principal stages in the spray-drying process (see Fig. 8.1): (1) feed atomization, (2) free moisture evaporation, (3) bound moisture evaporation, and (4) product recovery (air cleaning).

As the feed is in the fluid state, generally, at least 1 kg of liquid (usually, water) must be evaporated per kilogram of product. The limitation on solids content is the feed rheology, which determines both the pumpability and the spraying behavior.

Drying in the Process Industry, First Edition. C.M. van 't Land.
© 2012 John Wiley & Sons, Inc. Published 2012 by John Wiley & Sons, Inc.

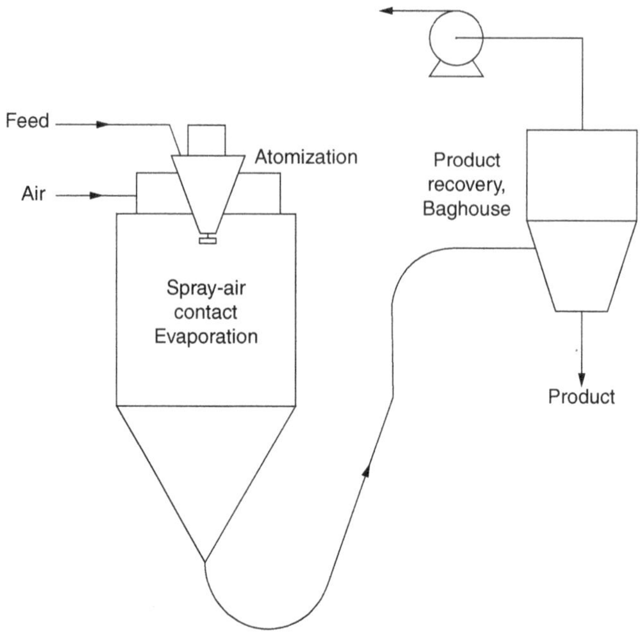

Figure 8.1 Process stages of spray drying.

Thus, the variable costs are high and the reasons for using spray drying must be discussed. These reasons are:

1. The conventional sequence of concentration by evaporation, crystallization, liquid–solid separation, and drying is too complicated.
2. The quality of the spray-dried material is better than the quality of the material made conventionally. The use of spray drying in ceramic manufacturing is an example. A spray-dried pressbody has a controllable particle-size distribution, consists of spherical particles, and is, because it does not contain fines, ideal for pressing operations.
3. Proper mixing of the feed is important. The past use of spray drying in the manufacture of detergents is an example. The spray dryer feed contained a large number of ingredients that had to be mixed well.
4. Spray drying can function as a shaping step. The manufacture of fluid cracking catalysts is an example. The spray dryer feed is a suspension of the active ingredients in a waterglass solution. Spray drying produces spherical particles of approximately 80 μm, which consist of a matrix of waterglass, in which the active ingredients are embedded. These particles are subsequently re-slurried, separated and leached on a belt filter to remove alkali, and the final drying occurs in a flash dryer.
5. Spray drying does not need a mother liquor purge. Process control is thus relatively easy.

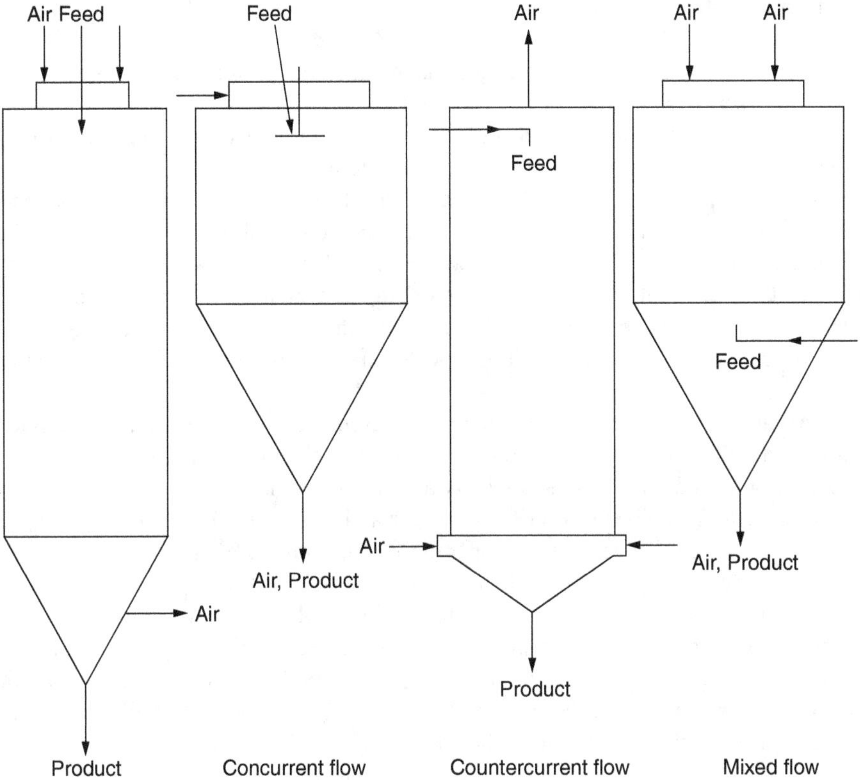

Figure 8.2 Product and air flow modes.

The spray dryer's short residence time (about 25 s for both the product and the gas) is a bonus for materials that cannot be exposed to high-temperature atmospheres for long periods.

Typical products produced by means of spray drying are milk powder, coffee powder, ceramic materials, detergents, and pigments.

Spray drying may proceed concurrently, countercurrently, or as a mixed-flow process (see Fig. 8.2). Both concurrent spray drying and the mixed-flow process occur more often than countercurrent spray drying.

Concurrent drying exposes the droplets to the highest air temperature and, hence, rapid evaporation results. Initially, however, the droplet is at the adiabatic saturation temperature and the mode is therefore suitable for thermolabile materials such as milk powder. The initial rapid evaporation may affect the powder properties, such as the bulk density, strongly.

Countercurrent drying exposes particles that are almost dry to the highest temperatures, and hence very dry products can be produced. The mode is suitable for relatively heavy particles that are thermally stable. The spray drying of detergents is an example.

The mixed-flow process shown in Figure 8.2 is one possibility for this mode. This type of design is sometimes called a *fountain spray dryer.* Passing the product from the top to the bottom while the air is both added and withdrawn at the top is a further possibility (see also [1]).

It is impossible to both dry and cool in the same equipment; thus, cooling is often effected in a subsequent pneumatic transport system.

As a technique, spray drying dates back to the 1920s when it was first introduced into the milk and the detergent industries. With regard to scaling up, there is no limit, and present maximum feed rates are well over 100 $t{\cdot}h^{-1}$. However, this capacity does lead to very high capital costs when compared with other drying techniques. Carlisle Process Systems at Gorredijk in The Netherlands recently manufactured a spray dryer for the production of milk powder that has a diameter of 15.5 m for the cylindrical part.

There are three different devices that can be used to atomize the feed: (1) a single-fluid nozzle, (2) a wheel (rotary atomizer), and (3) a pneumatic nozzle.

The single-fluid nozzle is used widely and produces droplets of size in the range 100 to 250 μm. It can be used for feeds having a viscosity smaller than 0.4 $N{\cdot}s{\cdot}m^{-2}$. The feed is forced through a small opening by high pressure, causing its atomization. The spray bottle for indoor plants is the most common domestic example. The direction of the spray is usually vertical and hence parallel to the wall. Cocurrent and countercurrent spray dryers equipped with this atomizer usually have a cylindrical height exceeding the cylindrical diameter by a factor of 3 to 4. Building spray dryers tall and narrow minimizes the backmixing of gas. It is possible to build them that way because it is possible to direct the spray downward or upward deliberately. Atomization pressures are in the range 20 to 100 bar; however, higher pressures also occur. A typical nozzle pressure at the production of coffee powder is 30 to 35 bar. Increasing the atomization pressure of a given nozzle leads to increased production and decreased droplet size. Small nozzle diameters are required for the production of small droplets. However, the capacity of a small nozzle is also small. Hence, an (often large) number of nozzles is needed. A typical spray dryer for milk powder was equipped with 28 nozzles. The drawback of a large number of nozzles is the possibility of nozzle plugging. This phenomenon may lead to incrustations that can take fire. Still, single-fluid nozzles are often preferred over rotary atomizers because of the powder properties and the possibility to recycle fines. Of the three different devices to atomize the feed, the single-fluid nozzle requires the smallest amount of energy per metric tonne of feed.

Often, the rotary atomizer (see Fig. 8.3) is preferred for large feed streams (i.e., exceeding 5 $t{\cdot}h^{-1}$). It produces relatively small droplets (30 to 120 μm) and there is little tendency for clogging caused by the large flow ports. Thus, the rotary atomizer is suitable for suspensions. However, clogging has been experienced at the spray drying of titanium dioxide. Rotary atomization can be used for all feed viscosities; however, it must be possible to pump the feed to the wheel. The feed liquid is centrifugally accelerated to high velocity before being discharged into the drying chamber. On entering the drying chamber, the liquid sheet breaks up into droplets. Circumferential velocities of the rotary atomizer are in the range 100 to 150 $m{\cdot}s^{-1}$.

Figure 8.3 Modern rotary atomizer F100 meeting the hygienic requirements of food and dairy spray drying. (Courtesy of GEA Process Engineering A/S, Søborg, Denmark.)

Increasing the circumferential velocity leads to decreased droplet size. If, at these high velocities, a rotary atomizer breaks away from its support, a serious accident can happen. In actual fact, such an incident has happened, although without personal damage. The atomizer wheel passed through the walls of both the dryer and the building. Vibration monitors are usually installed to detect an out-of-balance situation before serious damage occurs. The direction of the spray is horizontal and the cylindrical diameter of the spray-drying chamber equals approximately its cylindrical height because the chamber must be wide to prevent wet material from hitting the wall. The energy consumption of the rotary atomizer per metric tonne of feed is in between the energy consumptions of the single-fluid nozzle and the pneumatic nozzle.

The pneumatic (or two-fluid) nozzle is used for test work, small drying operations, and sometimes with viscous feeds. The atomizing gas is usually air. The spray gun is the most common representative of this nozzle type. Like the rotary atomizer, the pneumatic nozzle is suitable for all feed viscosities. The air pressure drops are between 1 and 10 bar. The droplet size is a function of the air/feed ratio in $kg{\cdot}kg^{-1}$ mainly, which is usually between 0.1 and 1. Increasing this ratio decreases the droplet size. The corresponding average droplet size range is 400 to 50 μm. The clogging tendency is smaller than the clogging tendency of the single-fluid nozzle. The feed flow channels are relatively wide and the feed velocities relatively low because, unlike the other two atomization techniques, the atomization does not rely on high liquid velocities. Here, the atomization is caused by high air velocities. The pneumatic nozzle is suitable for the processing of suspensions. This type of atomization can be achieved with relatively simple equipment. The trade-off is the relatively high specific energy consumption, higher than that for the other two devices.

Spray dryers can be fired directly using natural gas. Firing them directly is sometimes not recommended when they are used to process food (e.g., milk powder). The nitrosamine issue is not directly responsible for this viewpoint, as low-NO_x burners are available. Coffee powder is produced in spray dryers with direct firing. Indirect heating by means of steam or electricity is an alternative to direct firing.

As a rule of thumb, for both quality and process safety reasons, the air-inlet temperatures of spray dryers producing organic powders do not exceed 200°C. The air-inlet temperatures of spray dryers producing inorganic materials are substantially higher, even as high as 800°C at the production of silica powder. Inlet temperatures should remain at least 20 K below the dry product's melting point to prevent sticking of the product to the hot dryer wall.

Many spray dryers have a built-in fluid-bed dryer. This permits the reduction of the volume of the drying chamber, as the powder, still containing some moisture, is postdried in the fluid bed. This posttreatment also permits the lowering of the exhaust-gas temperature because the fluid-bed dryer provides residence time.

The application of spray dryers at the incineration of refuse is a recent development. Flue gases from a refuse incinerator having a temperature of 245°C, for example, enter the drying chamber of a spray dryer. The feed is a suspension of gypsum particles in an aqueous solution of calcium chloride coming from an absorption column through which the cooled flue gases pass. The powder produced is a mixture of gypsum and calcium chloride. The flue gas is cooled from 245°C to 170°C.

8.2 SINGLE-FLUID NOZZLE

We begin this discussion with a simple case (see Fig. 8.4). The liquid to be atomized leaves the nozzle with a given velocity. The liquid is assumed to be incompressible and the flow is assumed to be without friction. Bernoulli's law is then applicable:

$$\tfrac{1}{2}{\cdot}\rho{\cdot}v_1^2 + \rho{\cdot}g{\cdot}h + p_1 = \tfrac{1}{2}{\cdot}\rho{\cdot}v_2^2 + \rho{\cdot}g{\cdot}h + p_2$$

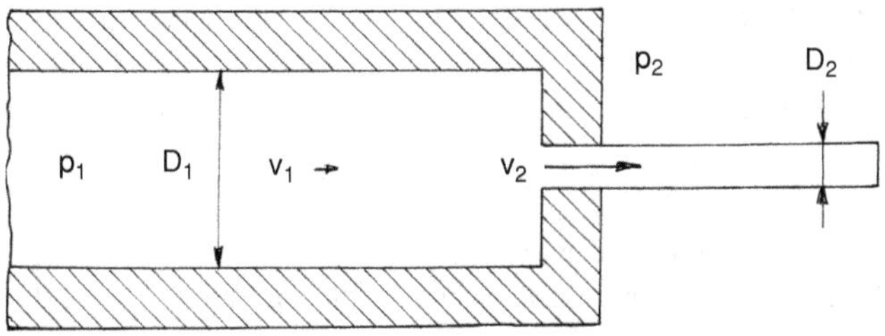

Figure 8.4 Simple single-fluid nozzle.

v_1 is much smaller than v_2 and can be neglected. It follows that

$$v_2 = \sqrt{\frac{2 \cdot \Delta p}{\rho}}$$

Because of friction losses, the actual velocity is smaller; for example,

$$v_2 = C \cdot \sqrt{\frac{2 \cdot \Delta p}{\rho}} \tag{8.1}$$

This simple expression indicates that the velocity is proportional to the square root of the pressure difference and inversely proportional to the square root of the specific mass. C is approximately 0.8 if the flow is turbulent, the hole length is between two and five times the hole diameter, and the hole is sharp and has a circular cross section.

Example 8.1 Water flows at ambient temperature and a pressure difference of 8 bar through a nozzle having a diameter of 3 mm and a length of 9 mm.

$$\rho = 1000\,\text{kg}\cdot\text{m}^{-3}$$
$$\mu = 0.001\,\text{N}\cdot\text{s}\cdot\text{m}^{-2}$$

The nozzle velocity is

$$v_2 = 0.8 \cdot \sqrt{\frac{2 \cdot 800{,}000}{1000}} = 32\,\text{m}\cdot\text{s}^{-1}$$

$$\text{Re} = \frac{\rho \cdot v_2 \cdot d}{\mu} = \frac{1000 \cdot 32 \cdot 0.003}{0.001} = 96{,}000$$

The flow is turbulent as Re > 2100. The flow through the nozzle is

$$\frac{\pi}{4} \cdot d^2 \cdot v_2 = \frac{\pi}{4} \cdot 0.003^2 \cdot 32 = 2.26 \cdot 10^{-4}\,\text{m}^3\cdot\text{s}^{-1} = 0.814\,\text{m}^3\cdot\text{h}^{-1}$$

Simple single-fluid nozzles, working with relative small pressure differences, generate relatively large droplets. A Sauter diameter $\bar{d}_p$ of 3.2 mm is calculated for the nozzle of this example [2]. See Chapter 5 for a discussion of the Sauter diameter.

Generally, a single small-diameter hole or ejector is not considered an atomizer. Real single-fluid nozzles have internals causing the formation of fine droplets.

Various nozzle types are used in spray-drying operations. Nozzles having hollow-cone spray patterns, symmetrical with respect to the nozzle axis, are used frequently. Conversion of pressure energy is carried out such that the resulting liquid motion is rotary. The liquid leaves the nozzle as a hollow cone that is broken up into droplets. Typical spray angles are 60° and 80°. The spray angle is the top angle of the cone. This nozzle type is suitable for cocurrent spray dryers and for mixed-flow dryers when the first phase of the droplet-air flow is cocurrent. Where droplet-air flow is under countercurrent conditions, use of a solid-cone spray can lead to chamber operation with reduced deposit formation. Solid-cone nozzles also produce small droplets. The discussion will be restricted to nozzles having hollow-cone spray patterns. Figure 8.5 depicts a typical design in which the liquid flows axially through the

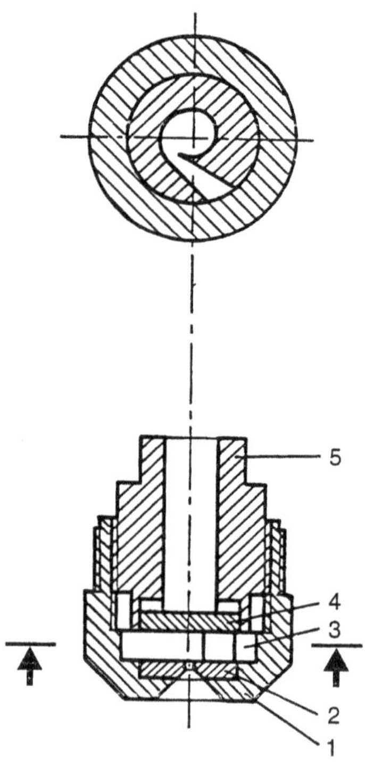

Figure 8.5 Single-fluid nozzle with swirl chamber. 1, Nozzle head; 2, orifice; 3, swirl chamber; 4, end plate; 5, screw pin.

Table 8.1

Nozzle Number	Spray Angle (deg)	ϕ_l (L·min^{-1})	Sauter Diameter (μm)
1	60	0.31	110
2	80	0.31	88
3	60	4.2	205
4	80	4.2	175
5	60	11.15	265
6	80	11.15	250

nozzle containing a swirl chamber. The liquid enters the swirl chamber tangentially and rotates in the swirl chamber. This rotation causes the liquid to leave the nozzle as a hollow cone. Generally also in this case, the liquid velocity is proportional to the square root of the pressure difference and inversely proportional to the square root of the specific mass. Lechler provided data on six different nozzles having swirl chambers and hollow-cone spray patterns [3]. The liquid sprayed was water. The experiments were carried out at ambient temperature. Table 8.1 contains the nozzle throughputs and the Sauter diameters of the water droplets at 10 bar pressure difference.

The nozzles were tested between 5 and 100 bar pressure difference. The Sauter diameters of the droplets produced by the larger nozzles are also larger. Increasing the spray angle of a given nozzle leads to droplet size reduction. The Sauter diameter of the droplets produced by the nozzles having a spray angle of 60° appeared to be proportional to $(\Delta p)^{-0.375}$. The Sauter diameter of the droplets from the nozzles having a spray angle of 80° appeared to be approximately proportional to $(\Delta \mathrm{p})^{-0.375}$. The viscosity of the liquid was not varied at these tests. Still, in actual practice, the viscosity may be in the range 10^{-3} to 0.4 N·s·m^{-2}. It is recommended to take into account that the Sauter diameter is a function of $\mu^{0.17}$ [4]. This means that, ceteris paribus, on increasing the viscosity from 10^{-3} to 0.4 N·s·m^{-2}, the Sauter diameter of the droplets increases by a factor of 2.8. The surface tension of the liquid was not varied either. As the system to be sprayed is usually aqueous, there is not too much variation in the surface tension of the spray dryer feeds.

Example 8.2

(a) The production and the Sauter diameter of the droplets produced by the Lechler No. 3 nozzle at a 50-bar pressure difference are to be calculated.

$$\phi_l = 4.2\sqrt{\frac{50}{10}} = 9.4\,\mathrm{L{\cdot}min^{-1}}$$

$$\bar{d}_p = \left(\frac{50}{10}\right)^{-0.375} \cdot 205 = 112\,\mu\mathrm{m}$$

(b) The calculation is also to be done for a Lechler No. 1 nozzle at a 35-bar pressure difference and for a liquid viscosity of 10^{-2} N·s·m^{-2}.

$$\phi_l = 0.31\sqrt{\frac{35}{10}} = 0.58\ \text{L·min}^{-1}$$

The implicit assumption is that the flow through the nozzle is still turbulent.

$$\bar{d}_p = \left(\frac{35}{10}\right)^{-0.375} \cdot \left(\frac{10^{-1}}{10^{-2}}\right)^{0.17} \cdot 110 = 102\ \mu\text{m}$$

The Lechler nozzles in Table 8.1 are equipped with a swirl chamber. Rotary motion is imparted by the use of a tangential flow entry into the swirl chamber. It is also possible to use slotted inserts or swirl inserts. The slotted inserts feature multiple feed inlets into the orifice, but each slot has a small cross-sectional area, and fine feed filtering is required to prevent nozzle plugging. Nozzles having a swirl chamber are, in this respect, less sensitive.

Highly abrasive or corrosive feeds can wear nozzles quickly and make frequent replacements necessary.

The range of pressure nozzles commercially available is extensive, so it will usually be possible to select a nozzle that is capable of producing the required mean droplet size and distribution for the application. Because there are many nozzle designs it is not possible to recommend a practical working formula.

Their low cost, ease of replacement of wear parts, and simple maintenance procedures are the advantages of single-fluid nozzles. The need for and maintenance of the high-pressure pumps, the need for feed straining, and the need for multiple-nozzle assemblies are the primary disadvantages.

Example 8.3 A total of 6 m^3·h^{-1} of spray dryer feed must be atomized by means of Lechler nozzles No. 5 at a pressure of 35 bar. The number of nozzles is to be calculated and the Sauter diameter of the droplets verified. The feed properties are

$$\rho = 1100\ \text{kg·m}^{-3}$$
$$\mu = 10^{-2}\ \text{N·s·m}^{-2}$$

The nozzle production is

$$\phi_l = \sqrt{\frac{35}{10}\cdot\frac{1000}{1100}} \cdot 11.15 = 19.9\,\text{L·min}^{-1} = 1.193\,\text{m}^3\text{·h}^{-1}$$

The feed atomization requires five nozzles.

$$\bar{d}_p = \left(\frac{35}{10}\right)^{-0.375} \cdot \left(\frac{10^{-2}}{10^{-3}}\right)^{0.17} \cdot 265 = 245\ \mu\text{m}$$

Single-fluid nozzles are considered less flexible than rotary atomizers, as both the nozzle production and the Sauter diameter of the droplets are dependent on the pressure. However, it is possible to improve the flexibility through interchangeable orifices and swirl inserts.

Example 8.4 The pressure of the nozzles in Example 8.3 increases to 40 bar. The production and the Sauter diameter of the droplets are both to be checked.

$$\phi_l = \sqrt{\frac{40}{35}}\cdot 1.193 = 1.275\,\text{m}^3\cdot\text{h}^{-1}$$

$$\bar{d}_p = \left(\frac{40}{35}\right)^{-0.375}\cdot 245 = 233\ \mu\text{m}$$

The theoretical power for nozzle operation is

$$P = \phi_l\cdot\Delta p \quad \text{W} \tag{8.2}$$

In practice, the power drawn from the electricity grid will be aproximately 1.25 times greater.

Example 8.5 A total of 6 $\text{m}^3\cdot\text{h}^{-1}$ of feed is atomized by means of single-fluid nozzles. The pressure difference is 50 bar and the feed specific mass is 1100 $\text{kg}\cdot\text{m}^{-3}$. We are asked to calculate the atomization energy consumption in $\text{kWh}\cdot\text{t}^{-1}$.

From equation (8.2):

$$P = \phi_l\cdot\Delta p = \frac{6}{3600}\cdot 5\cdot 10^6 = 8333\,\text{W}$$

The hourly production is $6\cdot 1100\cdot 10^{-3} = 6.6$ t.

The theoretical specific atomization energy consumption is

$$\frac{8.333}{6.6} = 1.26\,\text{kWh}\cdot\text{t}^{-1}$$

The practical specific energy consumption is $1.25\cdot 1.26 = 1.58$ $\text{kWh}\cdot\text{t}^{-1}$.

Single-fluid atomization is used almost exclusively in Europe for the production of milk powder. The possibility of recycling fines is an important reason for this production mode.

8.3 ROTARY ATOMIZER

The feed liquid is centrifugally accelerated to high velocity before being discharged into the drying chamber. A modern rotary atomizer is shown in Figure 8.3. Wheels

Table 8.2 Estimating Droplet Size from Rotary Atomizer Operation

Atomizer Operation		Exponent				Constant
Wheel Peripheral Speed ($m \cdot s^{-1}$)	Wheel Liquid Loading ($kg \cdot h^{-1} \cdot m^{-1}$)	*a*	*b*	*c*	*d*	K
85–115	250	0.24	0.82	0.60	0.24	1.40
85–180	250–1500	0.20	0.80	0.60	0.20	1.60
180–275	1000–3000	0.12	0.77	0.60	0.12	1.25
85–180	3000–60,000	0.12	0.80	0.60	0.12	1.20

Source: Courtesy of SprayDryConsult, Charlottenlund, Denmark.

with vanes or bushings are used to prevent slippage of the feed in the atomizer. Vanes can be either radial or curved, with the curved vanes being more suitable to accomplish deaeration, which can lead to higher bulk densities [5].

The weight-average droplet size of the atomized feed can be estimated from the relationship [5]

$$d_{50} = K \cdot \frac{M^a}{N^b \cdot d_1^c (n \cdot h)^d} \cdot 10^4 \text{ } \mu\text{m} \tag{8.3}$$

and

$$\begin{aligned} d_{95} &= 2 \cdot d_{50} (d_{50} < 60 \text{ } \mu\text{m}) \\ &= 2.5 \cdot d_{50} (60 \text{ } \mu\text{m} < d_{50} < 120 \text{ } \mu\text{m}) \end{aligned}$$

M is the feed rate in $kg \cdot h^{-1}$, N the wheel speed in min^{-1}, d_1 the wheel diameter in meters, n the number of vanes, bushings, pins, or holes, and h the height of a vane or pin and half the circumference of a circular bushing opening or drilled hole in meters. Table 8.2 can be used to select the power and the K values. Because the K value is a function of the feed, equation (8.3) should be used to establish order-of-magnitude values only. The table shows that the weight-average droplet size is relatively independent of the feed rate and the wetted perimeter since the values of exponents a and d are relatively low. The values of exponents b and c are not low, meaning that the weight-average droplet size is strongly affected by the circumferential speed, which makes sense. Equation (8.3) shows the flexibility of the rotary atomizer, as, to keep the weight-average droplet size constant, a smaller feed flow can be compensated by a lower rotational speed. It may give rise to surprise that the specific mass, the surface tension, and the viscosity of the feed are not in equation (8.3). The first two can be explained readily, as, the feed is usually aqueous, meaning that neither the specific masses nor the surface tensions of spray dryer feeds show great variations. The viscosities of spray dryer feeds, however, can vary by a factor of 10, 100, or even 1000. A power relation of 0.2 is considered realistic in industrial spray-drying operations, at least with low-viscosity liquids [5]. However, viscosity effects are not taken into account by equation (8.3). On scaling up, equation (8.3) should give the same d_{50} value.

Example 8.6

$$M = 1105.6\ \text{kg·h}^{-1}$$
$$N = 10{,}000\ \text{min}^{-1}$$
$$d_1 = 0.25\ \text{m}$$
$$n = 45$$
$$h = 0.03\ \text{m}$$
$$v = \frac{\pi \cdot 0.25 \cdot 10,000}{60} = 131\ \text{m·s}^{-1}$$
$$\text{Vane liquid loading: } \frac{1105.6}{45 \cdot 0.03} = 819\ \text{kg·h}^{-1}\text{·m}^{-1}$$

From Table 8.1 the exponents are $a = 0.2$, $b = 0.8$, $c = 0.6$, and, $d = 0.2$.

$$K = 1.60$$
$$d_{50} = \frac{1.60 \cdot 1105.6^{0.2}}{10{,}000^{0.8} \cdot 0.25^{0.6}(45 \cdot 0.03)^{0.2}} \cdot 10^4 = 89\ \mu\text{m}$$
$$d_{95} = 2.5 \cdot 89 = 222\ \mu\text{m}$$

The minimum dryer diameter should be 4 m if $d_{50} = 89\ \mu$m [6].
Marshall [7] gives the theoretically required power for a rotary atomizer:

$$P = 3.8 \cdot 10^{-10} \cdot M \cdot N^2 \left(2 \cdot d_1^2 - d_d^2\right) \quad \text{kW} \tag{8.4}$$

d_d is the distance between the feeding point and the axis. In practice, the result calculated should be doubled because of air ingress and friction.

Example 8.7 The power required for the rotary atomization of the previous example is calculated.

$$d_d = 0.1\ \text{m}$$
$$P = 3.8 \cdot 10^{-10} \cdot 1105.6 \cdot 10{,}000^2 (2 \cdot 0.25^2 - 0.1^2) = 4.8\ \text{kW}$$

The power required is $2 \cdot 4.8 = 9.6$ kW. A 15-kW motor should be chosen. The energy consumption in kWh per metric tonne of feed: $9.6/1.1056 = 8.7$.

8.4 PNEUMATIC NOZZLE

Atomizers utilizing a gas for atomization are called *pneumatic atomizers* or *two-fluid atomizers*. A pneumatic nozzle is shown in Figure 8.6. The depicted nozzle is called a *plain jet nozzle*. The pressurized gas flows to the nozzle through an annular channel, whereas the liquid flows through a central line. In some nozzle constructions, the gas leaves under an angle of 45° with the nozzle axis.

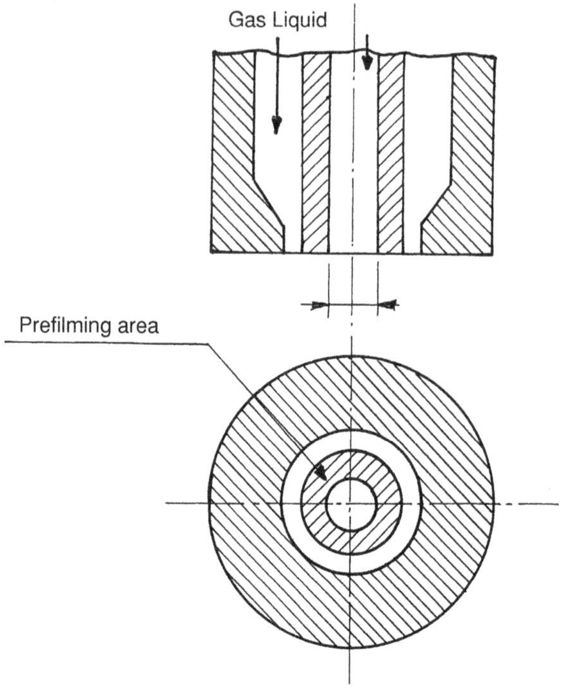

Figure 8.6 Pneumatic nozzle.

In single-fluid nozzles, the energy is supplied to the liquid by the flowing liquid itself. In rotary atomizers, the energy is supplied to the liquid by the rotation. The liquid obtains a high velocity in the latter two types of atomizers, whereas the air or other gas that receives the droplets is stationary. In pneumatic nozzles, the liquid velocity is relatively low until it meets the air or other gas having a high velocity. The atomization energy is now supplied by the atomizing gas.

Pneumatic atomizers do not need small holes for small flows, thus the plugging tendency is reduced. Hence, two-fluid nozzles are in principle suitable for the atomization of suspensions. For the same reason, they are in principle suitable for the atomization of viscous liquids.

Two types of pneumatic nozzles may be distinguished: with external mixing and with internal mixing. The two flows contact each other outside the nozzle in nozzles with external mixing. In nozzles with internal mixing, the contact occurs inside the nozzle. Nozzles with external mixing can be met more often than nozzles with internal mixing. The latter type can be used for clear liquids.

External mixing occurs in the pneumatic nozzle shown in Figure 8.6. It has a relatively simple construction.

The mass ratio air (gas)/liquid is an important independent variable influencing the droplet size distribution. Pneumatic nozzles operate with a mass ratio in the range 0.1 to 5, with the majority of applications below 1 [8].

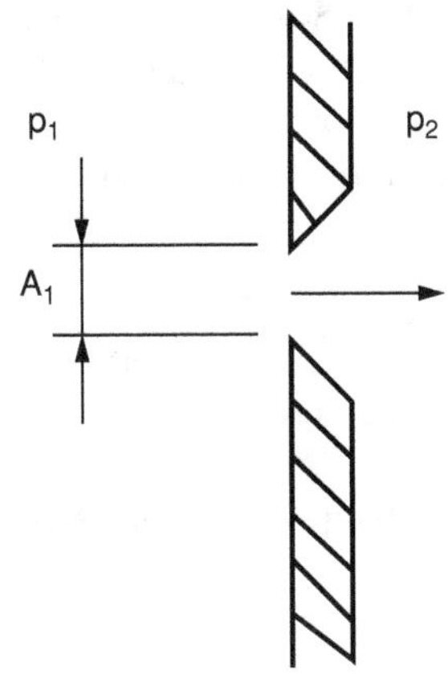

Figure 8.7 Flow of gas through an orifice.

The spray angle of the nozzle type pictured in Figure 8.6 is variable in the range 20 to 40°. The spray angle is the top angle of the sprayed cone. Generally, pneumatic nozzles produce solid-cone sprays.

The flow of air through an orifice is considered next [9] (see Fig. 8.7). It is assumed that the pressure loss due to the flow occurs in the orifice only. The air velocity increases on increasing the pressure difference $p_1 - p_2$. However, if $p_2 = 0.53{\cdot}p_1$ (for air), the velocity of sound is reached at the orifice. A further decrease in p_2 will not result in a greater orifice air velocity. This is because pressure waves travel with the velocity of sound, so when the downstream pressure equals $0.53{\cdot}p_1$ or has a lower value, the upstream pressure determines the flow completely. The expression for the airmass flow is

$$\phi_m = C_K{\cdot}A_C\sqrt{p_1{\cdot}\rho_l} \quad \text{kg}{\cdot}\text{s}^{-1} \tag{8.5}$$

The factor C_K is a function of $\gamma = c_p/c_v$. It is 0.684 for air. Furthermore, $A_C = C_C{\cdot}A_1$, where $C_C \approx 0.6$ for sharp-edged orifices and $C_C \approx 0.97$ for rounded orifices, and A_1 is the orifice cross-sectional area.

Example 8.8

(a) Air with a pressure of 4 bar absolute and a temperature of 20°C flows to the atmosphere through a sharp-edged orifice that has a diameter of 2 mm. The airmass flow is to be calculated.

$$\begin{aligned}\phi_m &= C_K{\cdot}C_C{\cdot}A_1\sqrt{p_1{\cdot}\rho_l} = 0.684{\cdot}0.6{\cdot}\tfrac{\pi}{4}{\cdot}0.002^2\sqrt{400{,}000{\cdot}4{\cdot}1.2}\\ &= 1.79{\cdot}10^{-3}\,\text{kg}{\cdot}\text{s}^{-1}\\ &= 6.43\,\text{kg}{\cdot}\text{h}^{-1}\end{aligned}$$

(b) Air with a pressure of 8 bar absolute and a temperature of 20°C flows through the pneumatic nozzle of part (a). The airmass flow is to be calculated.

$$\phi_m = 0.684 \cdot 0.6 \cdot \tfrac{\pi}{4} \cdot 0.002^2 \sqrt{800,000 \cdot 8 \cdot 1.2} = 3.57 \cdot 10^{-3} \quad \text{kg} \cdot \text{s}^{-1}$$
$$\phi_m = 12.86 \, \text{kg} \cdot \text{h}^{-1}$$

The airmass flow in part (b) is twice the airmass flow in part (a). The airmass flow is proportional to the upstream pressure.

Richter [10] gives an expression for the Sauter diameter of the droplets produced by a pneumatic nozzle:

$$\frac{\bar{d}_p}{D} \approx 0.48 \left(\frac{\sigma}{\rho_g \cdot D \cdot v_{\text{rel}}^2} \right)^{0.4} \left(1 + \frac{1}{\eta} \right)^{0.4} + 0.15 \left(\frac{\mu_l^2}{\sigma \cdot \rho_l \cdot D} \right)^{0.5} \left(1 + \frac{1}{\eta} \right) \qquad (8.6)$$

A definition of the Sauter diameter is given in Chapter 5.

The expression is applied by Richter [10] to a pneumatic nozzle according to Figure 8.6 as described in Example 8.9.

Example 8.9 A pneumatic nozzle (see Fig. 8.6) has an inner diameter of 2 mm. The gas/liquid mass ratio η is 1. The gas velocity is assumed to be 330 m·s^{-1}. The physical data of the gas and the liquid are:

$$\rho_l = 850 \text{ kg} \cdot \text{m}^{-3}$$
$$\rho_g = 2.5 \text{ kg} \cdot \text{m}^{-3}$$
$$\mu_l = 0.05 \text{ N} \cdot \text{s} \cdot \text{m}^{-2}$$
$$\sigma = 0.04 \text{ N} \cdot \text{m}^{-1}$$
$$\frac{\bar{d}_p}{D} \approx 0.48 \left(\frac{0.04}{2.5 \cdot 0.002 \cdot 330^2} \right)^{0.4} \left(1 + \frac{1}{1} \right)^{0.4} + 0.15 \left(\frac{0.05^2}{0.04 \cdot 850 \cdot 0.002} \right)^{0.5} \left(1 + \frac{1}{1} \right)$$
$$\approx 0.01406 + 0.05752 = 0.07158 \rightarrow \bar{d}_p \approx 143 \; \mu\text{m}$$

The contribution of the first part of the right-hand side is only 20%, whereas the contribution of the second part of the right-hand side is 80%. The second part is strongly affected by η. On decreasing η to 0.25, for example, the outcome is

$$\frac{\bar{d}_p}{D} \approx 0.02028 + 0.1438 = 0.1641 \rightarrow \bar{d}_p \approx 328 \; \mu\text{m}$$

Thus, equation (8.6) reflects that the average droplet size is strongly dependent on η.

Lechler [3] provided data concerning the Sauter diameter of droplets produced by a two-fluid nozzle. Air was used to atomize water. Air pressures of several bars were used. The water flows were several liters per minute. The data show a strong dependency on the air/water mass ratio. The latter independent variable varied between 0.1 and 0.9. The Sauter diameter of the droplets varied from 500 to 40 μm. The data also show an influence of the air pressure employed with smaller droplets produced for higher air pressures.

The theoretical energy for the adiabatic compression of 1 kg of air is

$$N = \frac{\gamma}{\gamma - 1} \cdot \frac{R \cdot T_1}{M} \left[\left(\frac{p_2}{p_1} \right)^{(\gamma-1)/\gamma} - 1 \right] \quad \text{J·kg}^{-1} \tag{8.7}$$

The compressor power is

$$P = \frac{\phi_m \cdot N}{\eta_{\text{ad}}} \quad \text{W} \tag{8.8}$$

in which η_{ad} is the adiabatic compression efficiency, which is, for example, 0.8.

Example 8.10 A total of 180 L·h^{-1} of water is atomized by 37.5 $\text{nm}^3\text{·h}^{-1}$ air, which has a pressure of 4.5 bar absolute. It is requested to calculate the energy consumption in kWh per metric tonne of water. The air temperature is 20°C and $\eta_{\text{ad}} = 0.8$.

$$N = \frac{1.4}{1.4{-}1} \cdot \frac{8314 \cdot 293}{29} \left[\left(\frac{4.5}{1} \right)^{(1.4-1)/1.4} - 1 \right] = 157{,}835 \text{ J·kg}^{-1}$$

$$P = \frac{37.5 \cdot 1.3 \cdot 157{,}835}{3600 \cdot 0.8} = 2672 \text{ W}$$

$$\text{kWh·t}^{-1}: \frac{2672}{0.180} = 14.8$$

8.5 PRODUCT QUALITY

One of the main characteristics of spray drying is the production of a spherical particle. In general, aqueous solutions of materials such as coffee, soap, gelatin, and water-soluble polymers (which form tough tenuous outer skins on drying) can be spray dried to hollow spherical particles. This is possible because of the formation of a case-hardened outer surface on the particle, which prevents liquid from the particle interior from reaching the surface. Because of high heat-transfer rates to the drops, the liquid at the particle's center evaporates, causing the outer shell to expand and form a hollow sphere. Sometimes the vapor generation rate within the particle is sufficient to blow a hole through the spherical shell's wall. Thus, the shape of the particles produced by spray drying is generally different from the shape of particles produced by crystallization processes. As a rule of thumb, the bulk density of

Figure 8.8 Spray-dried milk powder. (Courtesy of SPX Flow Technology Danmark A/S, Søborg, Denmark.)

crystallized materials such as vacuum-pan salt and sucrose, is 55% of the solid's specific mass. This also means that 55% by volume of a particle collection is taken up by solid material, and 45% by volume is occupied by air. As a rule of thumb, the bulk density of spray-dried materials such as milk powder and coffee powder is 55/2 = 27.5% of the solid's specific mass. Thus, in this case, 72.5% by volume of a particle collection is taken up by air. Figure 8.8 exhibits spray-dried milk powder.

Microscopy

Walton and Mumford [11] distinguish among agglomerated particles, particles having a skin, and crystallized particles. Agglomerated particles are mainly inorganic and are made up of elementary particles of size 1 to 10 μm. It is probable that agglomerated particles are crystalline; however, this is not directly clear from the photographs. X-ray diffractometry is a method to detect the crystallinity of a material. Particles having a skin are made in the production of milk powder, coffee powder, and other food products. The co-drying of egg and skimmed milk produces hollow spheres. It is probable that these materials are amorphous. Crystallized particles are inorganic and often hollow. Their crystalline nature is clear from photographs; trisodium orthophosphate is an example.

Moisture Content

Spray-dried materials are often not completely dry. Milk powder contains 2 to 4% by weight of water. The water content of egg powder is 8% by weight. Detergents can contain up to 10% by weight of water.

Bulk Density

A high bulk density means relatively little packing. However, a low bulk density is often not a serious drawback as long as the bulk density remains constant. The following independent variables affect the bulk density.

Air Temperature The air temperature affects the bulk density at cocurrent drying. A high air temperature causes a rapid evaporation and solidification, and this tends to lead to a low bulk density. Masters [12] has given an example concerning the spray drying of waterglass. The following table is valid for waterglass having a mass ratio SiO_2/Na_2O of 2 : 1.

Air Temperature (°C)	ρ_b (kg·m^{-3})
150	650
180	500
250	250
500	70–100

Particle Size The adhesion forces dominate the gravity forces when the particles are smaller than 10 μm, and this tends to lead to low bulk densities. However, Freudig et al. [13] report a case where the bulk density decreases when the particles become larger. Agglomerated powder was produced from skimmed milk and screened into four fractions:

Particle size (μm)	<200	200–450	450–1000	>1000
Bulk density (kg·m^{-3})	440	300	210	170

The larger particles are more porous than the smaller particles.

Particle-Size Distribution Generally, a wide particle-size distribution leads to a relatively high bulk density, and vice versa. This is because the smaller particles accommodate themselves between the larger particles.

Using Steam Air ingress is often a cause of low bulk density. By atomizing the feed into superheated steam, this effect can be avoided. In a later stage, the droplet comes into contact with warm air, accomplishing the drying. The processing of casein products in dairy plants is an example.

Precharging of Drying Air Dolinsky [14] reports a 30% bulk density increase at the spray drying of detergents when the moisture content of the drying air is increased from 12 g to 150 g per kilogram of dry air. The cause of this effect is probably the decrease in the driving forces for heat and mass transfer.

Dissolving Carbon Dioxide or Nitrogen Carlisle Process Systems, a spray dryer manufacturer, developed the dissolution under pressure of carbon dioxide or nitrogen into a spray dryer feed. The gas is resorbed on atomizing the feed, and this leads to the formation of hollow spheres. The production of cappucino powder is a typical application. Cappucino powder is made by adding, at the production of milk powder, coffee powder to recycled milk powder.

Particle Size

The particle size of spray-dried materials was discussed in Sections 8.2 through 8.4.

Flow Characteristics

Measuring the angle of repose is one method to characterize the flow properties. Hard spherical particles having a narrow size distribution have a small angle of repose and good flow properties. At the production of ceramic powders, spray drying has replaced the previous production method at which the materials were separated in press filters, dried by means of other convective dryers, and milled. The reason is that at the next step, spray-dried material fills the presses perfectly. A spray-dried pressbody consists of spherical particles and does not contain fines. Pressbody made by the traditional wet process consists of irregularly shaped particles with a wide particle-size distribution.

Attrition Resistance

There is much variation concerning attrition resistance. Fluid cracking catalysts must be very attrition-resistant because they are, in the refineries, recycled at high speed between the reactor and the regenerator. The temperatures in these two pieces of equipment are 550 and 650°C, respectively. The spray dryer feed is made by suspending active ingredients, such as zeolites and kaolin, into a waterglass solution. At spray drying, the active ingredients are embedded into a hard waterglass matrix.

Color

Incrustations in the spray dryer may, as they crumble away, lead to colored specks in the product. The prevention of black specks is important at the spray drying of emulsion PVC and titanium dioxide, for example.

Flavor Retention

Spray drying is a heat treatment in the presence of water and can cause the decomposition or conversion of sensitive organic compounds that give a flavor. It can be detected by means of sensorial analysis. The production of coffee powder by means of spray drying can be compared to the production of coffee powder by means of freeze drying. The latter technique has a greater capability for saving flavors.

Dissolution

The dissolution or dispersibility of a spray-dried material (e.g., milk powder) is often an important property. Freudig et al. [13] report the dissolution behavior of skimmed milk powder. Agglomerated milk powder having an average particle size of 600 μm was studied. Experiments were also carried out with four particle-size fractions (i.e., >1000, 450–1000, 200–450, and <200 μm). The coarse material could be wetted more readily than the fine materials. The coarse material was more porous. The latter fact also caused lower bulk densities, as discussed previously. The addition of 1% by weight of lecithin promotes the dissolution of milk powder. The dissolution behavior is an important property of detergents as well.

Binding Agents

Binding agents can be used to strengthen the particles or to increase the particle size. At the production of fluid cracking catalysts (FCC), waterglass is used to obtain strong particles. Poly(vinyl alcohol) (PVA) is also used to obtain strong particles at the spray drying of ferrites. Ferrites are used for electrical applications. PVA can also be used to obtain larger particles. It was typically used at the spray drying of rubber accelerator slurries to obtain a more skin-safe (less dusty) material.

Attaining a Desired Product Quality

It is recommended to determine the independent process variable(s) affecting the property or properties concerned. The next step is to carry out a single-variable study with each of these process variables. It is also possible that a partial factorial experiment plan is required to sort out the effect of these variables. A factorial experiment plan is a plan for carrying out experiments laid out in a multidimensional matrix, usually at the upper and lower conditions for each variable.

8.6 HEAT OF CRYSTALLIZATION

Often, a considerable amount of material is crystallized during spray drying. Usually, heat is liberated at crystallization, although the reverse is also encountered, especially with solutes exhibiting an inverted solubility characteristic. The crystallization of hydrates is often an endothermic process.

The heat of crystallization is taken as being equal in magnitude but opposite in sign to the heat of solution. Although this simplification is not entirely correct, the error involved is small. Reported heat-of-solution values refer to dissolution in an infinite amount of solvent; consequently, the heat of dilution is neglected. Heats of solution are usually reported at 18 or 25°C.

Example 8.11 The heat of solution of sodium chloride is –0.93 kcal·mol^{-1} (the minus sign indicating an endothermic process). Therefore, the crystallization of

1 mol of NaCl liberates 0.93 kcal. Since the molecular weight of sodium chloride is 58.46 g·mol^{-1}, the crystallization of 100 kg of NaCl liberates

$$\frac{100 \cdot 10^3 \cdot 4.2 \cdot 0.93}{58.46} = 6681\ \text{kJ}$$

The evaporative crystallization of 100 kg of NaCl requires the evaporation of 300 kg of water. In a spray dryer, the heat liberated by the crystallization of 100 kg of NaCl will do for the evaporation of approximately 1 kg of water. Thus, the effect is minor.

It is also possible that a chemical reaction occurs during spray drying. In that case, the heat of reaction must be taken into account.

8.7 PRODUCT RECOVERY

Since the spray-dried product is invariably fine and the spent drying air from the dryer usually carries the total powder load, the selection of product recovery equipment is very important. One may distinguish between dry and wet product recovery. For dry product recovery, three principal choices exist: (1) dry cyclones (down to 5 to 10 μm), (2) bag filters (down to 0.1 to 1 μm), and (3) electrostatic precipitators (down to 0.5 to 1 μm).

A spray dryer is always equipped with a dry product-recovery system, cyclones and bag filters being the most popular choices. The average value of the gas face velocity can, for bag-filter design purposes, vary from 0.5 to 1.5 m·min^{-1}.

The application of a wet product-recovery system depends on specific requirements, wet scrubbers being used mainly for air cleaning.

8.8 PRODUCT TRANSPORTATION

Because of the small particle sizes being dealt with, pneumatic conveying is quite popular for spray-dried products. A typical value for the powder/air mass ratio is 4. It may be necessary to condition the air for the transport to prevent the powder from absorbing moisture from the transport air. Such conditioning can be relatively simple by cooling the air by means of a chilled-water unit to, for example, 5°C, whereby the water that is contained by the air condenses. The next step is reheating the air to, for example, 20°C. The relatively dry air can then be used for transport. As a rule of thumb, line velocities between 20 and 25 m·s^{-1} are chosen.

Because the spray-dried material is relatively fine, it is feasible to combine the transport and the cooling of the powder. The heat transfer from the small particles is very rapid (see Chapter 7). However, in that case, the airmass flow should probably be boosted to obtain a powder/air mass ratio of approximately 1. Of course, this will only promote the reliability of the pneumatic transport and will add to the cost of the pneumatic transport equipment. If the powder/air mass ratio is approximately 1, the pressure loss equals approximately the pressure loss for clean air.

8.9 DESIGN METHODS

The mass balance regarding the product is made first (see Chapter 4). The drying-gas exit temperature must be known to be able to calculate the heat transferred to the product flow, Q_{tot_1}. The relationship proposed between TA_{in} and TA_{out} is

$$TA_{out} = 88.39 \cdot {}^{10}\log TA_{in} - 112.35\ ^{\circ}C \tag{8.9}$$

Figure 8.9 depicts graphically points found in the literature [15–20] and observed for industrial dryers.

The data (65 sets of TA_{in}/TA_{out} values) were analyzed statistically. The aforementioned relationship holds (linear regression), with the correlation coefficient being 0.727. Apparently, there is a considerable spread in the data. A low TA_{out} value leads

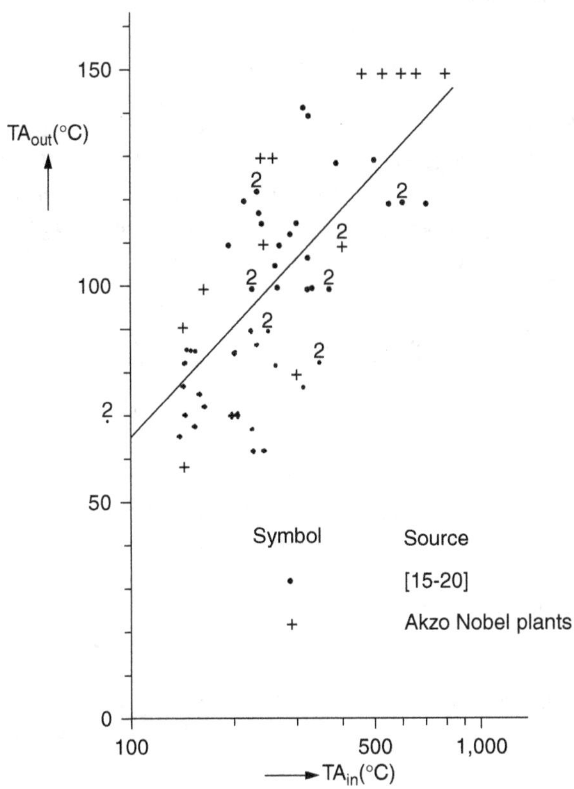

Figure 8.9 Relationship between air-inlet and air-outlet temperatures for spray dryers.

to good thermal efficiency; however, it is possible that the product still contains some moisture. A relatively high TA_{out} value favors a low residual-moisture content. It is possible to give spray-dried material a posttreatment in a fluid-bed dryer. In that case TA_{out} can be relatively low. Equation (8.9) predicts that $TA_{out} = 133.2°C$ when $TA_{in} = 600°C$ and $TA_{out} = 64.4°C$ when $TA_{in} = 100°C$.

To allow for the steady-state heat losses, Q_{tot_1} is multiplied by 1.25. Literature data on primary and secondary energy consumption of dryers are scarce (but see references 17 to 19). Plant data were collected, and these data together with the literature data led to the factor 1.25. The factor could be smaller for well-insulated dryers and even slightly higher when insulation is absent. The factor 1.25 does not differentiate between these two cases. Insulation is important because spray dryers have large areas. The load factors to arrive at the performance in the long run are 1.1 (full), 1.2 (normal), and 1.3 (low), respectively. These figures are based on a small number of literature data and data of five Akzo Nobel spray dryers. It must be pointed out, however, that important large operations are usually quite efficient. On the other hand, small minor equipment functioning intermittently exhibits high specific consumption data.

It is assumed that a conical-based spray dryer (60° cone) is selected. The volume of the dryer can be expressed as

$$V = 0.7854 \cdot D^2(H + 0.2886 \cdot D) \quad \text{m}^3 \tag{8.10}$$

where H is the height of the cylindrical part and D is the chamber diameter. A product residence time of 25 s is a good design value. It is assumed that this residence time prevails if the chamber's volume exceeds the exit gas flow in $m^3 \cdot s^{-1}$ by a factor of 25. The superficial air velocity is arrived at by dividing the exit gas flow by the chamber's cross-sectional area in m^2. Typical values for a spray dryer with rotary atomization are

Chamber Diameter (m)	Superficial Velocity ($m \cdot s^{-1}$)
4	0.20
6.5	0.35
9	0.50

The purpose is twofold: (1) to have a cylindrical height being approximately equal to the diameter, and (2) to have the residence time of about 25 s as a starting point.

Both concurrent flow spray dryers and mixed-flow spray dryers can be equipped with a fluid bed as part of the drying chamber cone. It is then possible to choose a residence time in the chamber of 15 s, for example. The spray dryer can thus become smaller. The fluid bed provides further processing options, such as agglomeration. The relatively large holdup of powder can be a drawback if the powder is combustible.

One further aspect has to be observed. Dryers with rotary atomization may suffer from wall incrustations if the diameter is too small and the droplets are too large. For example, for a mean size between 80 and 100 μm, a minimum chamber diameter of 4 m is required [6].

Typical values for the superficial velocity in nozzle towers are much higher and vary between 1 and several $m{\cdot}s^{-1}$. Here the cylindrical height normally exceeds the diameter by a factor of between 3 and 4.

A cocurrent, open-cycle spray dryer can be equipped witrh a fan to transport the air to the tower, possibly through a direct air heater. A Δp value of 2500 $N{\cdot}m^{-2}$ is a fair approximation. An indirect heater (e.g., a steam heater) would require about 500 $N{\cdot}m^{-2}$. An exhaust fan is often used to extract the gas–solids mixture from the spray chamber and can be located between the baghouse or the cyclone and a wet scrubber. A Δp value of 1500 $N{\cdot}m^{-2}$ is a good assumption. Another possibility is to have one exhaust fan draw the gases through the entire system. The underpressure in the spray chamber then usually exceeds the underpressure of a system equipped with two fans. A Δp value of 4000 $N{\cdot}m^{-2}$ appears to be a reasonable approximation of the actual state of affairs. An ingress air factor regarding the amount of fresh air is 1.1.

The miscellaneous secondary energy consumption is 10 kW.

Tests in a spray dryer having a diameter of, for example, 2 m should precede a design.

Example 8.12 A spray dryer for a foodstuff is to be designed.

Product

General: 500 $kg{\cdot}h^{-1}$ of product from the dryer
TP_{in}: 20°C
A_1: 55% water by weight (as fed)
A_2: 0.5% water by weight
c_s: 1.25 $kJ{\cdot}kg^{-1}{\cdot}K^{-1}$
Solubility solid: nil

Process

Ambient temperature: 10°C
TA_{in}: 205°C by indirect heat exchange with steam
c_p: 1.0 $kJ{\cdot}kg^{-1}{\cdot}K^{-1}$
Airflow: cocurrent
Atomization mode: vaned wheel

Steam data

Pressure: 23 bars absolute
Heat of vaporization: 1867 $kJ{\cdot}kg^{-1}$
Condensation temperature: 218.3°C

See Sections 4.13 and 4.15 for further data on water, steam, and air.

Mass balance ($kg{\cdot}h^{-1}$)

	In	Out
Solids	497.5	497.5
Water	+ 608.1	+ 2.5
	1105.6	500.0

$$\text{Evap} = 608.1 - 2.5 = 605.6$$

Net heat ($kJ{\cdot}h^{-1}$)

$$TA_{out} = 88.39{\cdot}^{10}\log 205 - 112.35 = 92.0°\text{C}$$
$$TP_{out} = 92.0 - 20 = 72°\text{C}$$

$$Q_1 = 605.6(2500 + 1.9{\cdot}92.0 - 4.2{\cdot}20) = 1{,}568{,}988$$
$$Q_2 = 497.5{\cdot}1.25(72 - 20) = 32{,}338$$
$$Q_3 = 2.5{\cdot}4.2(72 - 20) = +\ 546$$
$$Q_{tot_1} = 1{,}601{,}872$$
$$Q_{tot_2} = 1.25{\cdot}\frac{205 - 10}{205 - 92}{\cdot}1{,}601{,}872 = 3{,}455{,}365$$

The heat of crystallization of the solid is not taken into account.

Q_{tot_2} is the heat to be supplied to the drying air in the steam heater. See Section 8.9 for a discussion of the heat loss factor.

$$\text{kJ per kg of evaporated water: } \frac{3{,}455{,}365}{605.6} = 5706$$
$$\text{kg of steam per kg of evaporated water: } \frac{5706}{1876} = 3.0$$

Due to startup, shutdown, and cleaning, for example, the long-term consumption figure is probably a factor of 1.1 higher: $1.1{\cdot}5706 = 6277$ kJ per kilogram of evaporated water. The long-term consumption figure of kilograms of steam per kilogram of evaporated water is probably: $1.1{\cdot}3.0 = 3.3$.

Sizing drying-gas preparation unit

$$Q_{tot_2} = 3{,}455{,}365\ \text{kJ}{\cdot}\text{h}^{-1}$$

Buy a steam heater having a capacity of 3960 $MJ{\cdot}h^{-1}$ or 1100 kW (15% spare capacity).

The gas mass flow through the steam heater is

$$\frac{3{,}455{,}365}{1.0(205-10)} = 17{,}720\,\mathrm{kg{\cdot}h^{-1}}$$

$$R_A = \frac{355}{273+10} = 1.25\,\mathrm{kg{\cdot}m^{-3}}$$

$$\mathrm{Airflow}_1 = \frac{17{,}720}{1.25} = 14{,}176\,\mathrm{m^3{\cdot}h^{-1}}$$

$$\mathrm{Power}_{G1} = \frac{14{,}176{\cdot}2500}{3600{\cdot}1000{\cdot}0.5} = 19.7\,\mathrm{kW}$$

2500: pressure loss of the steam heater in N·m^{-2}

Install a fan with a motor of 25 kW.

Dryer sizing Take a superficial gas velocity of 0.3 m·s^{-1}. 17,720 kg·h^{-1} of air is passed on to the dryer. The quantity of air extracted from the dryer is 1.1·17,720 = 19,492 kg·h^{-1} (ingress air factor 1.1).

$$R_A = \frac{355}{273+92} = 0.973\,\mathrm{kg{\cdot}m^{-3}}\ \text{at}\ 92^\circ\mathrm{C}$$

$$\mathrm{Airflow}_2 = \frac{19{,}492}{0.973} = 20{,}033\,\mathrm{m^3{\cdot}h^{-1}}$$

$$\mathrm{Evap} = 605.6\,\mathrm{kg{\cdot}h^{-1}}$$

$$R_W = \frac{220}{273+92} = 0.603\,\mathrm{kg{\cdot}m^{-3}}$$

$$WA_{\mathrm{flow}} = \frac{605.6}{0.603} = 1004\,\mathrm{m^3{\cdot}h^{-1}}$$

$$\mathrm{Gasflow}_2 = 20{,}033 + 1004 = 21{,}037\,\mathrm{m^3{\cdot}h^{-1}}$$

$$D = \sqrt{\frac{4{\cdot}21{,}037}{\pi{\cdot}0.3{\cdot}3600}} = 4.98\,\mathrm{m}$$

Take 5.00 m and a residence time of 25 s. Bottom: 60° cone.

$$V = \frac{25{\cdot}21{,}037}{3600} = 146.1\,\mathrm{m^3}$$

$H = 6.0$ m [which can be calculated by means of equation (8.10)].

Droplet size distribution A weight average droplet size of 89 μm was calculated in Section 8.3.

It is not necessary to reconsider the chamber diameter because of the mean droplet size. A minimum chamber diameter of 4 m is needed for 89 μm [6].

Sizing exhaust-gas unit A bag filter is selected, face velocity 1 m·min^{-1}. The number of m^2 required is 21,037/60·1 = 350.6. Take 350 m^2.

$$\Delta p = 1500\,\text{N·m}^{-2}$$
$$\text{Power}_{G2} = \frac{21{,}037 \cdot 1500}{3600 \cdot 1000 \cdot 0.5} = 17.5\,\text{kW}$$

Take a fan with a motor of 25 kW.

Powder transport and cooling It is preferred to cool the powder during transport. A powder-to-air mass ratio of 1 is selected.

Line velocity: 25 m·s^{-1}.
Powder mass flow: 500 kg·h^{-1}
Airmass flow: 500 kg·h^{-1}

Calculation of the temperature of the cooled powder

$$500 \cdot 1.0(T - 10) = 500 \cdot 1.25(72 - T) \quad \text{kJ·h}^{-1}$$
$$T = 45°\text{C}$$

Calculation of the line diameter

$$\frac{500}{1.12} = 446\,\text{m}^3\text{·h}^{-1}$$
$$D = \sqrt{\frac{4 \cdot 446}{\pi \cdot 3600 \cdot 25}} = 0.079\,\text{m}$$

Take $D = 0.08$ m.

Calculation of the bag filter area

Face velocity: 1 m·min^{-1}

$$\frac{446}{60 \cdot 1} = 7.4\,\text{m}^2$$

Take 8 m^2.

Calculation of the fan power

Pressure loss: 2000 N·m^{-2}

$$\frac{446 \cdot 2{,}000}{3600 \cdot 1000 \cdot 0.5} = 0.5\,\text{kW}$$

Install a fan with a motor of 1 kW.

Survey

Dryer diameter: 5.0 m
Dryer cylindrical height: 6.0 m
Steam heater capacity: 1.1 MW
Long-term steam consumption per kilogram of evaporated water: 3.3 kg

Electricity Consumption

Air-supply fan:	19.7	
Exhaust fan:	17.5	
Solids-transport fan:	0.5	
Atomizer:	8.8	(8 kWh per metric t of feed)
Miscellaneous:	+ 10.0	
	56.5 kW	(93.3 kWh per t of evaporated water)

Weight-average droplet size: 89 μm
Main filter area: 350 m^2

REFERENCES

[1] Masters, K. (2002). *Spray Drying in Practice*, SprayDryConsult ApS, Charlottenlund, Denmark, p. 50.

[2] Richter, T. (2004). *Atomization of Liquids*, Expert Verlag, Renningen, Germany, pp. 35–37 (in German).

[3] Kircher, D. (1999). Private communication.

[4] Keey, R.B. (1999). Private communication.

[5] Reference [1], pp. 135–137.

[6] Masters, K. (1991). *Spray Drying Handbook,* Longman Scientific & Technical, Harlow, UK, p. 177.

[7] Marshall, W.R. (1954). *Atomization and Spray Drying*, American Institute of Chemical Engineers, New York, p. 39.

[8] Reference [1], p. 166.

[9] Beek, W.J., Muttzall, K.M.K., van Heuven, J.W. (1999). *Transport Phenomena*, Wiley, Chichester, UK, pp. 89–93.

[10] Reference [2], pp. 105–109.

[11] Walton, D.E., Mumford, C.J. (1999). Spray dried products-characterization of particle morphology. *Transactions of the Institution of Chemical Engineers*, *77*, A21–A38.

[12] Reference [1], pp. 408, 409.

[13] Freudig, B.S., Hogekamp, S., Schubert, H. (1999). Dispersion of powders in liquids in a stirred vessel. *Chemical Engineering and Processing*, *38*, 525–532.

[14] Dolinsky, A.A. (2001. High-temperature spray drying. *Drying Technology*, *19*, 785–806.

[15] Kamenkovich, V.V., Solov'eva, T.A., Goikhman, I.D., Pad'ko, S.I., Babenko, V.E. (1983). Features of the industrial process of spray drying an intermediate detergent composition. *The Soviet Chemical Industry*, *15*, 364–370.

[16] Krischer, O., Kast, W., Kröll, K. (1978). *Dryers and Drying Processes*, Springer-Verlag, Berlin, pp. 278, 279 (in German).

[17] Reference [6], pp. 489–688.

[18] Noden, D. (1972). Trend towards use of dispersion dryers. *Chemical and Process Engineering*, *53*, 48–52.

[19] Noden, D. (1974). Efficient energy utilization in drying. *Processing,* December, 25–27.

[20] Perry, R.H., Green, D.W. (2008). *Perry's Chemical Engineers' Handbook*, 8th ed., McGraw-Hill, New York, p. 12–96.

9

MISCELLANEOUS CONTINUOUS CONVECTIVE DRYERS AND CONVECTIVE BATCH DRYERS

The "big four"—fluid-bed, direct-heat rotary, flash, and spray dryers—were discussed in the preceding chapters. Some important convective-type dryers remain to be dealt with in this chapter.

The conveyor dryer is indispensable for the gentle drying of relatively coarse particulate material, the somewhat large particle size normally being obtained by a forming machine. The Wyssmont Turbo-Dryer is another gentle dryer for both fine and coarse particulate material. The Nara Media Slurry Dryer can process suspensions, using a fluidized auxiliary material (i.e., spherical particles) to accomplish drying. The particles are made of alumina (Al_2O_3) or zirconia (ZrO_2) and have a diameter of 2 mm, for example. This can be an alternative for a spray dryer. The Anhydro Spin Flash Dryer can dry pastes, filter cakes, and even sludges. The feed is disintegrated by means of an agitator in the drying chamber. The Hazemag Rapid Dryer is, in the ceramic industry, well established for the processing of plastic masses. It contains two agitators. The next convective drying method discussed is the simultaneous milling and drying of wet particulate material, which is a long-established practice. The milling is used not only to decrease the particle size, but also, in other applications, to preserve it by preventing particle agglomeration.

Fluid-bed and atmospheric tray dryers are examples of convective batch dryers; the fluid-bed dryer is the more powerful dryer. The atmospheric tray dryer is limited by relatively high personnel costs because of the amount of physical handling required in its operation. Convective batch dryers tend to be used for organic specialty chemicals and pharmaceuticals. When it is important to avoid human contact with the product at the stages of liquid–solid separation, leaching, and drying, it is possible to use a centrifuge-dryer.

Drying in the Process Industry, First Edition. C.M. van 't Land.
© 2012 John Wiley & Sons, Inc. Published 2012 by John Wiley & Sons, Inc.

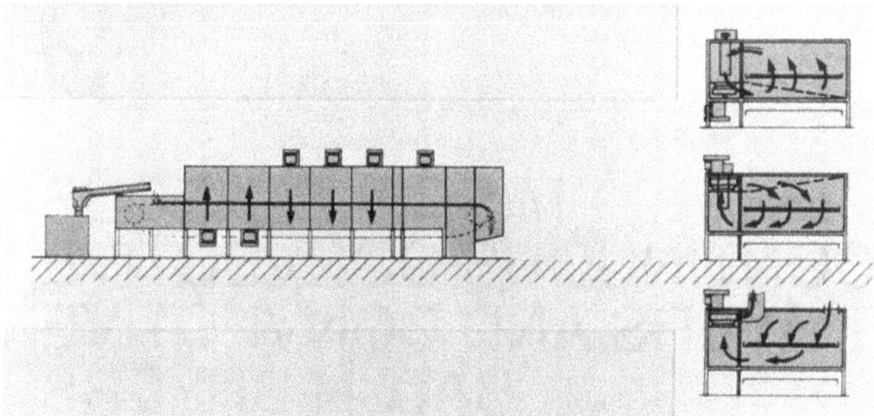

Figure 9.1 Single-conveyor dryer. (Courtesy of CPM Wolverine Proctor LLC, Horsham, PA.)

9.1 CONVEYOR DRYERS

A conveyor dryer is designed so that material is dried as it is continuously being transported horizontally on a perforated screen through which warm air is blown. The most widely used type is the single-conveyor dryer (see Fig. 9.1). The product is not moved in the dryer, which is an advantage for some products. It requires only a small installation height. Staging is employed when products with a high initial moisture content are to be dried (see Fig. 9.2). On drying these products, a large amount of shrinkage occurs (e.g., with the drying of onions), and repacking is necessary to promote material to air contact. Conveyors in series can have different speeds. Different drying conditions can be adjusted in the separate drying conveyors. Repacking also minimizes sticking to the conveyor.

Multiple-conveyor dryers are used when the zoning of air temperatures is not required and when, for quality reasons, drying must proceed with low air velocities and temperatures (e.g., for pastas; see Fig. 9.3). The drying time tends to become long in this case. Typically, the drying air flows through the product. For this dryer type, the reduced floor space is an advantage where only limited floor space is available. Product turnover, when being transferred from one conveyor to the next, is an additional feature of these dryers.

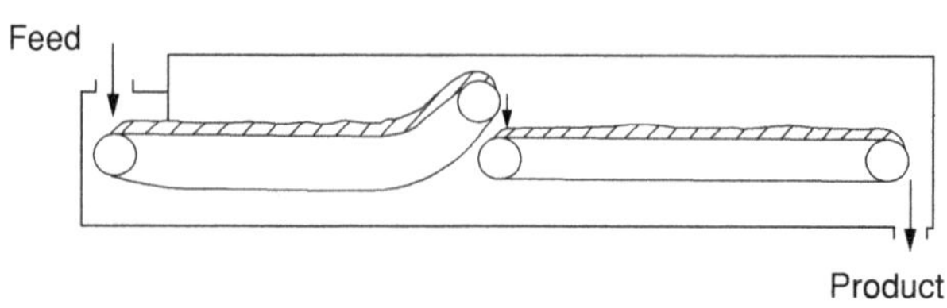

Figure 9.2 Two-stage conveyor dryer (seen schematically).

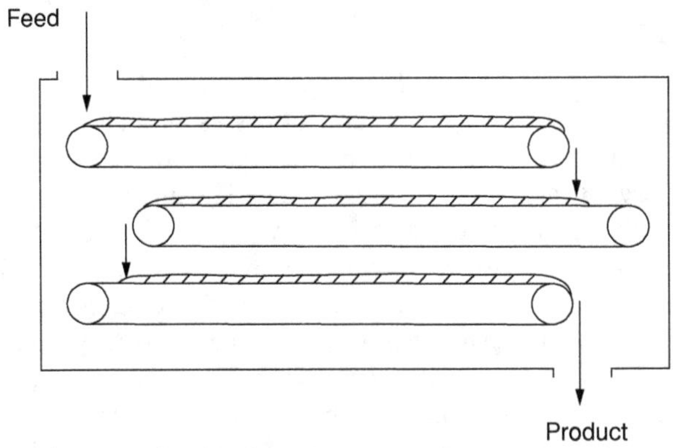

Figure 9.3 Multiple-conveyor dryer (seen schematically).

While being dried, the material is stationary, so the technique is suitable for friable products. The bed of wet material must be permeable, which means that the feed must be particulate. Since the required particle size is rather large (from 1 mm to several centimeters), feed preforming is often necessary (e.g., by an extruder). It is important to distribute the feed carefully (e.g., by means of an oscillating spreader) as there is no immediate opportunity to rearrange the bed of solids (compare, e.g., a fluid-bed dryer). Sticking of the particles to each other or to the apron cannot be tolerated.

Most conveyor dryers have zonal internal air recirculation. With zone or cell or a combination of cells, there is a fresh air makeup and a purge. Internal air recirculations pass the bed of material to be dried and go through a fan and a heat source (usually, a steam heater). Typically, the exhaust is controlled by the temperature in the loop (see Fig. 9.4).

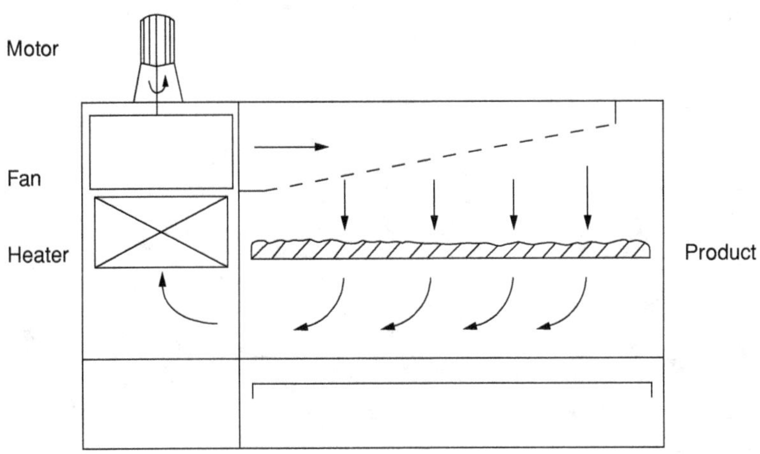

Figure 9.4 Cross-sectional view of a conveyor dryer cell.

A typical single-stage conveyor dryer consists of a number of almost identical cells in series between a feed part and a discharge section. An exhaust is active for a number of cells, and the same scheme applies for a makeup. It is conceivable to think of an airflow regime in which all warm air passes through once as in a fluid-bed dryer, which would make sense for a conveyor dryer, too, as long as the material is in the constant-drying-rate period. However, materials that are being processed in a conveyor dryer soon enter the period of drying with a decreasing rate because of the relatively large size of the particles. Operating on a once-through basis would then mean that exhaust air with a relatively low degree of saturation would be obtained. The relative humidity of the exhaust air can be increased by recycling drying air.

The last cell or last set of cells is often used to cool the material down to 40°C at least. Atmospheric air is used, which here passes through once.

A practice that is sometimes adopted to obtain a good product is to control not only the temperature in the recirculation loop, but also the relative humidity, by steam injection. An example is the drying of polyester chips, which if dried otherwise, are less suitable for a subsequent process step: in this case, dyeing.

Figure 9.5 shows a dryer that is installed and operating.

Layer Depth On the average, depth varies between 2 and 15 cm, with 4 and 6 cm being quite common values. The pressure loss of the drying air that flows through the bed is related to the air velocity and the layer depth. This fact provides one reason for working with relatively shallow layers, because pressure drops exceeding

Figure 9.5 Operating crouton conveyor dryer. (Courtesy of CPM Wolverine LLC, Horsham, PA.)

25 to 50 mm water gauge would lead to leaks or to bypassing (e.g., through the slit between the moving belt and the casing).

Air Velocity Practical air velocities vary between 0.5 and 2 $m \cdot s^{-1}$, with 1 $m \cdot s^{-1}$ being a common value. Often, an upward flow is chosen at the beginning of the drying period and a downward flow later. This regime promotes even drying and avoids dusting.

Temperature Most conveyor dryers have drying-air temperatures between 100 and 200°C, with even higher temperatures a possibility.

Size There is no practical limit with regard to the length of a conveyor dryer, and the width can vary between 0.5 and 3 m.

Conveyor Velocity Usually, the conveyor velocity is continuously adjustable between 0.1 and 2 $cm \cdot s^{-1}$.

Design Method The design of a conveyor dryer is based on the registration of drying curves. Representative wet material is deposited on a perforated test tray (e.g., $0.3 \cdot 0.3\ m^2$) with air blown through. Independent variables (no air recycle) include (1) air conditioning (temperature and relative humidity), (2) air velocity, and (3) layer depth.

Dependent variables (as functions of time) include (1) exhaust temperature, (2) pressure loss across the material, and (3) sample weight. These curves allow the selection of the correct air conditions, the air velocity, and the layer depth. These variables define the capacity per square meter and per hour. A scaling-up factor of $1/0.7 = 1.42$ is normally adopted. The reasons are that the material is, in practice, distributed more poorly and that the actual large-scale situation is comparable but not identical to the laboratory situation because of the zonal recirculation. In the laboratory, tests are performed on a once-through basis. Attention is also paid to product shrinkage, fines generation, fume or smoke generation, tendency to stick to the test tray, and cooling of the product.

Example 9.1

Starting points

Capacity: 1000 $kg \cdot h^{-1}$ out of the dryer
Initial moisture content: 0.163 $kg \cdot kg^{-1}$ (dry basis)
Final moisture content: less than 0.005 $kg \cdot kg^{-1}$
Physical form: 2-mm-diameter extruded pellets, length varies between 1 and 10 mm
Cooling: below 40°C required
Maximum product temperature: 80°C
Bulk density feed: 650 $kg \cdot m^{-3}$

The product is neither toxic nor corrosive, and the liquid to be evaporated is water. There is no generation of fines that might lead to a dust explosion. It is decided to carry out tray tests. The results of one of these tests is used for scaling up. The product is insoluble in water.

Calculation The drying curve shows a linear decrease of the moisture content with time. A test tray of $0.3{\cdot}0.3\ \mathrm{m}^2$ was utilized. Independent process variables that were selected are (1) air temperature 70°C (no adjustment of the relative humidity); (2) air velocity of $1\ \mathrm{m{\cdot}s^{-1}}$ (downward), and (3) layer depth of 3 cm.

Dependent process variables that were obtained include: (1) 8 min is required to come down to to $0.005\ \mathrm{kg{\cdot}kg^{-1}}$, (2) it is mandatory to raise the product temperature to 70°C; (3) the average temperature of the used air is 50°C; (4) the pressure drop of the bed is 24 to 30 mm water gauge; and (5) 3 min is required to cool the product to below 40°C with an air temperature of 20°C and a velocity of $1\ \mathrm{m{\cdot}s^{-1}}$. The product does not stick to the tray.

	Test Mass Balance (kg)	
	In	Out
Dry matter	1.515	1.515
Water	+ 0.247	+ 0.001
	1.762	1.516

$$\text{Drying capacity: } 1.516 \cdot \frac{60}{8} \cdot \frac{1}{0.09} = 126.33\ \mathrm{kg{\cdot}m^{-2}{\cdot}h^{-1}}$$

$$\text{Drying area required: } 1.42 \cdot \frac{1000}{126.33} = 11.2\ \mathrm{m}^2$$

$$\text{Cooling capacity: } 1.516 \cdot \frac{60}{3} \cdot \frac{1}{0.09} = 336.9\ \mathrm{kg{\cdot}m^{-2}{\cdot}h^{-1}}$$

$$\text{Cooling area required: } \frac{1000}{336.9} = 3.0\ \mathrm{m}^2$$

Conveyor dryers consist of standard sections. Probably, three sections would be required for drying and one for cooling. The heat efficiency of this dryer is relatively poor as the drying air is warmed from 20°C to 70°C; on average it only cools down to 50°C. By choosing the drying area 42% larger than the area calculated, 50°C might come down slightly in the plant as some air recycling is possible, the air velocity remaining at $1\ \mathrm{m{\cdot}s^{-1}}$. To allow recycling, the investment increases somewhat (more area, larger fans).

A tray test is a static test. It would be a bonus to carry out a second series of tests using a small continuous conveyor dryer. This makes it possible to confirm the results of the tray tests and shows the transport of the material to, through, and out of the dryer.

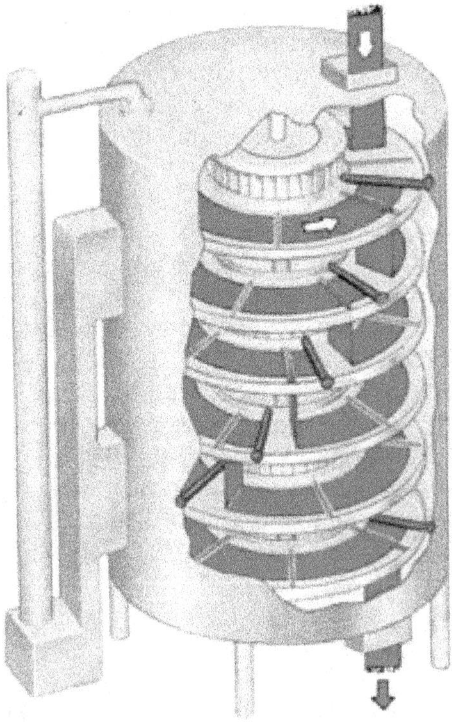

Figure 9.6 Wyssmont Turbo-Dryer. (Courtesy of Wyssmont Company, Fort Lee, NJ.)

9.2 WYSSMONT TURBO-DRYER

The Wyssmont Turbo-Dryer handles materials gently. Figure 9.6 shows this dryer schematically. It consists of a stack of slowly rotating circular trays. Material is fed onto the top tray. After one revolution the material is wiped onto the next lower tray, where it is leveled by a stationary leveler, and then, again after a second revolution, it is wiped to the next tray, where the operation is continued. Thus, the material is processed in plug flow with intermittent product redistribution. The trays are contained in an enclosure in which heated air or gas is circulated by internal fans. The fans are located in the center of the dryer. The name of the dryer is derived from these fans. Hot or warm drying air or drying gas is supplied at the dryer side and the exhaust gases are extracted at the dryer top. The dryer can operate with inert atmosphere recirculation with solvent recovery. Zonal adjustments of the temperature of the air or gas supply are possible. Thus, it is possible to cool the product on the lowest trays.

Typical Process Variables The dryer operates at atmospheric pressure. To avoid powder emissions, the pressure is slightly lower than the atmospheric pressure if water is evaporated and air is used as the drying medium. An inert gas (e.g., nitrogen) is used as the drying medium if a solvent is evaporated. The pressure is, then, to avoid air ingress and the risk of an air–vapor explosion, slightly higher than atmospheric pressure. The maximum air temperature is 650°C. Typically, a package unit having a diameter of 1.8 m contains 24 shelves. Using air of 150°C, this dryer can dry approximately 70 $kg \cdot h^{-1}$ of feed having a moisture content of 30% water by weight (wet basis). All moisture should be free moisture in that case.

Typical Applications The dryer is suitable for the handling of fragile materials such as crystals and pellets. Neither dust nor fines are generated. The material dried should not tend to incrustate, as the handling is gentle. It is possible to improve the product flow characteristics by backmixing.

Semitechnical Experiments and Scale-up Plug-flow drying with intermittent product redistribution is simulated in a cabinet dryer. Several kilograms of feed is sufficient for testing. The results of the cabinet dryer can be scaled up to industrial dryer sizes. Transport through the dryer must be tested in a pilot dryer. Package units have diameters in the range 1.2 to 3.6 m and heights in the range 1.5 to 4.8 m. The largest dryer has a diameter of 10.6 m and a height of 18.2 m. The manufacturer indicates a typical water evaporation capacity of 11.3 $t \cdot h^{-1}$ for this dryer.

9.3 NARA MEDIA SLURRY DRYER

The Nara Media Slurry Dryer can process suspensions (see Fig. 9.7). It is a dryer that uses an auxiliary material (i.e., spherical particles) to accomplish drying. The particles are made of alumina (Al_2O_3) or zirconia (ZrO_2) and have a diameter of 2 mm, for example. Essentially, the dryer consists of a circular fluidization chamber, in which, by means of warm air, the particles are fluidized. The feed is pumped into the bed and the suspension coats the particles. The moisture evaporation by the warm gas is the next step. Solids remaining on the surfaces of the spheres are removed by the collisions and friction of the spherical particles and are carried over by the fluidization gas. The product is separated from the gas stream by a cyclone and a filter. The particle size of the product is generally approximately equal to the particle size in the feed. Typically, the average particle size is in the range 1 to 10 μm.

Usually, water is evaporated and warm air is used as the drying gas. An inert drying gas (e.g., nitrogen) is recycled if a flammable solvent must be evaporated. There is a small overpressure in the drying chamber in that case. If solids being processed can, as a dust, explode with the oxygen in the air, an inert drying gas can also be used.

The process is controlled by keeping the warm gas flow and its temperature constant. Furthermore, the exhaust gas temperature is kept constant by adjusting the feed flow.

Typical Process Variables The maximum feed viscosity is 20,000 cP. The feed moisture contents are typically in the range 40 to 75% by weight, whereas the product moisture contents are typically in the range 0.1 to 2% by weight. The air-inlet

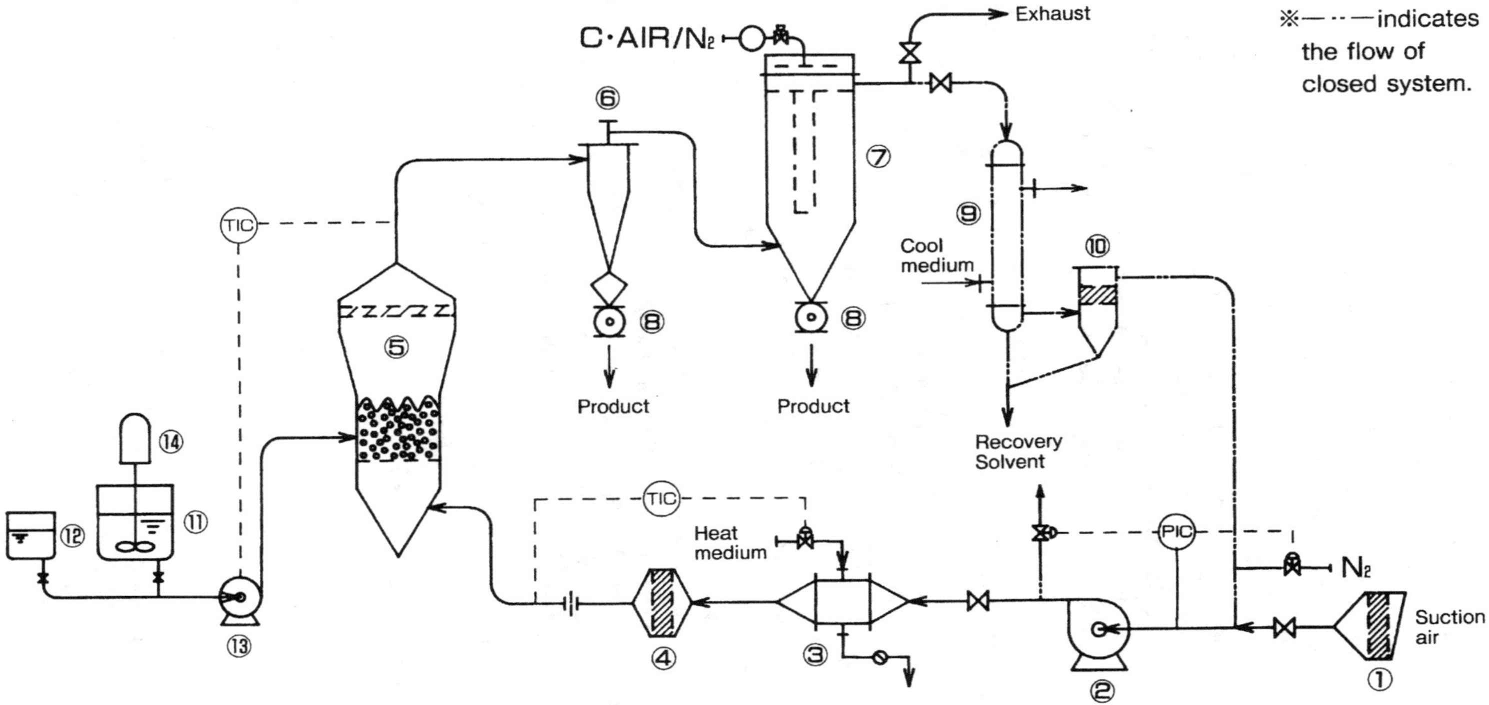

①Suction Filter
②Blower
③Hot Air Heater
④Hot Air Filter
⑤Media Slurry Drying Tower
⑥Cyclone
⑦Bag Filter
⑧Rotary Valve
⑨Condenser
⑩Demister
⑪Slurry Tank
⑫Cleaning Water Tank
⑬Slurry Pump
⑭Agitator

Figure 9.7 Nara Media Slurry Dryer. (Courtesy of Nara Machinery Co., Ltd, Frechen, Germany.)

temperature can vary from 50°C to a maximum of 250°C and the air velocity is at most $4\,m{\cdot}s^{-1}$.

Typical Applications Barium sulfate, polystyrene resin, and a ferrite (ferrites are complex salts of H_2FeO_4 that are used for electronic applications) are typical examples of products processed.

Semitechnical Experiments and Scale-up Semitechnical experiments are carried out in a piece of equipment having a diameter of 0.1 m. The results can be scaled up on the basis of the fluidization area. On keeping the air-inlet temperature and the fluidization air velocity constant, the scale-up factor is 100 for a dryer that has a diameter of 1 m. The largest dryer has a diameter of 1.5 m, and its maximum evaporative capacity is $1.5\,t{\cdot}h^{-1}$.

Comparison with a Spray Dryer

	Nara Media Slurry Dryer	Spray Dryer
Maximum air-inlet temperature (°C)	250	800
Feed	Suspensions	Suspensions and solutions
Maximum evaporative capacity ($t{\cdot}h^{-1}$)	1.5	No limit
Product particle size (μm)	1–10	50–250

The maximum air-inlet temperature of a spray dryer can be much higher than the maximum air-inlet temperature of the Nara Media Slurry Dryer because of the former's relatively large and simple construction.

For a given, relatively small evaporative capacity, and for equal air-inlet temperatures, the Slurry Dryer requires a smaller investment than that for a spray dryer. The reason is that the production intensity, measured in $t{\cdot}m^{-3}{\cdot}h^{-1}$, is then much greater for a Slurry Dryer than for a spray dryer. The variable costs are comparable.

9.4 ANHYDRO SPIN FLASH DRYER

The Anhydro Spin Flash Dryer can dry pastes, filter cakes, and even sludges. Essentially, the dryer consists of a cylindrical drying chamber into which the feed is conveyed and the warm or hot air is passed tangentially at the bottom. The feed is disintegrated by means of an agitator in the drying chamber (see Fig. 9.8). The tangential air inlet, together with the rotor action, causes a turbulent whirling airflow in the drying chamber. The turbulence prevents the agglomeration of the elementary particles during the drying process. The fine airborne powder particles pass through the classifier at the chamber top. The classifier prevents the larger particles from passing on to the filter. The larger particles tend to fall back to the agitator zone, are disintegrated, and continue drying.

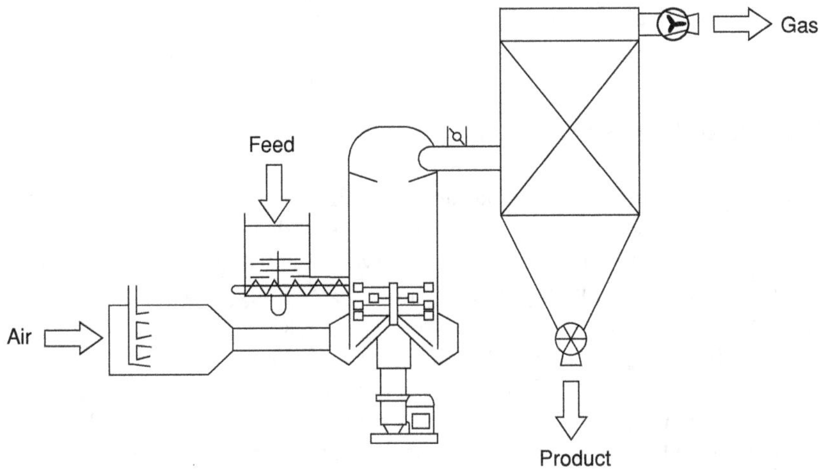

Figure 9.8 Anhydro Spin Flash Dryer. (Courtesy of Anhydro A/S, Søborg, Denmark.)

Usually, water is evaporated and warm or hot air is used as the drying gas. An inert drying gas (e.g., nitrogen) is recycled if a flammable solvent must be evaporated. A self-inertizing system can be taken if dust explosion hazards must be eliminated. Self-inertization is described in Chapter 14, as a method of safeguarding a drying operation. As an alternative to this system, it is possible to build the drying chamber pressure shock-resistant and to provide a bursting disk on the filter. Pressure shock-resistant means that the equipment can withstand the effects of a dust explosion; however, some parts must be replaced after the event.

The process is controlled by keeping the warm- or hot-air gas flow and its temperature constant. In addition, the exhaust gas temperature is kept constant by adjusting the feed flow.

Typical Process Variables Normally, the inlet-air temperature falls within the range 150 to 700°C. Drying chambers for inlet temperatures above 500°C feature special, heat-resistant steel air distributors. Filter cake moisture contents are typically in the range 30 to 55% by weight.

Typical Applications Calcium stearate, barium carbonate, and lactose are typical examples of products processed.

Semitechnical Experiments and Scale-up Semitechnical experiments are carried out in a piece of equipment having a diameter of 0.20 m. The results can be scaled up on the basis of the drying chamber diameter cross-sectional area. On keeping the inlet air temperature and the average upward air velocity in the drying chamber constant, the scale-up factor is 100 for a dryer having a diameter of 2 m.

The largest dryer has a diameter of 2.5 m and its maximum evaporative capacity is 7.8 $t \cdot h^{-1}$.

9.5 HAZEMAG RAPID DRYER

An example of an agitated continuous dryer, the Hazemag Rapid Dryer operates with a cocurrent flow of the feedstock and the drying gas, as depicted schematically in Figure 9.9. In Figure 9.10, an installed Rapid Dryer is shown that processes industrial sludge (the approximate dryer length is 6 m). The stationary dryer shell is provided with double pendulum self-sealing feed and dry-product connections, which may be coated with Teflon to prevent incrustations. A pair of rotating shafts carrying paddles disintegrates the feedstock and propels it through the dryer. A typical agitation system can be seen in Figure 9.11. The rotational directions are such that at the center a fountain is created. The shaft bearings near the feeding point can be water-cooled.

The material of construction of the dryer is steel, with the paddles, which are easily replaceable, being made of wear-resistant steel. However, stainless steel may also be used. Usually, the dryer is not insulated because in an insulated dryer the relatively high inlet temperatures (600 to 800°C) would lead to too-high temperatures for vital mechanical parts such as bearings.

Typical Process Variables The evaporative loads are in the range 50 to 25,000 $kg \cdot h^{-1}$. The spent drying gas characteristically exits at temperatures of 100 to 120°C. Typical peripheral speeds of the paddles are in the range 6 to 12 $m \cdot s^{-1}$.

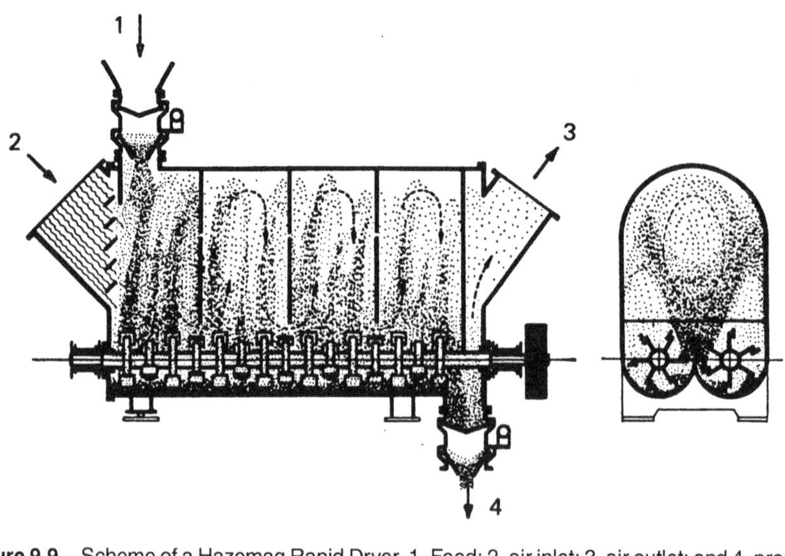

Figure 9.9 Scheme of a Hazemag Rapid Dryer. 1, Feed; 2, air inlet; 3, air outlet; and 4, product. (Courtesy of Hazemag & EPR GmbH, Dülmen, Germany.)

Figure 9.10 Hazemag Rapid Dryer used for industrial sludge. (Courtesy of Hazemag & EPR, Dülmen, Germany.)

Typical Applications The dryer is being used for mineral sludges, coal sludge, clay, and various filter cakes. The processing of plastic masses for the ceramics industry is particularly successful with this dryer. Akzo Nobel Chemicals used this dryer for the drying of fluorspar, natural calcium fluoride, which has a varying moisture content. The dry fluorspar reacted with sulfuric acid to obtain hydrogen fluoride, which was used to make propellants. This production was stopped because it was observed that this type of propellant attacks the ozone layer in the atmosphere.

Semitechnical Experiments and Scale-up Semitechnical experiments can be carried out in the smallest Rapid Dryer, which has an approximate length of 2.5 m. The amount of drying air increases proportionally with the capacity increase. The peripheral speed of the paddles is kept constant. The largest two-shaft dryer is about 9.1 m long, 4.5 m wide, and 8.7 m high.

Figure 9.11 View into the interior of a Hazemag Rapid Dryer. (Courtesy of Hazemag & EPR GmbH, Dülmen, Germany.)

9.6 COMBINED MILLING AND DRYING SYSTEM

The simultaneous milling and drying of coal in pulverized coal installations is a long-established practice, and the combination is also used in the food and chemical industry. Several companies are specialized in the manufacture of milling-drying systems. The combined milling and drying of coal is an example of a process step at which it is desired to both decrease the average particle size and to dry the material. In other industry branches, the operation is often used to preserve the average particle size. The turbulence within the mill due to the rotation then prevents agglomeration of the elementary particles in the feed during the drying step. The description will now focus on the combined milling and drying systems of Altenburger Maschinen Jäckering at Hamm, Germany. Their systems were originally introduced for the combined milling and drying of wheat and are now also used for other applications.

Figure 9.12 Combined milling and drying plant. (Courtesy of Altenburger Maschinen Jäckering GmbH, Hamm, Germany.)

A mill into which the feed is conveyed or pumped and the hot or warm air is passed is the first part of the system (see Fig. 9.12). The milling and the major part of the drying occur in the mill. The mill contains a rotor fitted with special blades and rotating at a high peripheral speed within a stator. The stator usually has a grooved surface and the gap between the rotor and the stator is narrow. Because of the centrifugal force, the solids are concentrated at the rotor's periphery. The air velocity within the eddies generated by the rotor between the rotor blades approaches the velocity of sound. The particles are milled or kept apart by the action of the air. The energy consumption of the mill is considerable, and within the mill it is converted into heat mainly. This heat supplies part of the water heat of evaporation. The remainder of the latter heat is supplied by the warm or hot air. Additional drying occurs in the flash drying tube, whereas solid–gas separation takes place in a filter. The air leaving the filter flows to a fan that supports the airflow through the installation. Therefore, there is a small underpressure in the entire system.

The process is usually controlled by keeping the exhaust gas temperature constant by adjusting the temperature of the warm- or hot-air gas flow.

The mill is built resistant to pressure shock if the material processed, together with the oxygen in the drying air, can give rise to a dust explosion. Being resistant to pressure shock means that the equipment can withstand the effects of a dust explosion; although some parts must be replaced after the event. However, Altenburg Maschinen Jäckering has, until the present day, not experienced a dust explosion in the mill. Dust explosions have occurred in the filter, so if a dust explosion can occur, the filter is provided with a rupture disk.

To improve the handling characteristics of the feed, recycling of milled and dried product is sometimes practised.

Typical Process Variables The maximum inlet air temperature is 300°C. Normally, outlet air temperatures fall within the range 60 to 100°C. The circumferential velocity of the rotor is in the range 80 to 140 $m \cdot s^{-1}$. A typical value for the tolerance between rotor and stator is 2 mm.

Typical Applications Wheat, methyl cellulose, and calcium carbonate (limestone) are typical products processed in the milling and drying systems described. The feed is pumped into the mill at the latter application. The average particle size of several micrometers is preserved in this case.

Semitechnical Experiments and Scale-up Semitechnical experiments are carried out in a small milling and drying system. The hot or warm airflow in the industrial mill, having the same temperature as the temperature used during the test, increases proportionally with the evaporation load. The large mill must be able to process the industrial feed and the larger airflow. The circumferential rotor velocity is kept constant to obtain the same milling effect. The upward gas velocity in the flash drying tube and its height should also be kept constant. The largest system has a maximum evaporative capability of 3.5 $t \cdot h^{-1}$. It can be equipped with a motor of up to 900 kW.

9.7 BATCH FLUID-BED DRYER

A batch fluid-bed dryer processes wet particulate material that can be dried by blowing warm air through its mass. There are similarities with the continuous fluid-bed dryers that are described in Chapter 5. The material to be dried is charged in a layer between 0.3 and 1 m thick into a container with a perforated supporting plate. The container is placed in a drying cabinet and airflow through the layer causes drying. Gradually, as the material gets drier, fluidization starts. The air leaves the dryer through filter bags (see Figs. 9.13 and 3.4).

The air velocities can be quite high in these dryers, and proper fluidization may not be achieved. This is not too much of a problem since the material cannot be carried over. It is common practice to blow air through the bed for awhile and then interrupt the gas flow to allow material on the filter cloth to fall down. This type of dryer is widely used for drying fine chemicals and pharmaceuticals. The evaporation of free moisture can be described with the aid of the humidity chart (see Fig. 4.2). The basic design equation is found as follows:

$$\begin{aligned} G_{H_2O} &= G_{\text{air}}(x_o - x_i) \qquad && \text{kg} \\ t_1 &= \frac{G_{\text{air}}}{(\pi/4)D^2 \cdot v_F \cdot \rho_F} \qquad && \text{s} \end{aligned} \tag{9.1}$$

Chapter 5 includes cases where the simple model was found to be applicable to drying from the initial moisture content down to the critical moisture content. Usually, at the end of the first drying step, the material still contains too much moisture. A reduction of the moisture content from a few tenths of a percent to almost nil is

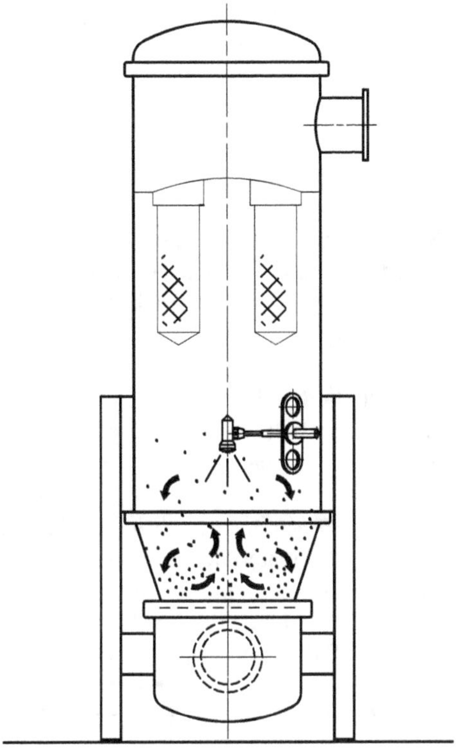

Figure 9.13 Batch fluid-bed dryer with a nozzle for granulation. (Courtesy of Glatt GmbH, Binzen, Germany.)

often still required. As the evaporation duty is usually small, the second drying step can be described as heating up the granular material, whereby thermal equilibrium is reached (i.e., the temperature of the fluidized bed equals that of the exit gas). Thus, the driving force for heat transfer decreases with the time elapsed as the heating air temperature is constant. $-d[TA_{\text{in}} - T_p(t)]/dt$ is proportional to $TA_{\text{in}} - T_p(t)$ in such a case. On integrating the corresponding differential equation, an expression containing a natural logarithm is obtained. A heat balance for a small time dt:

$$\frac{\pi}{4} \cdot D^2 \cdot v_F \cdot \rho_F \cdot c_p \cdot dt[TA_{\text{in}} - T_p(t)] = G_s \cdot c_s \cdot dT_p(t)$$

Rearranging and integrating between $t = 0$, $T_p(t = 0)$ and $t = t_2$, $T_p(t = t_2)$ yields

$$t_2 = \frac{G_s \cdot c_s}{(\pi/4) \cdot D^2 \cdot v_F \cdot \rho_F \cdot c_p} \ln \frac{TA_{\text{in}} - T_p(t = 0)}{TA_{\text{in}} - T_p(t = t_2)} \quad \text{s} \qquad (9.2)$$

More background concerning this design equation may be found in Section 5.3.

One may, however, also encounter substances where the situation is more complicated and the need to evaporate a considerable amount of bound moisture exists.

For example, this can be caused by the granule structure when there are crevices and/or pores. In cases of diffusion-limited evaporation, the temperature gradually increases and the moisture content gradually decreases. The time requirements must be established experimentally.

Cooling is carried out prior to packaging and can be dealt with in the same way as the heating at the evaporation of bound water.

Typical Process Variables Drying-gas temperatures seldom exceed 200°C. The drying gas is air or an inert gas (usually, nitrogen). A typical warm-air fluidization velocity is $1\,\text{m}\cdot\text{s}^{-1}$. It is possible to combine batch fluid-bed drying with fluid-bed granulation and fluid-bed coating.

Typical Applications Batch fluid-bed dryers are used in the pharmaceutical industry and to dry organic fine chemicals. The drying of solid peroxidicarbonates, a class of organic peroxides, is a typical application.

Semitechnical Experiments and Scale-up Small-scale trials can be carried out in dryers containing up to 45 L of material. Scale-up occurs by, first, verifying the applicability of equations (9.1) and (9.2) on a small scale. Second, the equations are applied to large-scale drying. Large batch fluid-bed dryers can contain up to $1.5\,\text{m}^3$ of material. The corresponding fan capacity is on the order of $12{,}000\,\text{m}^3\cdot\text{h}^{-1}$.

Example 9.2 A batch fluid-bed dryer for an organic peroxide is to be designed.

Product

General: batch size 120 kg of feed
TP_{in}: 18°C
TP_{out}: 20°C
A_1: 30% water by weight (wet basis)
A_2: 0.0% water by weight
Critical moisture content: 0.1% water by weight
c_s: $1.2\,\text{kJ}\cdot\text{kg}^{-1}\cdot\text{K}^{-1}$
Solubility solid: nil

Process

Ambient temperature: 10°C
Ambient relative humidity: 50%
TA_{in}: 40°C by indirect heat exchange with warm water
Maximum product temperature: 35°C
Fluidization velocity: $0.7\,\text{m}\cdot\text{s}^{-1}$ under the distribution plate
c_p: $1.0\,\text{kJ}\cdot\text{kg}^{-1}\cdot\text{K}^{-1}$

Dryer

D: 0.9 m

Mass balance (kg)

	In	Out
Solids	84.0	84.0
Water	+ 36.0	+ 0.1
	120.0	84.1

$$\text{Evap} = 36.0 - 0.1 = 35.9$$

First drying period $x_o - x_i = 0.013 - 0.004 = 0.009$ kg of water per kilogram of dry air. See the humidity chart. The adiabatic saturation temperature is 18°C.

$$G_{\text{air}} = \frac{35.9}{0.009} = 3988.9 \text{ kg}$$

Note: This is dry air. The moisture content of the drying air is very low (0.004 kg of water per kilogram of drying air). Hence, the amount of drying air excluding the moisture is set equal to the amount of drying air, including the moisture.

The pressure under the distribution plate is approximately 104,000 N·m^{-2}.

$$\rho_g = \frac{355}{273+40} \cdot \frac{104{,}000}{101{,}300} = 1.164 \text{ kg·m}^{-3}$$

$$V = \frac{3988.9}{1.164} = 3426.9 \text{ m}^3$$

$$t_1 = \frac{3426.9}{(\pi/4) \cdot 0.9^2 \cdot 0.7} = 7695.4 \text{ s (2 h and 8.25 min)}$$

Second drying period From equation (9.2) we obtain:

$$t_2 = \frac{G_s \cdot c_s}{(\pi/4) D^2 \cdot v_F \cdot \rho_F \cdot c_p} \ln \frac{TA_{\text{in}} - T_p(t=0)}{TA_{\text{in}} - T_p(t=t_2)} \quad \text{s}$$
$$= \frac{84 \cdot 1.2}{(\pi/4)\, 0.9^2 \cdot 0.7 \cdot 1.164 \cdot 1.0} \ln \frac{40-18}{40-35} = 288 \text{ s (4.8 min)}$$

Cooling period

$$\rho_g = \frac{355}{273+10} \cdot \frac{104{,}000}{101{,}300} = 1.288 \text{ kg·m}^{-3}$$

$$t_3 = \frac{84 \cdot 1.2}{(\pi/4)\, 0.9^2 \cdot 0.7 \cdot 1.288 \cdot 1.0} \ln \frac{10-35}{10-20} = 161 \text{ s (2.7 min)}$$

Review

First drying period: 2 h and 8.25 min
Second drying period: 4.8 min
Cooling period: 2.7 min
Loading/unloading: 15 min
Spare: 15 min
Cycle time: 2 h and 45.75 min

9.8 ATMOSPHERIC TRAY DRYER

An atmospheric tray dryer is shaped as an enclosed, insulated housing in which trays containing the feed are placed (see Fig. 9.14). The material to be dried is placed

Figure 9.14 Atmospheric tray dryer. (Courtesy of CPM Wolverine Proctor LLC, Horsham, PA.)

in layers that are between 1 and 5 cm thick on trays that are then loaded on trucks which are wheeled into drying cabinets. Alternatively, trays with material are placed manually into drying cabinets. After closing the doors, the air is heated and circulated through the dryer. There is a purge of warm air containing the evaporated moisture, and fresh makeup air is introduced. The loading and unloading of the trays require considerable manual labor, and although automation is possible in principle, this is one aspect that in many instances has caused the replacement of tray dryers with other systems (e.g., fluid-bed dryers).

A further aspect to be considered is the relatively low output of tray dryers when measured in $kg \cdot h^{-1} \cdot m^{-3}$. A drying cycle of between 5 and 40 h is not unusual in tray dryers. Fluid-bed dryers are superior in this respect. The final drawback is the nonuniform airflow in tray dryers, which can result in local overheating even though in other parts of the dryer wet product can still be found. Because of product quality or process safety requirements, a restriction is often placed on the drying-air temperature. Large individual items can be stacked in piles or placed on shelves, which demonstrates that the tray dryer can accept a wide range of feed physical forms (e.g., filter cakes, pastes, sludges). Sometimes, however, the product must be converted into a free-flowing powder by means of a screen or other device.

The most common practice is to circulate the gas between the trays (i.e., the dryers operate with cross-airflow). An alternative is to circulate the air through the material. In the case of coarse particulate material, this leads to higher specific evaporation rates in kilograms per m^2 tray area per hour. Atmospheric tray dryers are used for the drying of pharmaceuticals, pigments, and fine chemicals.

Normally, a maximum air temperature of approximately 200°C is employed, with the air being heated indirectly by means of steam. An average recirculation figure of 80% is used. Cross-airflow velocities can be as high as $3\ m \cdot s^{-1}$, but a value of 1 to $1.5\ m \cdot s^{-1}$ is customary. Through-circulation equipment typically uses an air velocity of 0.5 to $1\ m \cdot s^{-1}$.

Typical performance data vary between 0.1 and 1.5 kg of water evaporated per m^2 tray area per hour, with the higher values being obtained at high drying temperature, and vice versa (e.g., 0.2 at 50°C and 1.0 at 150°C). Usually, steam consumption varies between 1.5 and 3.5 kg of steam per kilogram of evaporated water. Normally, tray dryers are from several cubic meters to several tens of cubic meters.

The results of small-scale drying trials can be readily scaled up. It is essential that the temperature, relative humidity, and velocity of the air accomplishing the drying are kept constant. The amount of drying air per kilogram of product must also be kept constant.

Example 9.3 A pharmaceutical product is to be dried with air of 95°C and subsequently cooled to 40 to 50°C. A tray dryer with 29 trays is available; the area per tray is $1\ m^2$. Approximate tray dryer dimensions: width 1 m, depth 1.5 m, and height 2 m. The initial moisture content (wet basis) is 32% by weight. The bulk density of wet product is $600\ kg \cdot m^{-3}$.

Take a layer thickness of 2 cm. The feed charge is $29 \cdot 1 \cdot 0.02 \cdot 600 = 348$ kg.

	Mass Balance Product (kg/batch)	
	In	Out
Solids	237	237
Water	+111	+ 0
	348	237

Evaporation: 111 kg per batch

A specific evaporation capacity of 0.6 $kg \cdot m^{-2} \cdot h^{-1}$ is assumed.

Drying time: $\dfrac{111}{29 \cdot 0.6} = 6.4$ h

Cooling to 40 to 50°C: 2 h

Cycle time: $6.4 + 2 = 8.4$ h

Capacity: $\dfrac{237}{8.4} = 28.2\ kg \cdot h^{-1}$

9.9 CENTRIFUGE–DRYER

Liquid–solid separation, leaching, and drying are hermetically sealed from the environment in a centrifuge–dryer. This can be important to avoid human contact with the product. The centrifuge–dryer FSD (FIMA Suspension Dryer) is described. The machine is shown in Figure 2.3. The slurry to be separated flows through the hollow shaft into the rotating drum. During centrifuging, a cake builds up on the drum wall while most of the liquid disappears through a filter cloth and the perforated drum wall. The rotating drum actually consists of two drums: an inner, conical perforated drum and an outer, cylindrical solid drum.

One or more cake washes can be carried out after the initial liquid–solid separation. Again, most of the wash liquid disappears through the perforated drum wall covered with a filter cloth.

There are centrifuge cakes that at the solid–liquid separation step, suffer from compaction and sometimes even become impermeable. In those instances, the cake can be kept loose by passing in gas pulses from below and from outside the drum into the drum. Gas having a pressure of 6 bar gauge is throttled through nozzles. The manufacturer calls this method *counter pulse technology*. The pressure in the drum is 0.2 bar gauge then. The rotational speed at the liquid–solid separation steps can be chosen at will.

Drying is the next phase. Drying occurs convectively, and in many instances nitrogen obtained from a liquid nitrogen storage is warmed to 70°C, for example. There are alternatives for the drying step.

First, it is possible to pass warm nitrogen or air into the centrifuge drum through the hollow shaft. The warm gas flows through the cake. This mode is called *fixed-bed drying*. There are centrifuge cakes that in fixed-bed drying, suffer from compaction and sometimes even become impermeable. In those instances, the cake can, as in the centrifugation step, be kept loose by passing in warm-gas pulses from below and from outside the drum into the drum. Warm gas, having a pressure of 6 bar gauge is throttled through nozzles. This warm gas accomplishes drying. The radial acceleration necessary to loosen the cake against the centrifugal force is, for example, only $4 \cdot g$. To shorten the drying step, in addition to the gas pulses, it is possible to pass warm drying gas into the drum through the hollow shaft.

Second, it is possible, while the drum has a rotational speed of $5\ \text{min}^{-1}$, for example, to dislodge the cake from the drum by means of warm nitrogen or air pulses. The gas, which has a pressure of 6 bar gauge, is throttled through nozzles and enters the drum at the lower drum parts. The drum pressure is 0.2 bar gauge. The material collects in the lower drum parts and is dried by the gas pulses. That mode is called *fluid-bed drying*. The latter mode is unsuitable for materials that, under the influence of the rotation, suffer from balling. A variant of this method is to dislodge part of the cake while the drum is stationary. The dislodging is continued after the drum has rotated over a relatively small angle.

The temperature of the gas is 70°C, for example, at both 6 and 0.2 bar gauge. At the throttling of an ideal gas, which is an isenthalpic process, the temperature remains unchanged. The energy released by the gas is used to overcome the friction in the narrow passage.

The gas cools down during drying. The conical drum wall acts as a dust filter. At the same time, the exhaust gas cools the perforated drum wall. The drum wall cannot attain the temperature of the drying gas due to this effect. This is important for thermally sensitive materials. Postdrying is part of the drying process.

The temperature of the leaving gas can be used to control the product's final moisture content.

The fourth and final phase (the first three phases being mother liquor separation, cake washing, and drying) is discharge of the drum contents. The drum pressure is atmospheric in this phase. The rotational speed of the drum is raised when the drying step is complete. The material slides down the conical drum, leaves the machine, and by means of an integrated pneumatic system can subsequently be conveyed to a mixer, for example. The material movement in the drum can be supported by a gas flow.

It is possible to pass the nitrogen, used for drying, through an incinerator to burn gaseous organic components.

Typical Process Variables The maximum radial acceleration is approximately 600·g. Generally, peeler centrifuges can operate with greater radial accelerations (e.g., 1000·g or even 1500·g). Greater radial accelerations are not required because the centrifugation step is, in the same machine, followed by a drying step. The drum's rotational speed is continuously adjustable. The range of FSD machines can accommodate 20 to 800 L of cake.

The maximum temperature of the drying gas is in the range 130 to 140°C. A typical mass flow is 800 kg of warm, dry gas per square meter of drum area per hour. The hole size in the perforated drum varies from 5 to 18 μm.

Typical Applications The centrifuge–dryer is used primarily in the pharmaceutical and chemical industries. The processing of aspirin is a typical example. Antibiotics, penicillin, insulin, and steroids are additional examples.

Semitechnical Experiments and Scale-up Pilot-plant experiments can be carried out with a machine having a cake volume of 20 L, and the results can be scaled up. A normal cake thickness for the 20-L machine is 6 cm, and 20 cm can be accomodated in a 800-L centrifuge. The dependence of the filtration rate on the cake thickness can be checked in the small machine, and the results can be extrapolated. Assessment of the correct centrifugation time is not critical, as it is usually shorter than the drying time. The drying gas mass flow per kilogram of solid material should be kept constant.

Example 9.4 The drying time in a centrifuge–dryer is to be calculated. A slurry containing a particulate solid suspended in water is processed in an FSD-1300 centrifuge–dryer. This machine can accommodate a maximum cake volume of 530 L. The drum's diameter, length, and filtration area are, respectively, 1300 mm, 800 mm, and 3.27 m^2. A dry airflow of 2500 $kg{\cdot}h^{-1}$ at 70°C is used to dry the solid.

Liquid–solid separation and washing take 1 h. The cake remaining in the centrifuge has a mass of 200 kg and contains 50 kg of water. The time needed to evaporate this moisture will be calculated. All moisture is free moisture. The drying process occurs at a pressure of 120,000 $N{\cdot}m^{-2}$.

Water properties

$$\Delta H = 2{,}433{,}000\ \mathrm{J{\cdot}kg^{-1}}\ \text{at } 29^\circ\mathrm{C}$$
$$P_{sat} = 861\ \mathrm{N{\cdot}m^{-2}}\ \text{at } 5^\circ\mathrm{C}$$
$$P_{sat} = 3986\ \mathrm{N{\cdot}m^{-2}}\ \text{at } 29^\circ\mathrm{C}$$

Calculation An adiabatic saturation temperature of 29°C is assumed.

$$\begin{array}{lll} \text{At } 29^\circ\mathrm{C}: & P_{sat} = \quad 3{,}986 & \quad 3.3 \\ & P_{air} = +\ 116{,}014 & +\ 96.7 \\ \hline & P_{tot} = \quad 120{,}000\ \mathrm{N{\cdot}m^{-2}} & \quad 100.0\% \end{array}$$

$$\begin{array}{lll} \text{At } 5^\circ\mathrm{C}: & P_{sat} = \quad 861 & \quad 0.7 \\ & P_{air} = +\ 119{,}139 & +\ 99.3 \\ \hline & P_{tot} = \quad 120{,}000\ \mathrm{N{\cdot}m^{-2}} & \quad 100.0\% \end{array}$$

2500 kg of dry air is 86,207 mol·h^{-1}.

$$\text{Air preloading at 5°C: } 86{,}207 \cdot \frac{0.7}{99.3} \cdot 0.018 = 10.9 \text{ kg of } H_2O \text{ per hour}$$

$$\text{Air loading at 29°C: } 86{,}207 \cdot \frac{3.3}{96.7} \cdot 0.018 = 53.0 \text{ kg of } H_2O \text{ per hour}$$

Drying rate 53.0 – 10.9 = 42.1 kg of H_2O per hour

Heat balance check: 42.1·2,433,000 ≈ 2500·1000(70 – 29)

Conclusion: The assumption of 29°C was correct and the drying time is approximately 71 min.

Notes About 67 min can be calculated for drying at atmospheric pressure. This can be verified by repeating the calculation for a system pressure of 101,300 N·m^{-2}. The adiabatic saturation temperature is 27°C in this case, as can be read from the humidity chart. Generally, the convective drying process becomes less effective when the system pressure increases. This is because the water partial pressure is a function of the temperature only, whereas the air partial pressure is variable. The background is that air is a gas, whereas water is a vapor. Pressures higher than atmospheric pressure are effective to avoid air ingress. The example given concerns water evaporation, and the heat is supplied convectively by warm air. It is also possible to evaporate *i*-propanol, for example, by using warm nitrogen.

10

ATMOSPHERIC CONTACT DRYERS

A series of contact dryers is covered in this chapter, their fundamental bonus being their relatively modest energy consumption. In this respect, they are more efficient than convective dryers, but they all have a limited capacity. Scaling up is not effective because the heat-transfer area per unit of volume decreases. For example, equipment having triple the volume has only double the area. Continuous plate dryers, agitated contact dryers (both mildly and vigorously agitated), vertical thin-film dryers, drum dryers, steam-tube dryers, and spiral conveyor dryers are all considered in this chapter. The batchwise operating agitated atmospheric dryer is also treated.

Some atmospheric contact dryers, such as the plate dryer and the agitated atmospheric dryer, can also operate under vacuum. Vacuum drying is discussed separately in Chapter 11.

10.1 PLATE DRYERS

The plate dryer is a continuous contact dryer depicted schematically in Figure 10.1 (see also Fig. 3.12). The manufacturer is Andritz KMPT at Vierkirchen in Germany. The feedstock enters at the top of the dryer via a rotating lock. The product is conveyed in a spiral pattern across stationary plates by rakes on a vertical rotating shaft. The plates are heated or cooled by a medium. Even-numbered plates have a larger diameter than odd-numbered plates. Dried product also leaves the dryer via a rotating lock.

Drying in the Process Industry, First Edition. C.M. van 't Land.
© 2012 John Wiley & Sons, Inc. Published 2012 by John Wiley & Sons, Inc.

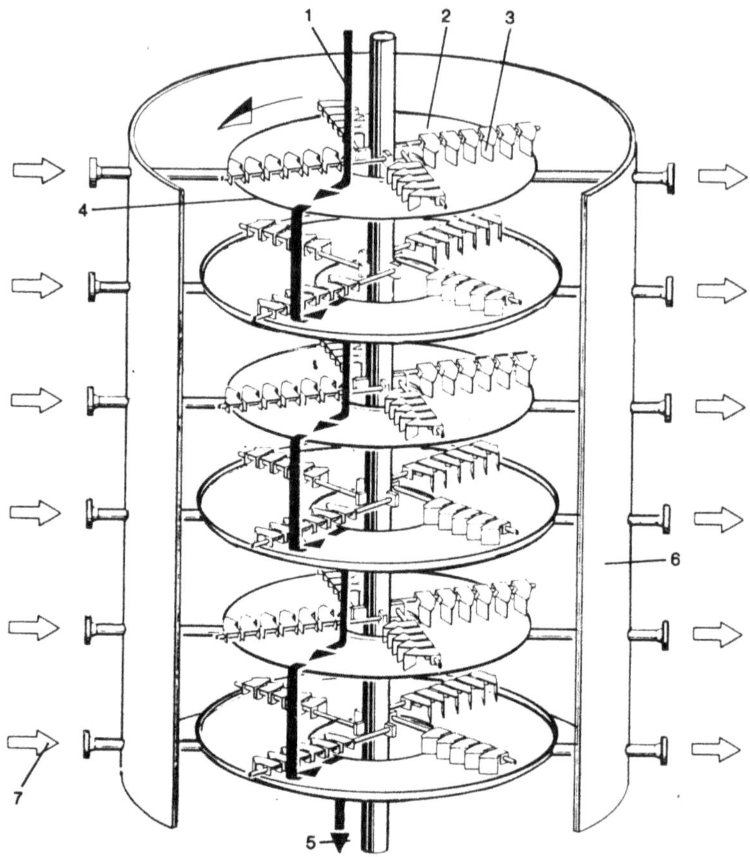

Figure 10.1 Plate dryer (seen schematically). 1, Wet product; 2, plate; 3, rabble mechanism; 4, plate rim; 5, dry product; 6, housing; 7, heating or cooling medium. (Courtesy of Andritz KMPT GmbH, Vierkirchen, Germany.)

Both this arrangement and positioning of the plows attached to the radial arms enable the processed material to take the path that is indicated. Note that this setup guarantees plug flow. Each plate or group of plates can be heated or cooled individually. Evaporated water or solvent is removed by strip gas (air or nitrogen) in cross-flow. The strip gas flows, horizontally between the plates, from the dryer center radially to the periphery. Thus, the gas velocity is minimum at the periphery, giving minimum dust entrainment. The combined gas flows flow upward along the dryer wall. To prevent emissions, the dryer is operated slightly below atmospheric pressure when water is evaporated. The dryer is operated slightly above atmospheric pressure (20 to 50 mbar gauge nitrogen atmosphere) when a solvent is evaporated. This is to eliminate the possibility of a gas explosion.

It is also possible to dry continuously under vacuum. The relatively high price of the vacuum equipment may be more than compensated by a faster evaporation

Figure 10.2 Plate dryer GTT 27/29 (29 plates, nominal diameter 2.7 m). (Courtesy of Andritz KMPT GmbH, Vierkirchen, Germany.)

rate leading to a smaller dryer. The vacuum-tight locks for the product flow are an additional cost item. A large-scale plate dryer is depicted in Figure 10.2.

Typical Process Variables With the use of thermal oil, the plate temperature can be as high as 300°C. The shaft rotates typically at 2 to 6 rpm. A characteristic figure is the product turnover number, which indicates the number of times that a specific particle is "reshoveled" during its stay in the dryer. (This is elucidated in Example 10.1.) An important aspect is the low gas velocity, typically between 0.01 and 0.1 $m \cdot s^{-1}$, leading to little, if any, dust carryover. The size of the gas cleaning equipment is hence quite modest. The dryer can provide residence time, which is often required for the removal of firmly bound moisture (5 to 300 min). It is usual to work with shallow product layers (i.e., 10 to 15 mm).

Typical Applications The dryer is applied in cases where a gentle treatment of the material dried is important. The feed should be free-flowing. The particle size can be up to 5 mm. It is also an advantage that the product can be dried and cooled in the same piece of equipment. The removal of methanol from rubber accelerators in a closed system is a typical example. The methanol vapor is entrained by nitrogen.

Semitechnical Experiments and Scale-up Experiments with a 0.1- or 0.15-m^2 batch plate dryer are usually adequate for a check of the feasibility. Essentially, this small dryer is one plate with an agitator. Semitechnical experiments with a continuous plate dryer having an area of 3.5 m^2 is the next step.

There are two series of industrial plate dryers. The nominal diameter of the larger plate of the dryers of the first series is 2 m, whereas the nominal diameter of the larger plate of the second series is 2.7 m. The largest dryer of the second series has 40 plates and a nominal area of 200 m^2.

Example 10.1

Starting points

Capacity: 350 $kg{\cdot}h^{-1}$
Initial moisture content: 0.25 $kg{\cdot}kg^{-1}$ (dry basis)
Final moisture content: 0.00 $kg{\cdot}kg^{-1}$
Feed physical form: wet powder
Cooling: under 40°C required
Maximum product temperature: 100°C

The product is a skin irritant, and the liquid to be evaporated is water. The solid material is water insoluble. It is decided to carry out tests in a laboratory plate dryer having one plate and an area of 0.1 m^2. The gas used is air. The agitator has two arms, and each arm carries two plows.

Calculation Independent process variables that were selected are (1) plate temperature 95°C, (2) air temperature 95°C, (3) air velocity 0.2 $m{\cdot}s^{-1}$, (4) agitator rotational speed 5 min^{-1}, and (5) initial feed layer thickness 10 mm.

The drying time was 37 min and the final product temperature was 85°C. The product neither stuck to the plate nor to the agitator. The product turnover number is $2{\cdot}5{\cdot}37 = 370$.

	Test Mass Balance (g)	
	In	Out
Dry matter	492	492
Water	+ 123	+ 0
	615	492

Based on this test, the required industrial drying area would be

$$0.1 \cdot \frac{37}{60} \cdot \frac{350}{0.492} = 43.9\,\mathrm{m}^2$$

With an allowance for losses and cooling, a 50-m^2 dryer would be a preliminary choice. A Model 20 dryer with 20 plates has a nominal area of 50 m^2. As the large dryer agitator has four arms, the rotational speed could be 2.5 min^{-1}. Thus, about the same product turnover number would be obtained.

10.2 MILDLY AGITATED CONTACT DRYERS (PADDLE DRYERS)

Figure 10.3 exhibits the functioning of typical paddle dryers. The wet feed enters at the left and exits at the right. The agitators (typically, two), which consist of shafts and paddles, and the jacket are heated, and the moisture evaporated is entrained by a small amount of carrier gas entering at both ends and sucked off centrally. The mass to be dried is agitated, which causes the product to flow through the dryer by breaking down the angle of repose.

The discussion will now focus on the GMF Paddle Dryer. The dryer is installed with a small slope (1 to 5°), and the particles flow by gravity from the feed end to the opposite end of the trough, where the product is discharged over a weir. Evaporation of both water and solvents is possible, and heating, cooling, and calcining can also be carried out. An aspect is that contact of the product with a warm or hot wall is required for heat transfer to occur. Because the amount of entraining gas is relatively small, the gas–solid separation is simple.

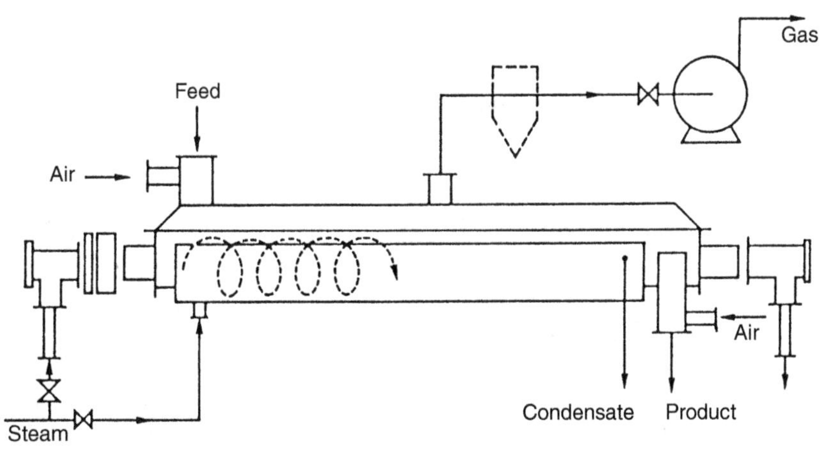

Figure 10.3 GMF Paddle Dryer (schematically). (Courtesy of GMF-Gouda Processing Solutions, Waddinxveen, The Netherlands.)

Typical Process Variables A maximum steam pressure of 28 bar (condensation temperature 229°C) can be employed. With thermal fluids, temperatures of approximately 300°C can be used. The smallest dryer has a heat-exchanging area of 1.5 m^2; the largest offers 295 m^2. Typical rotational speeds are between 5 and 30 min^{-1}. The hold-up volumes range from 10 L to 32.7 m^3. The height of the weir at the end of the dryer is adjustable. Residence times are in the range 30 to 180 min. Heat transfer coefficients can vary from approximately 100 to about 550 $W{\cdot}m^{-2}{\cdot}K^{-1}$. The higher values are found for products that have surface moisture adhering to them. The lower values are applicable for cooling and heating relatively dry particulate materials. The rotational speed has an influence on the overall heat transfer coefficient.

Typical Applications The drying of moderately wet, fine, and light materials is a typical application: for example, gypsum, pigments, dyestuffs, poly(ethylene terephthalate), municipal sludges, and oily sludges. The products are gently handled in this dryer.

Semitechnical Experiments and Scale-up Semitechnical experiments are typically carried out in a GMF Paddle Dryer that has an area of 2.5 m^2. If scaling up is carried out on the basis of the area, the residence time increases. This is because, quite generally, the volume of a piece of equipment increases proportionally with D^3, whereas the surface increases proportionally with D^2. Hence, $V/S \sim D$. This effect is relevant for the jacket area, however, it is less important for the agitator area. An installed Komline–Sanderson paddle dryer is shown in Figure 10.4.

Example 10.2 The results of a small-scale drying test are scaled up. The dryer is heated by steam. Air is used as strip gas. The stationary test lasted 75 min. Incrustations did not occur. Cooling of the dried product must be considered separately. Industrially, 600 $kg{\cdot}h^{-1}$ from the dryer is in focus.

Product

General: 90 $kg{\cdot}h^{-1}$ of a wet organic material to the dryer
TP_{in}: 12°C
TP_{out}: 100°C
Product thermal stability: good
Maximum product temperature: 110°C
d_{50}: 30 μm
A_1: 22.5% water by weight (as fed)
A_2: 0.2% water by weight
c_s: 1.9 $kJ{\cdot}kg^{-1}{\cdot}K^{-1}$
ρ_b (wet): 600 $kg{\cdot}m^{-3}$
ρ_b (dry): 500 $kg{\cdot}m^{-3}$
Solubility solid: nil

Figure 10.4 Paddle dryer drying biosolids for use as fertilizer. (Courtesy of Komline-Sanderson Engineering Corporation, Peapack, NJ.)

Dryer

Type: GMF Paddle Dryer GPD 1.6W2.5
Jacket area: 0.8 m^2
Agitator area: 1.7 m^2
Effective volume: 0.07 m^3
N (12.5–62 min^{-1}): 30 min^{-1}
d: 160 mm
Slope: 3.5°
Weir height: 80%

Process

Pressure: atmospheric
Product temperature: 100°C

Jacket and agitator temperature: 125°C
Dryer filling degree: 95%
Strip air temperature: 110°C

See Sections 4.13 and 4.15 for data on water, steam, and air.

Mass balance (kg·h^{-1})

	In	Out
Solids	69.75	69.75
Water	+20.25	+ 0.14
	90.0	69.89

$$\text{Evap} = 20.25 - 0.14 = 20.11$$

Net heat (kJ·h^{-1})

Process flow heat requirement

$$\begin{aligned}
Q_1 &= 20.11(2500 + 1.9{\cdot}100 - 4.2{\cdot}12) = 53{,}082 \\
Q_2 &= 69.75{\cdot}1.0(100 - 12) = \quad 6{,}138 \\
Q_3 &= 0.14{\cdot}4.2(100 - 12) = \quad + \underline{\quad 52} \\
Q_{\text{tot}_1} &= \quad 59{,}272
\end{aligned}$$

Calculation of the heat transfer coefficient

$$U = \frac{59{,}272{\cdot}1000}{3600{\cdot}2.5(125 - 100)} = 263.4\ \text{W}{\cdot}\text{m}^{-2}{\cdot}\text{K}^{-1}$$

This is a relatively high value. Replacing the steam heating by heating with warm water would reduce the heat transfer coefficient to approximately 230. The water in the boundary layer boils.

Residence time (min)

Flow entering the dryer: $\frac{90}{600} = 0.150\ \text{m}^3{\cdot}\text{h}^{-1}$

Flow leaving the dryer: $\frac{69.89}{500} = 0.140\ \text{m}^3{\cdot}\text{h}^{-1}$

Average flow through the dryer: $0.145\ \text{m}^3{\cdot}\text{h}^{-1}$

Residence time: $\frac{0.07}{0.145}{\cdot}60 = 29.0$ min

Scaling up It is possible to obtain an idea of the scaling-up consequences from the screening test. Real scale-up would require more experiments. To test the transport through the dryer thoroughly, the time of the tests should also be longer. The scaling-up factor is $600/69.89 = 8.58$. The dryer area is $2.5 \cdot 8.58 = 21.5\ m^2$.

Note that it is assumed that the heat-exchanging capabilities of the jacket and the agitator are equal. This is not entirely true, as the agitator rotates and the jacket is stationary.

Take a GPD 6W25 having a heat-exchanging area of 24.9 m^2. The ratio agitator area/jacket area remains constant. The effective volume of this dryer is 1.51 m^3. The residence time on a large scale becomes

$$\frac{1.51}{8.58 \cdot 0.145} \cdot 60 = 72.8\ \text{min}$$

It is possible to keep the product at 100°C for more than 70 min. It should be possible to vary the rotational speed of the industrial dryer. It should be ensured that dry product cannot reach a temperature higher than 110°C in the dryer (e.g., by switching to cooling water if a disturbance occurs).

The jacket steam consumption is (steam pressure 2.5 bar absolute, heat of evaporation 2185 $kJ \cdot kg^{-1}$)

$$59{,}272 \cdot 8.58 \cdot \frac{1.1}{2185} = 256.0\ kg \cdot h^{-1}$$

Note that a factor 1.1 is used to allow for heat losses.

Air-heater steam consumption

Industrial evaporation: $8.58 \cdot 20.11 = 172.5\ kg \cdot h^{-1}$
Airflow: 1,500 $m^3 \cdot h^{-1}$ (10°C, 60% RH)
Air specific mass: 1.25 $kg \cdot m^{-3}$

	Airflow In ($kg \cdot h^{-1}$)	Evaporation ($kg \cdot h^{-1}$)	Airflow Out ($kg \cdot h^{-1}$)
Dry air	1865.7	0.0	1865.7
Water vapor	+ 9.3	+ 172.5	+ 181.8
	1875.0	172.5	2047.5

The dew point of the exhaust gas is 52°C. See the humidity chart. The steam requirement is

$$\frac{1.1 \cdot 1500 \cdot 1.25 \cdot 1.0(110-10)}{2185} = 94.4\ kg \cdot h^{-1}$$

The steam consumption of this drying operation will be $256.0 + 94.4 = 350.4$ $kg \cdot h^{-1}$ (say, 350 $kg \cdot h^{-1}$), or approximately 2 kg of steam per kilogram of water evaporated.

10.3 VIGOROUSLY AGITATED CONTACT DRYERS

A further type of continuous contact dryer is built as a horizontal jacketed pipe with an agitator. Air travels concurrently or countercurrently with the product through this dryer (see Fig. 10.5). The heat is supplied by means of the jacket and the air; however, the jacket contribution is usually more important than the convective contribution. The product travels along the wall from the inlet to the outlet. A criterion to be observed here is the Froude number (Fr), which gives the ratio between the radial acceleration and the acceleration that is due to gravity. The radial acceleration is v^2/R, with v being the circumferential velocity and R the radius. The acceleration that is due to gravity is 9.81 m·s^{-2}.

$$\mathrm{Fr} = \frac{v^2}{R \cdot 9.81}$$

If Fr > 10, the product moves in a circular path along the wall. The heat transfer coefficient can hardly be increased in this region by rotating faster. For all Froude numbers exceeding 10, the heat transfer coefficient is dependent on (1) the mass fed

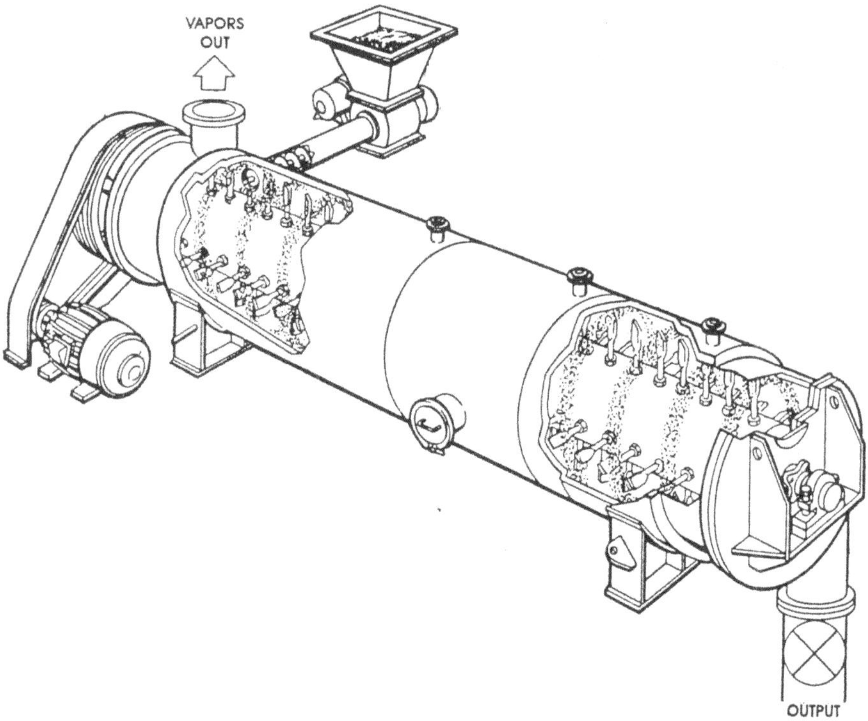

Figure 10.5 Solidaire (seen schematically). (Courtesy of Bepex International LLC, Minneapolis, MN.)

per unit of time and per meter of circumference, (2) the moisture content and the material properties, and (3) the paddle position, which is adjustable, in combination with the rotational speed. The specific dryer load, the paddle position, and the rotational speed influence the degree to which the heat transfer area is utilized.

Typical Process Variables Maximum heating-medium and drying-air temperatures are up to 500°C. A high heating-medium temperature can be obtained by using molten salt. The gas exit velocity is typically 5 $m{\cdot}s^{-1}$ for concurrent operation and 0.5 $m{\cdot}s^{-1}$ for the countercurrent mode. A typical dryer has hundreds of paddles moving with a tip speed between 10 and 20 $m{\cdot}s^{-1}$. Residence times between 1 and 15 min are possible. Relatively short residence times are registered on operating concurrently, and relatively long times are obtained for the countercurrent mode. Typical heat transfer coefficients vary from about 100 to approximately 400 $W{\cdot}m^{-2}{\cdot}K^{-1}$ (e.g., 100 for dry PVC powder and 200 to 250 for a normal drying problem).

Typical Applications Sludge drying is a typical application. The dryer is also suitable for processing cakes that contain organic solid material. Feeds that exhibit a tendency to cake in flash dryers can often be processed in a vigorously agitated contact dryer. Generally, the wetter the feed, the less suitable this type of dryer is. The moisture content of the feed can be reduced by recycling dried material. The vigorous agitation may lead to a reduction in the particle size of the material. In addition, the dryer is unsuitable for abrasive materials. An option is to combine a vigorously agitated contact dryer with a dryer providing residence time. Such a combination is used in the drying of polyolefins. Figure 10.6 exhibits a vigorously agitated contact

Figure 10.6 Vigorously agitated contact dryer in an Akzo Nobel Chemicals plant (area 3.8 m^2, stirrer speed 500 min^{-1}).

dryer installed in an Akzo Nobel Chemicals plant. Generally, this dryer type is used more often than the vertical version (see Section 10.3).

Semitechnical Experiments and Scale-up The smallest Solidaire dryer (from Bepex International LLC) has a diameter of 8 in. and a barrel length of 4 ft. This dryer is also used for testing. Further testing can be done on a 16 in. × 10 ft dryer. The largest dryer is an 88 in. × 50 ft dryer.

Example 10.3 The results of a small-scale drying test are scaled up. The dryer is heated by steam. Hot air is also passed through the dryer. The stationary test lasted several hours. Incrustations did not occur. The cooling of the dried product must be considerred separately. Industrially, 750 kg·h^{-1} is in focus.

Product

General: 186.55 kg·h^{-1} of a filter cake consisting of an organic material and water
TP_{in}: 20°C
TP_{out}: 75°C
Product thermal stability: stable up to 160°C (browning at 180°C)
A_1: 20.0% by weight (as fed)
A_2: 0.5% by weight
c_s: 1.25 kJ·kg^{-1}·K^{-1}
Solubility solid: slight, strong caking tendency

Dryer

Type: solidaire
D: 0.25 m
L: 2.3 m
N (0 – 1300 min^{-1}): 1300 min^{-1}

Process

Pressure: atmospheric
Airflow: 720 kg·h^{-1}
TA_{in}: 170°C (wet-bulb temperature 42°C)
TA_{out}: 100°C

See Sections 4.13 and 4.15 for data on water, steam, and air.

Mass balance (kg·h^{-1})

	In	Out
Solids	149.25	149.25
Water	+ 37.30	+ 0.75
	186.55	150.00

$$\text{Evap} = 37.30 - 0.75 = 36.55$$

Net heat (kJ·h^{-1})

Process flow heat requirement

$$Q_1 = 36.55(2500 + 1.9 \cdot 100 - 4.2 \cdot 20) = 95{,}249$$
$$Q_2 = 149.25 \cdot 1.25(75 - 20) = 10{,}261$$
$$Q_3 = 0.75 \cdot 4.2(75 - 20) = + \quad 173$$
$$Q_{\mathrm{tot}_1} = 105{,}683$$

Air contribution

$720 \cdot 1.0(170 - 100) = 50{,}400$ kJ·h^{-1}
Spent on the process: approximately $0.7 \cdot 50{,}400 = 35{,}280$ kJ·h^{-1}
$0.3 \cdot 50{,}400 = 15{,}120$ kJ·h^{-1} is lost

Jacket contribution

$105{,}683 - 35{,}280 = 70{,}403$ kJ·h^{-1}

$$(\Delta T)_m = \frac{(160 - 42) - (160 - 75)}{\ln[(160 - 42)/(160 - 75)]} = 100.6\,\mathrm{K}$$

Calculation of the jacket heat transfer coefficient

$$U = \frac{70{,}403 \cdot 1000}{3600 \cdot 1.5 \cdot 100.6} = 129.6\,\mathrm{W \cdot m^{-2} \cdot K^{-1}}$$

The method of calculating U by using the logarithmic mean temperature difference can be criticized.

Stirring

$$v = \frac{\pi \cdot D \cdot n}{60} = 17.0\,\mathrm{m \cdot s^{-1}}$$

This tip speed prevents the occurrence of incrustations.

$$\mathrm{Fr} = \frac{v^2}{R{\cdot}9.81} = \frac{17.0^2}{0.125{\cdot}9.81} = 236$$

The product moves from the inlet to the outlet along the dryer wall.

Particle size The vigorously agitated contact dryer decreased the weight average particle size by a factor of 3 to 4. This was not a problem.

Scaling up The scale-up factor is 5. Thus, on maintaining the jacket temperature and the drying-air temperature, a dryer having an area of $5{\cdot}1.5 = 7.5\ \mathrm{m}^2$ would be a good choice. However, a choice must be made between a dryer having an area of $6.6\ \mathrm{m}^2$ and a considerably larger dryer. The dryer having an area of $6.6\ \mathrm{m}^2$ has a diameter of 0.6 m. The maximum rotational speed of the agitator of this dryer is 600 min^{-1}. The dryer having an area of $6.6\ \mathrm{m}^2$ is chosen. v and Fr are, for $n = 600\ \mathrm{min}^{-1}$, $18.8\ \mathrm{m{\cdot}s^{-1}}$ and 121. Thus, incrustations will not occur and the product will also move from the inlet to the outlet along the dryer wall. The agitator rotational speed is variable. It is also possible to adjust the paddle position. The industrial airflow is five times larger than the airflow at the scale-up experiment. If need be, it will be possible to increase this airflow, which has a temperature of 170°C, by 25%. The scale-up was successful.

10.4 VERTICAL THIN-FILM DRYERS

A vertical thin-film dryer is a vigorously agitated contact dryer that can convert a suspension or a solution into a dry powder (see Fig. 10.7). This equipment originated from the thin-film evaporator that is used to concentrate solutions. The design is sturdy and can process pastes and powders (see Fig. 10.8). The power consumed during the processing of the paste—the intermediate stage between fluid and solid—can be quite high.

The fluid is fed at the top and flows down. The agitator rotates at speeds corresponding with Fr numbers that are greater than 10, which means that the product moves in a spiral-shaped path along the wall (see Section 10.3). The agitator elements are not rigid but have a hinge halfway between the shaft and the wall. Thus, a wall-scraping action is exerted that is in itself safeguarded against damage.

Typical Process Variables Normally, the pressure in the dryer can be chosen from 1 torr up to 1 bar. Special constructions are adequate for higher pressures. The maximum temperature is 195°C when using steam and 350°C when using a thermal oil. Circumferential rotor velocities are in the range 8 to 10 $\mathrm{m{\cdot}s^{-1}}$. The clearance between the wall and the agitator elements is 0.3 to 0.5 mm.

Typical residence times vary from 10 to 20 s per meter of dryer length. Normal feed rates can be recalculated to a figure in the range 0.8 to 1.2 kg of feed per

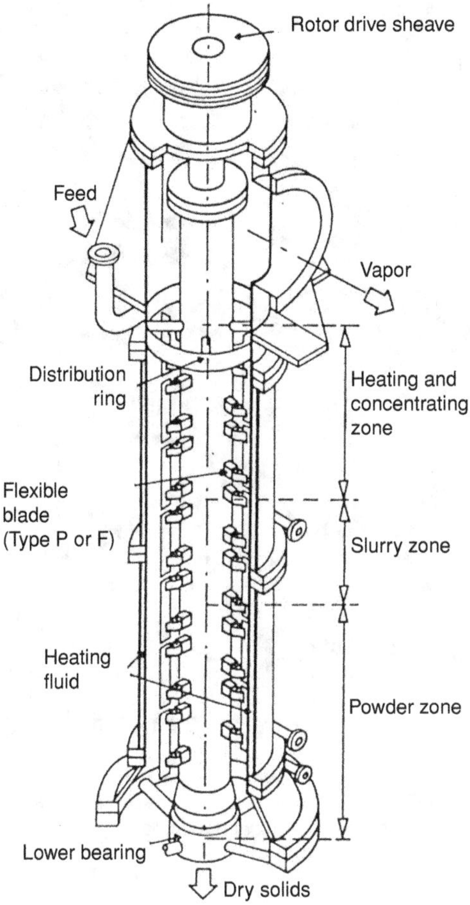

Figure 10.7 Vertical thin-film dryer. (Courtesy of Buss-SMS-Canzler GmbH, Butzbach, Germany.)

centimeter of circumference per hour. Because of the relatively short residence time, it is not too easy to obtain final moisture contents lower than 0.1% by weight.

Typical Applications Recovering solvents, crystallizing and drying of salts from solutions, and obtaining pigments from aqueous slurries are typical applications. Generally, this dryer type is used less often than the horizontal version (see Section 10.3).

Semitechnical Experiments and Scale-up Semitechnical test results can be scaled up by a maximum factor of 50. The kilograms of feed processed per hour and per square meter of heat transfer area is the basis. Commercial dryer sizes are in the range 0.5 to 60 m^2.

Figure 10.8 Vertical thin-film dryer equipped with vacuum locks for powder extraction. (Courtesy of Buss-SMS-Canzler GmbH, Butzbach, Germany.)

Remark If it is necessary to reach final moisture contents as low as 0.01% by weight, it could be advantageous to combine the vertical thin-film dryer with the vigorously agitated contact dryer (see Fig. 10.9). The residence time can be stretched up to 10 $\text{min}\cdot\text{m}^{-1}$ in horizontal dryer length. Typically, the horizontal dryer's process variables are comparable numerically to the process variables mentioned for the vertical thin-film dryer.

When it is necessary for safety or hygienic reasons to combine several unit operations in one piece of equipment, the combination of the two dryer types is also attractive.

10.5 DRUM DRYERS

A drum dryer processes materials effecting a thermal solid–liquid separation by applying the feed onto one or several slowly rotating steam-heated drums and removing the product by means of a knife. Solutions, suspensions, and pastes can be fed, and

Figure 10.9 Combination of a vertical thin-film dryer and a vigorously agitated horizontal contact dryer. (Courtesy of Buss-SMS-Canzler GmbH, Butzbach, Germany.)

powders, flakes, or chips can be obtained. Evaporated water is extracted through a vapor hood. It is also possible to evaporate solvents and, to keep the temperature down, to operate under a vacuum.

The drum dryer was introduced into the process industry approximately 120 years ago. Development started with the double-drum dryer, featuring the feed flowing into the nip between the drums (see Figs. 10.10 and 10.11). The double-drum dryer is still being used widely; however, because pressure on the rolls is then considerable, the dryer is less suitable for viscous feedstocks. This fact led to the introduction (in 1945) of the top-feed single-drum dryer (see Figs. 10.12 and 10.13), which can handle viscous feedstocks. Furthermore, for special applications, it is possible to apply the feed by means of dipping, splashing, spraying, or a bottom-feed roll.

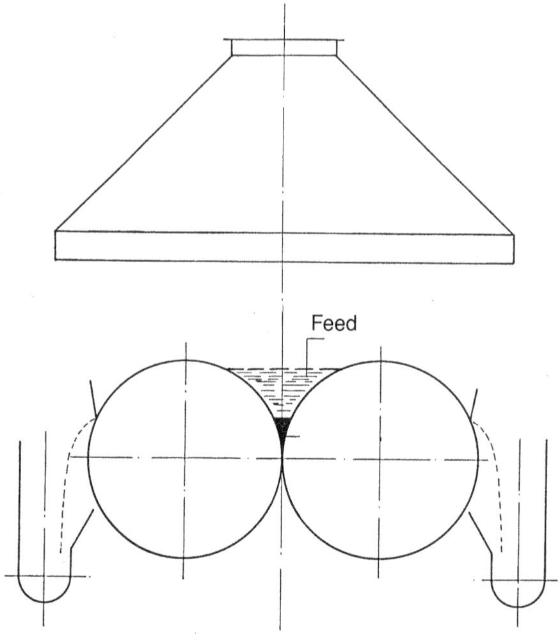

Figure 10.10 Double-drum dryer for low-viscosity solutions. (Courtesy of GMF-Gouda Processing Solutions, Waddinxveen, The Netherlands.)

Figure 10.11 Installed double-drum dryer. (Courtesy of GMF-Gouda Processing Solutions, Waddinxveen, The Netherlands.)

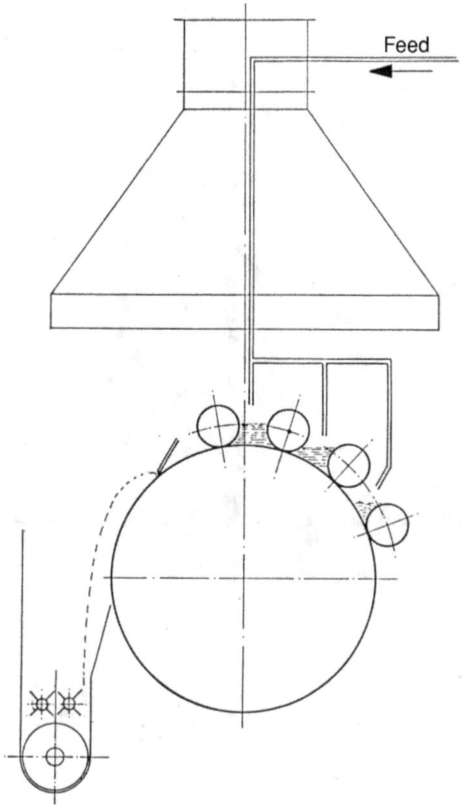

Figure 10.12 Single-drum dryer with applicator rolls for the processing of pastes and highly viscous products. (Courtesy of GMF-Gouda Processing Solutions, Waddinxveen, The Netherlands.)

The drum-drying system has limitations. Specifically, materials that do not adhere to metal cannot be processed, materials that cannot stand the temperature (considering the exposure time) are unsuitable, and materials that are or get too viscous cannot be handled.

Typical Process Variables The maximum steam pressure normally is 12 bar (condensation temperature 187°C). The rotational speeds are in the range 1 to 30 min^{-1}, and hence the residence times vary from 2 s to 1 min. Temperature-sensitive materials such as concentrated milk and gelatin are processed at high speeds. In a double-drum drying system, the product layer's thickness is a function of the distance between the drums and can be selected from the range 100 to 400 μm.

Typical Applications Drum dryers are being used in those instances where conventional methods to separate solids and liquids (e.g., filtration and convective

Figure 10.13 Installed single-drum dryer. (Courtesy of GMF-Gouda Processing Solutions, Waddinxveen, The Netherlands.)

drying) might fail. Typical materials dried on a double-drum dryer include baby food, devitalized yeast, glues, and dairy products; on a single-drum dryer, typical materials are starch, instant potatoes, dyestuffs, and gelatins.

Semitechnical Experiments and Scale-up Semitechnical experiments are typically carried out on a single- or double-drum dryer, the drums having both a diameter and a length of 0.5 m. Generally speaking, scaling up is based on the mass flow per unit area. A suitable temperature and feasible feeding arrangement are selected. The residence time is fixed by means of the rotational speed.

Drum dryers can be built up to a diameter of 2.2 m and a length of 6.2 m. Usually, the product capacities are in the range 5 to 30 $kg \cdot m^{-2} \cdot h^{-1}$. The maximum capacity of a single-drum dryer is thus approximately 1 $t \cdot h^{-1}$. The drum surface can be chrome-hardened to allow the processing of corrosive feedstocks.

10.6 STEAM-TUBE DRYERS

Indirect rotary dryers can be subdivided into low- and high-temperature devices. The former are usually heated by steam; the latter are fired by either oil or natural gas.

The steam-tube dryer is one of the most common types of indirect rotary dryers. It is a contact dryer in the form of a slowly rotating, almost horizontal cylinder, with heat-transfer tubes along the circumference. The material to be dried enters at one end and exits at the other end. The shell is inclined at a slight angle (1 to 5°). Steam is introduced into the tubes through a manifold located at the discharge end of the

Figure 10.14 Central discharge assembly on an 8×60 ft rotary steam-tube dryer. (Courtesy of Swenson Technology, Inc., Monee, IL.)

cylinder. The manifold is connected to a rotary steam joint that admits the steam and drains the condensate continuously. The steam tubes together with the radial flights serve to agitate the dryer contents. The rotating shell is sealed into the two end breechings. Various constructions exist.

There is a natural-draft stack to remove air and vapors. The purge passes through a cyclone, a settling chamber, or a wet scrubber to separate any entrained solids. A steam-tube dryer's central discharge assembly is depicted in Figure 10.14.

Typical Process Variables The steam pressure is normally in the range 4 to 10 bar; however, the pressure may be as high as 40 bar. Usually, there is a slight underpressure in the dryer. As many as three staggered, concentric rows of tubes are possible. The diameter of the tubes of the inner rows can be smaller than the diameter of the outermost row. The pitch-to-diameter ratio of the tubes must be sufficiently large (e.g., 2 : 1) to ensure that the solid material can flow freely around them. For sticky materials, normally a single row of tubes is used. Dried product may be back-mixed to improve the product flow properties. In addition, the shell can be equipped with rappers. They are intended to remove material that builds up on the tubes and between the tubes and shell because this reduces heat transfer and therefore capacity. The length/diameter ratio can vary from 4 : 1 to 10 : 1. The circumferential velocity is normally in the range 0.3 to 1 $m{\cdot}s^{-1}$. Percentages by volume of material vary from 10 to 20. The heat transfer coefficients are in the range 30 to 90 $W{\cdot}m^{-2}{\cdot}K^{-1}$.

Typical Applications The dryer is useful for fine dusty particles that contain bound moisture and would be entrained by the gas in a convective dryer. The drying of silica particles made by the reaction between waterglass and sulfuric acid and separated in a plate-and-frame filter press is a typical example of this application. Steam-tube dryers are also useful for drying products that should not come into contact with combustion products or be exposed to large quantities of air.

Semitechnical Experiments and Scale-up Semitechnical experiments are, for example, carried out in a dryer having a 12-in. diameter and a length of 8 ft. It is possible to have eight tubes with a diameter of 1 in. The largest dryer has a 12-ft diameter and a length of 90 ft. It has four rows of tubes and a heat-exchanging area of 19,567 ft^2 (1817.8 m^2).

Example 10.4 The results of a small-scale drying test are scaled up. The dryer is heated by steam. Air is used as a strip gas. The stationary test lasted several hours. Incrustations did not occur. The cooling of the dried product must be considered separately. Industrially, 1000 kg·h^{-1} from the dryer is in focus.

Product

General: 7.92 kg·h^{-1} of a wet inorganic material to the dryer

TP_{in}: 40°C

TP_{out}: 120°C

Product thermal stability: good

Maximum product temperature: 300°C

A_1: 75% water by weight (as fed)

A_2: 1% water by weight

Solubility solid: nil

Dryer

D: 12 in.

L: 8 ft

Number of tubes: 8

Tube diameter: 1 in.

N (5 – 30 min^{-1}): 20 min^{-1}

Slope: 3°

Process

Steam tube temperature: 169°C

Dryer filling degree: 1/3

Strip air temperature: 110°C

See Section 4.13 for water data.

Mass balance (kg·h^{-1})

	In	Out
Solids	1.98	1.98
Water	+5.94	+0.02
	7.92	2.00

$$\text{Evap} = 5.94 - 0.02 = 5.92$$

Net heat (kJ·h^{-1})

Process flow heat requirement

$$\begin{aligned}
Q_1 &= 5.92(2500 + 1.9{\cdot}100 - 4.2{\cdot}40) = 14{,}930.2\\
Q_2 &= 1.98{\cdot}0.75(120 - 40) = \quad 118.8\\
Q_3 &= 0.02{\cdot}4.2(120 - 40) = \quad +\ 6.7\\
Q_{\text{tot}_1} &= \quad 15{,}055.7
\end{aligned}$$

Calculation of the heat transfer coefficient Assume that the heat flow conducted through the tube wall is 1.2·15,055.7 = 18,066.8 kJ·h^{-1} (heat losses and air heat-up).

$$U = \frac{18{,}066.8{\cdot}1000}{(169 - 100)1.56{\cdot}3600} = 46.6\,\text{W}{\cdot}\text{m}^{-2}{\cdot}\text{K}^{-1}$$

Note: The full area is taken in the calculation. However, only a fraction (possibly one-third) of this area is in contact with the product. So, related to the area actually used for heat transfer, the heat transfer coefficient is higher.

Scaling up The scaling-up factor is 500. The process flow heat requirement is hence 500·15,055.7 = 7,527,850 kJ·h^{-1}. Multiply by 1.1 to allow for air heating up and heat losses: 1.1·7,527,850 = 8,280,635 kJ·h^{-1}. The factor 1.1 is smaller than the factor 1.2 used before because the large dryer has a smaller area/volume ratio than that of the small dryer.

The required heat transfer area is

$$\frac{8{,}280{,}635{\cdot}1000}{(169 - 100)46.6{\cdot}3600} = 715.4\,\text{m}^2$$

This area is provided by a dryer having a diameter of 8 ft and a length of 80 ft. It contains, for example, three rows of tubes having a diameter of 4.5 in. The three rows of tubes contain respectively 24, 30, and 36 tubes.

The assumption is that the heat transfer coefficient in the large dryer is equal to the heat transfer coefficient in the small dryer. This is the case if the fraction of the heat transfer area of the large dryer that is in contact with the product is equal to the fraction of the heat transfer area of the small dryer that is in contact with the product.

The size of the motor for the dryer rotation is 21.9 kW and the rotational speed is 3 min^{-1}. The dryer's slope is 3°. A relatively small airflow, having a temperature of 110°C, removes the water vapor. Entrained solid particles are retained by a settling chamber.

10.7 SPIRAL CONVEYOR DRYERS

See Figure 10.15. The wet particulate material is fed continuously to the dryer bottom. The dryer is vibrated to convey the product from the dryer bottom to the dryer top. The vibration motors causing the vibration can be mounted either at the dryer bottom or at the dryer top. The top mounting permits feeding at almost floor level; however, it is mechanically slightly more complicated than the bottom mounting. The spiral is heated indirectly by means of steam or a thermal oil. A carrier gas flows, concurrently with the product, through the dryer. It entrains the evaporated moisture. Nitrogen is used if a solvent is evaporated. The product is handled gently in this dryer type. The dryer occupies little floor space, and the drying operation is combined with vertical transport.

Typical Process Variables The data given apply for the spiral conveyor dryer of Vibra Maschinenfabrik Schultheis at Offenbach in Germany. The dryer is operated at atmospheric pressure. The medium temperature and pressure are determined by the strength of the flexible connections. The present maximum steam pressure is 5 bar gauge (condensation temperature 158°C). The maximum thermal oil temperature applied so far is 110°C. Further optimization is possible. It is possible to vary the amplitude of the vibration continuously.

Typical Applications The gentle drying of fragile, rather coarse organic crystals containing a low percentage by weight of a solvent is a typical application. It must be possible to convey the material by means of vibration. Elevating the product to a higher plant level enables the subsequent feeding thereof to, screening equipment, for example.

Semitechnical Experiments and Scale-up Small-scale experiments are carried out in a small spiral conveyor dryer. The basis of the scale-up is the amount of material processed per hour and per square meter. The temperature of the heating medium is kept constant. The layer thickness is also kept constant. The residence time is maintained by adjusting the vibration. The diameter of the spiral conveyor dryer

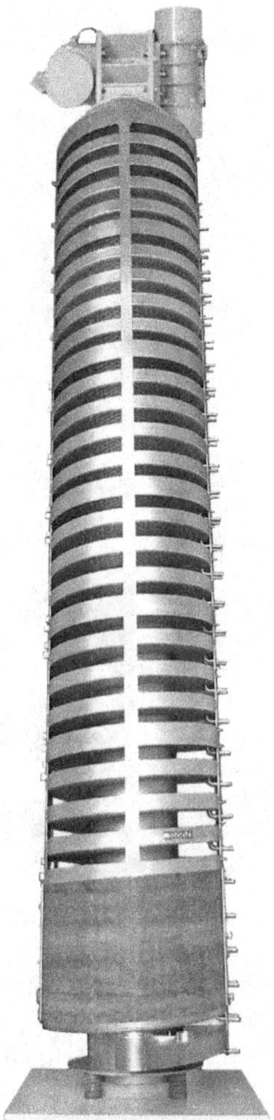

Figure 10.15 Spiral conveyor dryer. (Courtesy of Vibra Maschinenfabrik Schultheis GmbH & Co., Offenbach am Main, Germany.)

of Vibra Maschinenfabrik Schultheis varies from 550 to 1400 mm. The maximum height of their spiral conveyor is 8 m.

10.8 AGITATED ATMOSPHERIC BATCH DRYERS

In an agitated atmospheric dryer, a batch of wet particulate material can be dried at atmospheric pressure by heating the agitated material indirectly so as to cause evaporation of the moisture. When dealing with solutions, if the moisture is water and if the drying is to occur at a reasonable rate, the product temperature must be

Figure 10.16 Mechanical pan dryer. (Courtesy of Mitchell Dryers Ltd, Carlisle, UK.)

100°C or higher. Atmospheric dryers are still found in small production systems because of their simplicity.

Generally, one may distinguish between vertical and horizontal agitated dryers. The simplest vertical dryer is the pan-type dryer with top-driven paddle in which material cannot get trapped in seals or bearings (see Fig. 10.16). The horizontal agitated dryers usually consist of a jacketed horizontal cylinder. The agitators are often radial paddles, but scroll blades are also found. Ancillary equipment may simply consist of a vapor duct and chutes for charging and discharging material.

Typical Process Variables

GENERAL Typical jacket temperatures are in the range 100 to 200°C. Low-speed agitation is practised (several rotations per minute). Initial moisture contents are in the range 30 to 80% by weight (wet basis). Usual drying times vary from several hours up to more than 24 h.

VERTICAL DRYERS Vertical dryers can vary in size with a diameter of up to several meters and a diameter/depth ratio varying between 3 and 6. A usual filling ratio is 40%.

HORIZONTAL DRYERS The length/diameter ratio varies between 2.5 and 5. The largest dryers have a capacity of several tenfolds of cubic meters. The filling ratio should normally be between 0.3 and 0.6.

Typical Applications Generally, this type of dryer is best suited to the drying of materials that are friable and that do not pass through an extremely viscous stage during the drying process. Often, the feed has the consistency of a paste. It is particularly suitable for the drying of toxic products, where total enclosure is necessary.

Semitechnical Experiments and Scale-up Experiments can be carried out in a small dryer, and the drying time can be established. On maintaining the filling ratio, the drying time of the industrial dryer increases proportionally with the dryer diameter (see Section 10.2 for the explanation). The agitator can be scaled up by keeping $W \cdot kg^{-1}$ constant. It should, to this end, be possible to vary the agitator rotational speed.

11

VACUUM DRYING

Vacuum drying, freeze-drying, and vacuum pumps are considered in this chapter. For reasons of clarity, the discussion is first restricted to the removal of water. In vacuum drying, water is evaporated while having a temperature higher than 0°C. In freeze-drying, frozen water (i.e., ice) is sublimed while having a subzero temperature. In principle, this distinction can also be made when solvents are being dealt with. Vacuum drying then relates to the evaporation of a liquid, and freeze-drying relates to the sublimation of a frozen liquid.

Vacuum drying is not used as widely as atmospheric drying, but for certain applications, such as in handling materials that are heat-sensitive or difficult to dry, in applications requiring recovery of the volatile component, and for hazardous materials, it offers particular advantages. The application of a vacuum affects both the rate of drying and the temperature at which it occurs. The reduction of the boiling point of the volatile component enables the material dealt with to remain at lower temperatures for the early part of the drying cycle; later, when drying is controlled by diffusion, the temperature may rise to approach that of the heating medium. However, at that stage of the drying process, the material is generally less susceptible to thermal damage.

A general disadvantage of vacuum dryers is their relatively high cost per unit of capacity compared with atmospheric dryers. This is partly caused by their limited throughput, up to typical values of 500 $kg{\cdot}h^{-1}$, but the main reason lies in the increased manufacturing costs due to the provision of heating jackets and to stricter design codes, requiring thicker walls. Vacuum drying receives attention in Section 11.1.

Drying in the Process Industry, First Edition. C.M. van 't Land.
© 2012 John Wiley & Sons, Inc. Published 2012 by John Wiley & Sons, Inc.

Freeze-drying is also termed *lyophilization* (from the Greek "made solvent loving"). Two broad categories of materials are freeze-dried: pharmaceutical materials and food. The justification for employing lyophilization is that the nature of the product is hardly altered. For example, denaturation (the decomposition of proteins) and loss of flavor occur in hot-air drying of food but are prevented from occurring in freeze-drying. Both spray-dried and freeze-dried coffee are available on the market, and the tastes can be compared. Normally, freeze-dried products can be rehydrated almost perfectly.

Lyophilization is usually carried out under vacuum because the process proceeds more economically at low pressures than convectively at, for example, atmospheric pressure. The reason is water's very low saturated vapor pressure: for example, 0.9 mbar at $-20°C$. This would mean that for a certain degree of moisture removal, for convective dying the dry airflow would have to be more than 1000 times greater in volume than the evaporated moisture flow. In some instances, even the dry airflow would have to be replaced by a dry inert gas flow to avoid oxidation. A trivial example of convective freeze-drying at atmospheric pressure is wet laundry drying outside on a cold winter day. Vacuum equipment manufacturers are often also freeze-drying equipment suppliers.

The successive steps of systematic freeze-drying are (1) preparation and pretreatment, (2) prefreezing to solidify the water, (3) primary drying at which the ice is sublimed under vacuum, (4) secondary drying to desorb residual moisture under high vacuum, and (5) packing after the vacuum has been broken with a dry inert gas. Freeze-drying is treated in Section 11.2.

Vacuum pumps are dealt with in Section 11.3. It is useful to distinguish between pumps without moving parts and pumps with moving parts. Steam ejectors and air ejectors are pumps without moving parts. Air ejectors are, at vacuum drying, often used in combination with liquid ring pumps. The latter type of pump has a moving part, a rotor. In many instances, the liquid of the liquid ring is water. The category of pumps having moving parts can be subdivided into liquid ring pumps, oil-sealed pumps, and pumps running dry. These three subcategories of vacuum pumps concern all three pumps in which a volume filled with gas is cyclically isolated from the inlet, the gas being transferred to an outlet. The liquid ring pump is widely used at vacuum drying. Pumps of this type produce nearly isothermal compression and can handle dry gases or mixtures of vapors and gases. Due to the isothermal compression, it is possible to handle explosive gases or gases subject to polymerization. The pumping of vapor-laden gases results in condensation in the pump, giving an enhanced capacity. Furthermore, oil-sealed pumps are used frequently. Two types can be distinguished: two-vane pumps and rotary piston pumps. These pumps are used widely for freeze-drying. However, the pumping of corrosive or inflammable materials, sometimes combined with abrasive dust, complicates the operation of oil-sealed pumps and can reduce their normal long life and performance. It also sometimes happened that oil flowed back and caused system contamination. These two aspects led to the development of pumps that did not need liquid sealing, called *dry vacuum pumps*. The Roots pump, the claw pump, and the screw pump are used for vacuum drying. In actual drying plants, combinations of vacuum pumps are met.

11.1 VACUUM DRYING

Continuous Vacuum Band Drying

Figure 11.1 shows a continuous vacuum band dryer. It is used mainly in the food and pharmaceutical sectors. The band dryer consists of a vacuum chamber, housing a number of conveying bands that pass over heated or cooled plates. The drying of liquid or pasty products and the drying of solid products are different (see Fig. 11.2). The drying of liquid or pasty products proceeds in somewhat simpler fashion than the drying of solid products. At the drying of these products, oscillating nozzles or fixed nozzles are used to distribute the feed continuously from a common manifold, distributing it onto the bands operating in parallel. Thus, the feed requirements are that it has to be pumpable; however, it should not flow through a woven band mesh or over the band edge. Each band runs across a sequence of plates, usually heated by steam or hot water and controlled independently so that different sections of the dryer can be operated at different temperatures. As the product passes through the dryer, the water it contains evaporates. It is possible to cool the final plate so that the product hardens sufficiently for proper discharge. Typically in the food industry, the cake is thermoplastic. The cooled product cake is detached at the deflection rollers. It falls through a breaker at the exit. The product, which typically has a particle size of 2 cm, leaves the dryer via an automatic double-hopper system that is operated intermittently.

If the feed has been preheated to a temperature above the saturation temperature corresponding to the absolute pressure in the dryer, vapor flashes off as it leaves the feed nozzles. Many materials then expand to form a cellular, honeycomb cake structure which makes them readily soluble or dispersible.

Figure 11.1 Continuous vacuum band dryer. (Courtesy of Bucher Processtech AG, Niederweningen, Switzerland.)

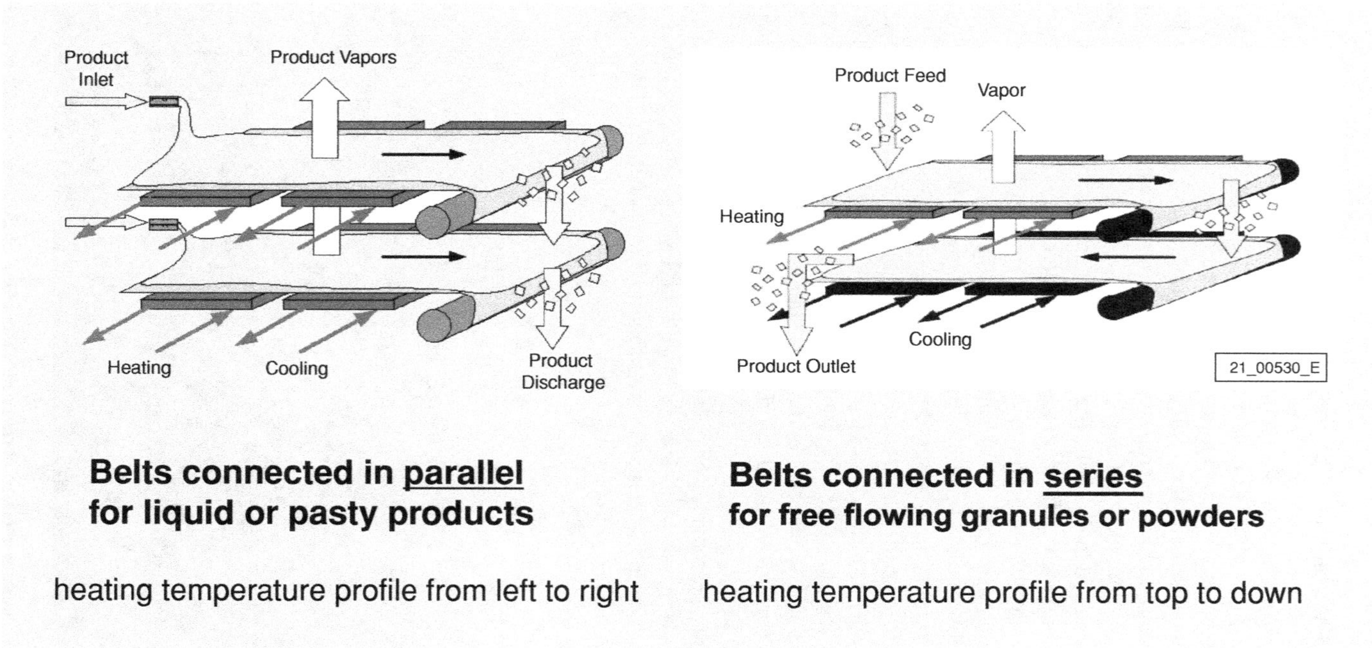

Figure 11.2 Feeding liquids and solids to a continuous vacuum band dryer. (Courtesy of Bucher Processtech AG, Niederweningen, Switzerland.)

At the drying of solid products, a double-hopper system is used to feed the wet particulate material. The particles are charged onto the upper band only; the bands are run in opposite directions so that at the end of the upper band the product falls down onto the next lower band and is conveyed back. The lumpy material runs back and forth through the dryer until it is discharged from the lowest band. The temperature of the plates of the second band can be different from the temperature of the first band. The plates of the lowest band can be cooled.

A combination of an air ejector and a water ring pump, or dry vacuum pumps, are used to create the vacuum in the dryer.

Typical Process Variables At the drying of liquid or pasty products, the feed solids content is typically in the range 50 to 90% by weight. In many cases the feed is prepared at lower solids concentrations to enable thorough mixing of the constituents. It is then concentrated by evaporation to attain a high feed concentration consistent with dryer performance. The throughput of the dryer increases with feed concentration. The feed viscosity is typically in the range $3 \cdot 10^4$ to $3 \cdot 10^5$ cP. Hot plate temperatures are in the range 20 to 165°C. The absolute pressure in the dryer is in the range 10 to 100 mbar, 15 mbar being a typical value. The water evaporation capability of the vacuum band dryer varies typically from 0.5 to 2 $kg \cdot m^{-2} \cdot h^{-1}$. Residence times are in the range 15 to 75 min.

Typical Applications Powders for breakfast drinks, powder of malt extract, instant soup, chocolate crumb, and fruit and vegetable powders are typical applications in the food industry. The production of plant extract powders is a typical application in the pharmaceutical industry.

Semitechnical Experiments and Scale-up Pilot-plant vacuum band dryers typically have a plate area of several square meters. The specific production in $kg \cdot m^{-2} \cdot h^{-1}$ is the basis for the scale-up. The largest vacuum band dryer for the processing of solid feeds at Bucher Processtech in Niederweningen, Switzerland has a plate area of 203 m^2. The largest vacuum band dryer for the processing of liquid feed at Bucher Processtech has a plate area of 225 m^2.

Batchwise Vacuum Drying

Cone Dryer with Screw Figure 11.3 illustrates a drying plant based on a cone dryer with screw (Fig. 3.3 is a photograph of such a dryer.) The dryer contains an agitator, which is an Archimedean screw rotating around its own axis and pumping upward. It is also moved over the cone's area by a rotating arm. The screw can pump down when discharging the equipment. The vacuum system consists of a condenser with a receiver, a vacuum pump, and a second condenser with a receiver. A filter, retaining entrained solid particles, is placed between the dryer and the vacuum system. The dryer and the filter can be heated or cooled by means of a thermal fluid, which is heated or cooled itself by means of media that are available locally.

The evaporation of free moisture occurs while the product temperature is constant and, ideally, this temperature is the boiling point at the prevailing pressure. The

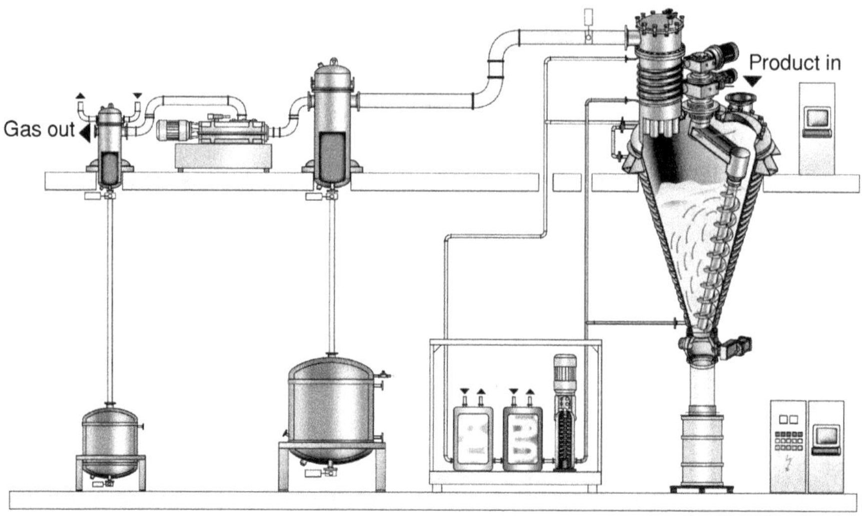

Figure 11.3 Cone dryer with screw. (Courtesy of Hosokawa Micron B.V., Doetinchem, The Netherlands.)

transport of heat from the metal wall to the wet particles provides the main resistance to heat transfer. Because this resistance is high, the water temperature drop on flowing through the jacket is usually 1 to 2 K.

When the amount of bound moisture is small, the thermal process can be described as a heating up of the granular material.

$$U \cdot A[T_i - T_p(t)]dt = G_s \cdot c_s \cdot dT_p(t)$$

Integrating and rearranging yields

$$U = \frac{G_s \cdot c_s}{t_2 \cdot A} \cdot \ln \frac{T_i - T_p(t=0)}{T_i - T_p(t=t_2)} \qquad \mathrm{W \cdot m^{-2} \cdot K^{-1}} \tag{11.1}$$

Thus, a U-value can be found for scaling-up calculations. However, if the amount of bound moisture cannot be neglected, the thermal process is more complicated. In this case the drying time should be multiplied by the cube root of the volume ratio.

The cone dryer with screw has several advantages compared with other types of agitated vacuum dryers:

1. A top-driven stirrer without a bottom bearing precludes the penetration of the material being dried into a bearing or packing.
2. The largest diameter is at the top, and this keeps the gas velocity low, thus minimizing powder entrainment.
3. The headspace serves as a freeboard, allowing the settling of particles.

It is useful to distinguish among three different drying periods. Free moisture is removed in the first drying period. Bound moisture is removed in the second drying period. The third period is meant to attain equilibrium between the product's moisture content and the prevailing conditions of temperature and pressure. If the amount of bound moisture is small (e.g., less than 0.5% by weight), the second and third drying periods coincide.

Typical Process Variables The maximum working temperature is 250°C, whereas maximum and minimum working pressures are 10 bar and 1 mbar, respectively. Typical rotational speeds of the Archimedean screw are in the range 30 to 60 min^{-1}. The rotational speed of the arm is much lower, it is typically in the range 0.5 to 5 min^{-1}. The drying of a filter cake in this dryer type has been tested [1, 2]. The dryer feed was a filter cake having a tendency to form incrustations, but ultimately it became a free-flowing powder. Drying the product from 60% by weight to 0.1% by weight of water occurred with an average heat transfer coefficient of 70 $\text{W}\cdot\text{m}^{-2}\cdot\text{K}^{-1}$. The published data on heat transfer coefficients exhibit a large scatter [3,4]. This research emphasizes the necessity to rely on test results. Typically, the heat transfer coefficients are 100 to 150 $\text{W}\cdot\text{m}^{-2}\cdot\text{K}^{-1}$ at the start of the first drying period and decrease to 25 to 50 $\text{W}\cdot\text{m}^{-2}\cdot\text{K}^{-1}$ at the end of this period.

The two screw rotations cause a power dissipation in the dryer contents. Its effect on the drying process should be checked. This can be done by noting the power consumptions during the drying and comparing them to the power consumptions when the dryer is empty.

Typical Applications The drying of pharmaceuticals and fine chemicals are typical applications.

Semitechnical Experiments and Scale-up Test work is typically carried out in a cone dryer having an effective volume of 300 L. If a substantial amount of heat is exchanged, scaling up proceeds, by means of the multiplication of the drying time by the cube root of the volume ratio.

It is then assumed that (1) the degree of filling on a volume basis is equal for the two scales, (2) there is geometric similarity, (3) the heating medium temperatures are equal, (4) the vacuum conditions are equal, and (5) the mixing conditions are equal.

The largest conical dryer with screw made by Hosokawa Micron in Doetinchem, The Netherlands had a volume of 22 m^3. This dryer was equipped with a cantilevered screw.

Example 11.1 The results of a small-scale drying test are scaled up. The dryer is warmed by warm water. Incrustations did not occur. Industrially, a 1000-L dryer is in focus.

Product

General: batch size 151.2 kg of a wet powder
TP_{in}: 19°C
TP_{out}: 45°C
Maximum product temperature: 45°C
A_1: 8.3% water by weight (wet basis)
A_2: $< 0.1\%$ water by weight
Critical moisture content: 2.0% by weight (wet basis)
c_s: 1.5 kJ·kg^{-1}·K^{-1}
Solubility solid: nil

Dryer

Type: cone dryer with screw
Effective dryer volume: 300 L
Warmed area: 2.1 m^2
Wetted area: 1.5 m^2
Arm rotational speed: 3.5 min^{-1}
Screw rotational speed: 50 min^{-1}
Screw diameter: 0.32 m

Process

Pressure: 20 mbar
Temperature entering warm water: 46°C
Temperature leaving warm water: 44°C
Vapor temperature in the first drying period: 20.5°C

First drying period This period lasted 144 min. Water boils at 17.5°C at 20 mbar.

	Mass Balance (kg)	
	In	Out
Solids	138.7	138.7
Water	+ 12.5	+ 2.8
	151.2	141.5

Product temperature according to [5]: $\dfrac{20.5 + 17.5}{2} = 19°\text{C}$

Calculation of the heat transfer coefficient The heat of evaporation at 19°C is 2466 kJ·kg^{-1}·K^{-1}.

$$U = \frac{(12.5 - 2.8)2.466{\cdot}10^3}{(45 - 19)1.5{\cdot}144{\cdot}60} = 71.0\,\mathrm{W{\cdot}m^{-2}{\cdot}K^{-1}}$$

This heat transfer coefficient is an average for the first drying period. The coefficient will initially have been higher and finally, have been lower.

Heat exchanged: (12.5 – 2.8)2466 = 23,920 kJ

Second drying period This period lasted 160 min. The product temperature gradually rose to 40°C and the moisture content fell to 0.15% by weight.

Mass balance (kg)

	In	Out
Solids	138.7	138.7
Water	+ 2.8	+ 0.2
	141.5	138.9

Heat balance (kJ)

$$\begin{aligned} Q_1 &= 2.6(2500 + 1.9{\cdot}30 - 4.2{\cdot}19) = 6{,}441 \\ Q_2 &= 138.7{\cdot}1.5(40 - 19) = 4{,}369 \\ Q_3 &= 0.2{\cdot}4.2(40 - 19) = +\ 18 \\ Q_{\mathrm{tot}_1} &= 10{,}828 \end{aligned}$$

The specific heats of water and steam are 4.2 and 1.9 kJ·kg^{-1}·K^{-1}, respectively. The amount of heat exchanged in this period is approximately half the amount of heat exchanged in the first period.

Third drying period This period lasted 45 min. The product temperature rose quickly to 45°C, and the material was kept at this temperature for 30 min. The final moisture content was below 0.1% by weight.

Scaling up

$$\text{Batch size industrial dryer: } \frac{1000}{300}{\cdot}151.2 = 504\ \mathrm{kg}$$

The specific heat exchanging area in m^2·m^{-3} decreases on scaling up. The factor is $(1000/300)^{1/3} = 1.49$. See Table E11.1.

Table E11.1 Processing Times (min)

	300 L	1000 L
Charging and homogenizing	30	60
First drying period	144	1.49·144 = 215
Second drying period	160	1.49·160 = 238
Third drying period	45	45
Discharging	15	30
Spare	+ 15	+ 30
	409	618

The rotational speeds of the screw and the arm can be varied. Thus, it is possible to obtain the test heat transfer coefficients. The screw diameter of the industrial dryer is 0.305 m. The scale-up was succesful. Incrustation did not occur during the tests or in the plant.

Final remark The volume of the industrial dryer is more than three times the volume of the test dryer. However, the capacity of the industrial dryer (44.9 $kg{\cdot}h^{-1}$) is hardly double the capacity of the test dryer (20.3 $kg{\cdot}h^{-1}$). The cause is the decrease in the specific heat exchanging area in $m^2{\cdot}m^{-3}$. Here, the effect is pronounced because the heat is supplied by the jacket. If heat is supplied by dryer internals, as in other dryer types, the effect is less pronounced.

Example 11.2 A 100-kg quantity of powdery copolymer of ethylene and vinyl acetate is warmed up from 20°C to 43°C in 45 min. The 300-L cone dryer with screw is utilized, and 2.1 m^2 is covered with powder. Water of 60°C flows through the jacket. $c_s = 2800\ J{\cdot}kg^{-1}{\cdot}K^{-1}$.

$$U = \frac{100{\cdot}2800}{2.1{\cdot}45{\cdot}60} \cdot \ln \frac{60 - 20}{60 - 43} = 42.3\ W{\cdot}m^{-2} \cdot K^{-1}$$

Filtrodry Filter–Dryer The Filtrodry filter–dryer FPP/XD, an agitated pressure Nutsche filter–dryer made by the Italian company 3V Cogeim in Dalmine, Italy, is a typical piece of equipment for combined liquid–solid separation and drying. Figure 11.4 shows a cross section of the filter–dryer. The piece of equipment is used primarily in the pharmaceutical and fine chemical industry. The good working hygiene is an advantage of the filter–dryer. The processing proceeds batchwise. The slurry is fed to a closed cylindrical vessel with a horizontal screen at the bottom. The liquid passes, driven by gas pressure, through the screen while the wet solid is retained. The filter–dryer is equipped with an agitator. Washing is possible by mixing the filter cake with the wash liquid or by passing the liquid through the stationary cake. Rings for the distribution of wash liquid can be seen in Figure 11.4.

Drying is accomplished by transferring heat from the hollow stirrer, jacket, and bottom to the vessel contents. A heating medium (e.g., water of 70°C) is circulated through the stirrer and the half-pipe jackets welded to the cylindrical body and the

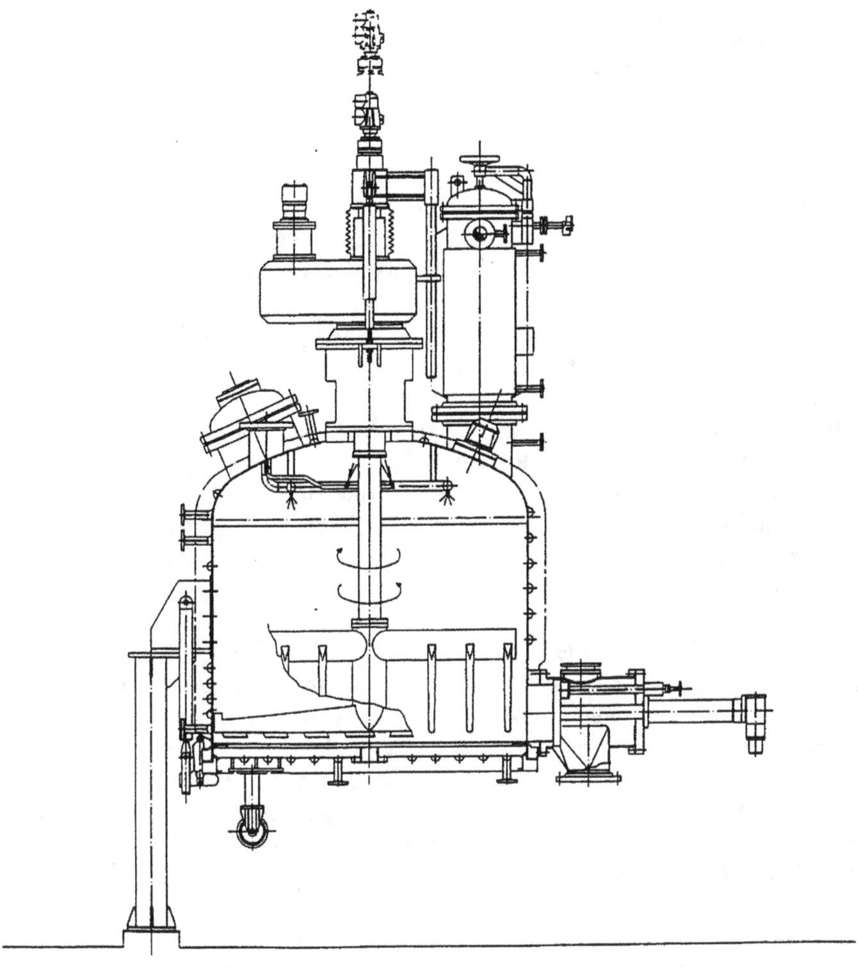

Figure 11.4 Filter–dryer. (Courtesy of 3V Cogeim SRL, Dalmine, Italy.)

bottom. The vapors generated in the filter–dryer pass through a dust filter and are evacuated.

Basically, the apparatus consists of a pressure vessel with a cylindrical body, dished top head, and flat bottom. The agitator, which can move up and down, consists of two elements at 90° to each other. The lower element achieves axial mixing, cake smoothing prior to washing, and product discharge through the side hatch. The vertical hollow blades of the upper element provide heating area. It is, depending on the process requirements, possible to choose from several blade shapes.

A side hatch with plug valve for the product discharge is flanged to the cylindrical body. The plug valve is flush with the inside of the vessel. It is hydraulically driven.

Typical Process Variables Normally, the maximum pressure is 4 bar gauge; however, it is possible to obtain filter–dryers that can be used at higher pressures.

A typical gas pressure at the filtration step is 500 mbar gauge. A typical cake thickness after filtration is 30 cm. The agitator circumferential velocity can be varied from 0.15 $m \cdot s^{-1}$ to 1.5 $m \cdot s^{-1}$, for example. The flat bottom contains up to five gauze layers, one of which is active, whereas the other layers act as supporting layers.

The heat-exchanging efficiencies of the jacket area and the bottom area are, respectively, 60% and 10% of the efficiency of the stirrer area. The clearance between the agitator and the filter bottom is 5 to 6 mm.

Metal gauze, typically having 5-μm holes, is used for the filtration of dust in the top dome. As a rule of thumb, the dust filtration area is equal to the liquid filtration area.

Recently, it has sometimes been possible to shorten the drying time by combining the usual indirect heating with microwave heating.

Typical Applications As the piece of equipment has a completely sealed construction, it allows the processing of toxic materials without environmental pollution or contact with the operator. The advantage of a filter–dryer is that liquid–solid separation, solid washing, and drying are carried out in one piece of equipment, so that transfers between these process steps are avoided. Water or an organic solvent (e.g., ethanol) can be evaporated.

Semitechnical Experiments and Scale-up Semitechnical experiments are carried out in a filter–dryer having a diameter of 600 mm. The results can be scaled up, as we show in an example. The drying time in contact dryers often increases on scaling up, as the jacket heat-exchanging area per kilogram processed is inversely proportional to the linear dryer dimension. The filter–dryer of 3V Cogeim does not suffer from this effect as the stirrer heat-exchanging area, the most important heat-exchanging area, is approximately proportional to the filter-dryer volume. The largest filter–dryer has a diameter of 4000 mm.

Example 11.3 The processing of an aqueous slurry containing an API (active pharmaceutical ingredient) within, at that time, Akzo Nobel Pharma is in focus. The results of a small-scale filtration and drying test are scaled up. Filtration and washing last 45 min. At the drying step, the filter–dryer is warmed by warm water. All moisture is free moisture. Incrustation did not occur. Industrially, a filter–dryer having a diameter of 1800 mm is in focus.

Product

General: batch size 48.4 kg of a wet powder

TP_{in}: 20°C

A_1: 64.7% water by weight (wet basis)

A_2: $< 0.1\%$ water by weight

Layer thickness: 23 cm

Solubility solid: nil

Dryer

D: 600 mm
N (10–60 min^{-1}): 10

Process

p: 63 mbar
Temperature entering warm water: 71°C
Temperature leaving warm water: 69°C
Layer thickness after drying: 11.5 cm
Drying time: 13 h

Calculation of the stirrer heat transfer coefficient The stirrer heat transfer area decreases during the drying step from 0.25 m^2 to 0.15 m^2. The average value is 0.20 m^2. The jacket heat transfer area decreases during the drying step from 0.40 m^2 to 0.20 m^2. The average value is 0.30 m^2.

At 63 mbar, water boils at 37°C. The heat of evaporation at this temperature is 2412 $kJ{\cdot}kg^{-1}$.

$$U = \frac{0.647 \cdot 48.4 \cdot 2,412,000}{13{\cdot}3600(70-37)(0.1{\cdot}0.28 + 0.6{\cdot}0.30 + 0.20)} = 120\,W{\cdot}m^{-2}{\cdot}K^{-1}$$

Initially, the heat transfer coefficient will possibly have been as high as 200 $W{\cdot}m^{-2}{\cdot}K^{-1}$ and, at the end of the drying step, have decreased to 25 to 50 $W{\cdot}m^{-2}{\cdot}K^{-1}$. The time-average bottom heat transfer coefficient is $0.1{\cdot}120 = 12$ $W{\cdot}m^{-2}{\cdot}K^{-1}$. The time-average jacket heat transfer coefficient is $0.6{\cdot}120 = 72$ $W{\cdot}m^{-2}{\cdot}K^{-1}$.

The temperature of the circulating warm water is almost constant. This is because the resistance to heat transfer is concentrated primarily in the step from the warm wall to the process material.

Scale up

Product

General: batch size 568.2 kg of the wet powder
Cake thickness: 30 cm

Dryer

D: 1800 mm
N (1.6–16 min^{-1}): 3.3: min^{-1}

Process

> p: 42 mbar
> Warm water temperature: 70°C
> Layer thickness after drying: 15 cm

Calculation A time of 1.5 h is taken for filtration and washing. The stirrer heat transfer area decreases during the drying step from 2.8 m^2 to 2.0 m^2. The average value is 2.4 m^2. The jacket heat transfer area decreases during the drying step from 1.7 m^2 to 0.85 m^2. The average value is 1.3 m^2. At 42 mbar, water boils at 30°C. The heat of evaporation at that temperature is 2430 $kJ{\cdot}kg^{-1}$.

It is assumed that by keeping the stirrer circumferential velocity constant, the experimental heat transfer coefficients will also be found on a large-scale. The large-scale circumferential velocity can, if necessary, be adjusted. Thus, it is possible to calculate the industrial drying time as follows:

$$\tau = \frac{0.647{\cdot}568.2{\cdot}2,430,000}{3600(70-30)(12{\cdot}2.54 + 72{\cdot}1.3 + 120{\cdot}2.4)} = 15.1\,\text{h}$$

The actual drying time is approximately equal to this time. Thus, the large-scale drying time is comparable to the experimental drying time. On scaling up, the specific drying area in $m^2{\cdot}kg^{-1}$ hardly decreases and the driving force is slightly greater.

Tumble Dryers In a vacuum tumble dryer, wet particulate material can be dried by indirect heat transfer to the material while the charge is in a gentle rolling and folding motion that is due to gravity. This definition distinguishes the tumble dryer from systems having flights, vanes, plows, or scrapers that induce mobility. Unlike a vacuum tray dryer, the tumble dryer has the potential to produce a finely divided dry mix rather than a hard cake.

The resistance to heat transfer is located almost exclusively in the step from the wall to the product. The heating medium temperature decreases very little between the inlet and the outlet. Sometimes the tumbling action can cause balling; in such instances, the tumbler is not really suitable.

Typical Process Variables The jacket temperature can range up to 300 to 350°C, and the vacuum in the dryer can be as low as 1 mbar. The rotational speed is in the range 5 to 30 min^{-1}. Drying cycles can be up to 20 h. Overall heat transfer coefficients of between 85 and 115 $W{\cdot}m^{-2}{\cdot}K^{-1}$ can be found for solids being dried, whereas 6 to 12 $W{\cdot}m^{-2}{\cdot}K^{-1}$ is applicable for fairly dry solids [5].

Typical Applications The drying of polyester and nylon chips to a final moisture content of less than 0.01% by weight are typical applications. If nylon is in contact with air at temperatures higher than 40°C, oxidation occurs. Thus, prior to discharge,

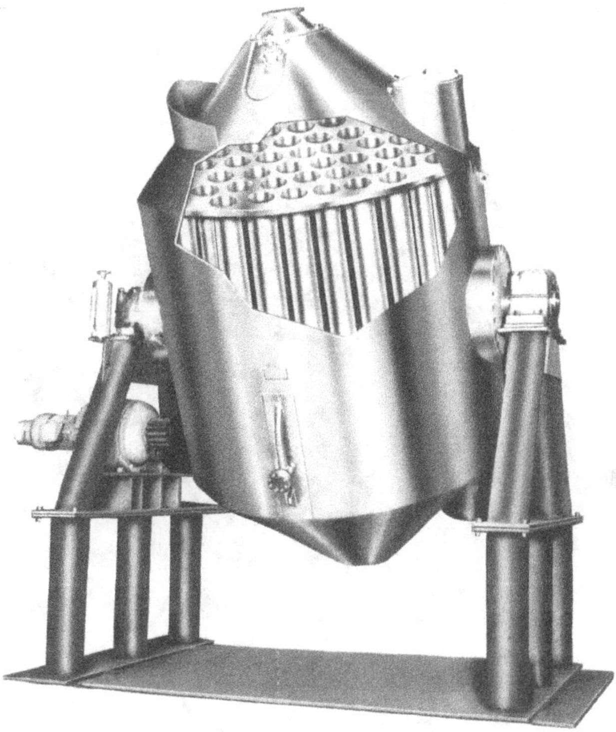

Figure 11.5 Cutaway showing 6-in.-diameter tubes in a Patterson-Kelley Tubular Vacuum Dryer (300 ft^3). (Courtesy of Patterson-Kelley/Harsco, East Stroudsburg, PA.)

indirect cooling is used to bring the maximum temperature of 110°C (nylon 6) down to less than 40°C.

Semitechnical Experiments and Scale-up Testing is usually carried out in dryers having volumes of up to several hundred liters. On scaling up, the number of units of heat-transfer area per unit of volume is decisive, assuming that the process conditions (vacuum, feed, mixing, and heating medium temperature) are the same for each size. To illustrate this, on increasing the volume by a factor of 3, the area becomes approximately twice the small-scale area (see also the earlier section on the cone dryer with screw). The largest dryers have volumes of up to 30 to 40 m^3, the batch usually occupying up to 60% of the dryer volume. Figure 11.5 illustrates an advanced tumbler. The tubes provide the heat-exchanging area and the maximum size made so far is 8.5 m^3.

Vacuum Tray Dryers A vacuum tray dryer is a contact dryer shaped like a chamber in which are placed trays containing the solid to be dried (see Fig. 11.6). Much of what was said concerning the atmospheric tray dryer is equally applicable here.

Figure 11.6 Vacuum tray dryer. (Courtesy of Mitchell Dryers Ltd, Carlisle, England.)

The heat supply is by means of warm water or other suitable heat exchanging medium, which is circulated through the metal parts on which the trays rest. The evolved vapors are absorbed by the vacuum unit. Understandably, this type of dryer has a low capacity. Specific hourly evaporation figures vary between 0.1 and 0.2 kg per square meter of tray area. Relatively high values are found for heating plate temperatures of 150°C.

Vacuum tray dryers are used fairly extensively for temperature-sensitive or easily oxidized materials. Industrial sizes of up to several cubic meters are available.

11.2 FREEZE-DRYING

Lyophilization of pharmaceutical materials is more important than the freeze-drying of food. For pharmaceutical materials, the method often cannot be replaced by a different type of drying and, furthermore, they are often quite expensive. In this field, one deals with unstable delicate biological materials (e.g., human and animal vaccines, blood proteins, complex antibiotics, and vitamins). Industrial freeze-drying was introduced for the preservation of blood plasma during World War II. Freeze-drying of pharmaceutical materials usually occurs batchwise, because of the necessity to adhere to good manufacturing practices.

Often, food is dried continuously in an intermittent way. In the food field, the freeze-drying of coffee and soup mixes has practical significance. Two aspects prevent the method from extensive use: the drying method is very expensive, and an efficient frozen-food distribution system exists in Japan and the Western world.

Freeze-Drying of Pharmaceutical Materials

Preparation and Pretreatment Pharmaceutical freeze-drying may comprise sterilization, filtration, or the addition of excipients. Excipients may protect the material during freezing or drying or may aid in the reconstitution. Alternatively, they serve other purposes.

Prefreezing Typically, the material to be lyophilized is an aqueous solution containing 10% by weight of dissolved material. The solution is contained in vials, ampoules, or bottles. Automated vial loading/unloading systems are now in common use for industrial freeze dryers in order to reduce the risk of contamination from humans and to make the process more efficient. Loading–unloading systems can be fixed (i.e., dedicated to one freeze-dryer) or flexible (they serve different freeze-dryers). Cooling below 0°C leads to ice crystallization. After complete solidification, the product typically contains ice and a glass. The glass contains the material that was originally dissolved in the solution, and water. Figure 11.7 depicts an industrial freeze-dryer schematically. Prefreezing, primary drying, secondary drying, and packing can all be carried out in the same drying chamber. The containers are placed on shelves and, during the prefreezing stage, a cooling medium circulates through the shelves. Cooling medium temperatures can be as low as –70°C or even –75°C. A typical vial diameter is 23 mm, and a typical filling height is 1 to 2 cm. The cabinet pressure at the prefreezing stage is atmospheric. Two important aspects regarding the prefreezing are the minimum temperature that must be attained and the freezing rate.

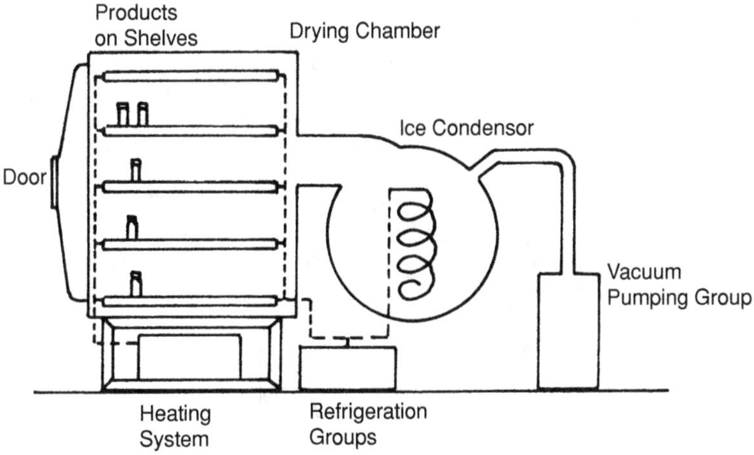

Figure 11.7 Industrial freeze-dryer for pharmaceuticals. (Courtesy of IMA Edwards Freeze Drying Solutions, Dongen, The Netherlands.)

The *freezing rate* is usually expressed in K·min^{-1}; 1 K·min^{-1} is a typical freezing rate. The freezing rate is defined as the rate at which the solution temperature decreases when placed on a shelf having a temperature of −50°C, for example. The prefreezing step influences the quality of the freeze-dried product. The freezing rate is a function of:

1. The cooling medium's temperature
2. The vial or ampoule material and dimensions
3. The filling height

A low freezing rate implies the formation of relatively large ice crystals, and vice versa. Relatively large pores are formed on subliming large ice crystals. This facilitates the vapor transport through the material processed and promotes reconstitution. Mean pore diameters in freeze-dried solutions were measured [6]. Values in the range 1 to 4 μm were found. The freezing rate varied between 25 and 3 K·min^{-1}.

The phase diagram of the pharmaceutical material can exhibit the existence of polymorphology and metastable phases. The fact that organic compounds can appear in various crystalline forms is known as *polymorphology*. Generally, the less stable forms are more soluble than the stable forms. Thus, the reconstitution attainable depends on the morphology obtained on freeze-drying. Measurement of the electrical resistance of a sample as a function of the temperature is a convenient tool for the investigation of prefreezing. The measuring cell is shown in Figure 11.8. The cell functions with liquid nitrogen that evaporates. A typical record is shown in Figure 11.9. The steep parts of the curve point to the formation of ice crystals [7].

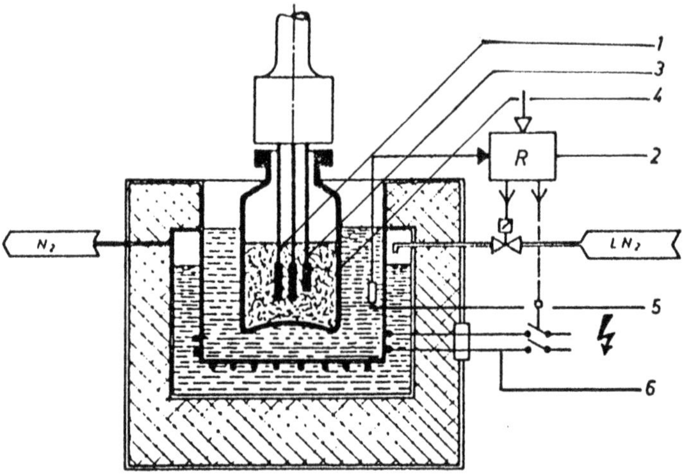

Figure 11.8 Measuring the resistance as a function of the temperature. 1, Platinum electrodes; 2, temperature controller; 3, resistance thermometer; 4, product sample; 5, heat transfer medium; 6, heating system.

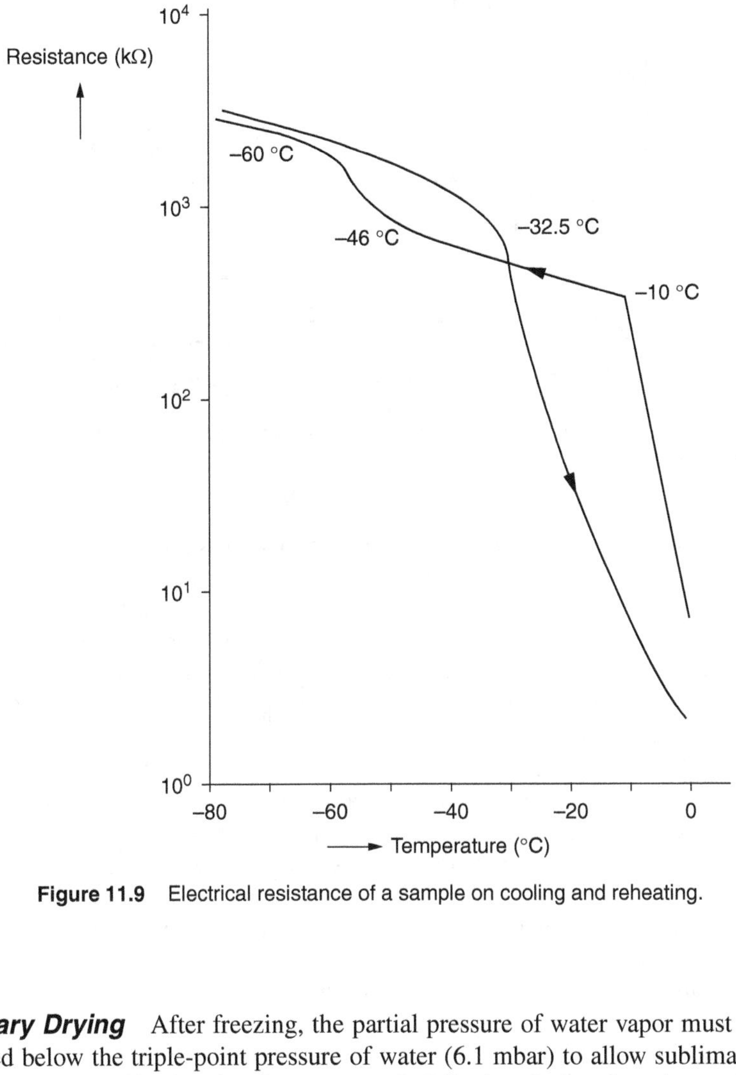

Figure 11.9 Electrical resistance of a sample on cooling and reheating.

Primary Drying After freezing, the partial pressure of water vapor must be decreased below the triple-point pressure of water (6.1 mbar) to allow sublimation to take place. Typically, the chamber pressure is 0.5 mbar during the primary drying. The material temperature would, without a controlled heat input, fall until drying came virtually to a standstill. It is thus necessary to supply heat to the product support shelves. For example, in 2 h the shelf temperature is increased linearly to 35°C. Subsequently, this temperature may be kept constant during the primary and secondary drying steps. A chamber pressure of 0.5 mbar, for example, is kept constant by means of pressure control. The saturated vapor pressure of ice at –27°C is 0.5 mbar. When the chamber pressure is 0.5 mbar, the saturated vapor pressure at the ice's surface will be 0.55 mbar, for example, because of the pressure loss experienced by the vapor flow. This means that the ice surface temperature is –26°C (see Fig. 11.10).

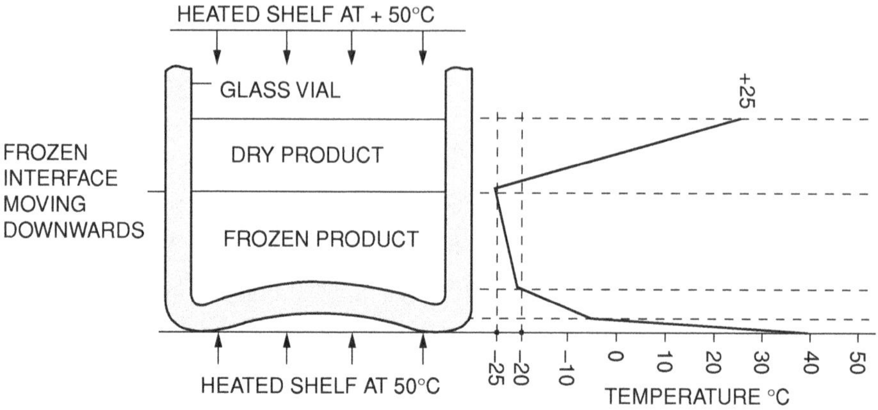

Figure 11.10 Temperature profile of the first drying period. (Courtesy of IMA Edwards Freeze Drying Solutions, Dongen, The Netherlands.)

A very low chamber pressure during primary drying is not necessarily a bonus, as the sublimation front temperature then is also low, as is the ice-saturated vapor pressure. Thus, the driving force for drying is small and drying lasts long. Generally, the primary drying rate increases when the chamber pressure increases. However, the temperature of the subliming ice front also increases when the chamber pressure increases. The temperature of the frozen product must be kept at least 5 to 7 K below the temperature at which the ice front starts to melt. Thus, it is important to select a combination of chamber pressure, shelf temperature, and vial characteristics that provides a satisfactory drying rate while product melting is avoided. A good combination can be selected by means of small-scale experiments.

A typical drying time for a 1-cm thickness of a "simple" material is between 10 and 20 h with favorable ice-crystal structure and optimized temperature and pressure conditions in the drying chamber. The drying time varies approximately with thickness to the power 1.5 [8].

The sublimed ice resolidifies on the cold metal surface of an ice condenser, which is located between the drying chamber and the vacuum pumps.

The pressure rise method is a reliable endpoint detection method. It comprises an isolation of the drying chamber from the vacuum pumps. Subsequently, the pressure rise in the drying chamber is measured. If the pressure rise is more than 3% of the original value, the primary drying is continued. The first observation is that the pressure measured after the isolation is the subliming ice front's saturated vapor pressure, and therefore the ice front's temperature is known (barometric temperature measurement). The second observation is that a relatively large pressure rise in the isolated drying chamber points to a relatively large vapor flow (associated with a certain pressure loss). The resorbed vapor flow decreases when the material gets "dry".

IMA in Bologna, Italy sell an instrument that measures the vapor flow from the drying chamber to the condenser directly by means of a laser technique. The instrument can also be used as an endpoint detection method.

Secondary Drying Secondary drying is the removal of bound moisture, which may be water of crystallization, randomly dispersed water in a glassy material, intracellular, or absorbed water. The bound moisture is removed by heating the product, typically for about one-third of the time for the primary drying time.

The secondary drying is initiated as soon as the pressure rise method detects that during primary drying, the pressure rise is less than 3% of the original value. The shelf temperature is kept at 35°C, for example, and the vacuum is lowered rapidly to, say, 0.07 mbar. Again, the endpoint is detected by measuring the pressure rise on isolation of the drying chamber or by measuring the vapor flow directly.

Like prefreezing, secondary drying has an influence on the quality of the freeze-dried product. Typically, the product possesses, after the secondary drying step, a porous structure that facilitates reconstitution. *Collapse*, a phenomenon by which the particles are solid after secondary drying, making reconstitution difficult, is sometimes noted. Typically, the moisture content is 2 to 3% by weight after secondary drying.

Packing Closing the vials or bottles occurs in the drying plant after the vacuum has been broken with a dry inert gas. Most freeze-dried materials are hygroscopic and must be stored in sealed containers.

Scales Figure 11.11 shows a small-scale freeze-drying plant. It can also be used as a pilot freeze-dryer to establish experimentally the conditions for a large freeze-dryer. Lyophilizers with shelf areas between 2 and 20 m^2 are considered "medium-scale". Figure 11.12 shows a medium-scale lyophilizer of 20 m^2 shelf area, large enough to process some 36,000 bottles of 23-mm diameter or 400 kg of wet material. Figure 11.13 shows a small-scale industrial freeze-dryer having a shelf area of 30 m^2.

From 20 m^2 shelf area upward in size are considered large-scale production units. The largest units in current use in the pharmaceutical industry go up to about 100 m^2.

Freeze-Drying of Food

Preparation and Pretreatment The materials to be dried should be in a physical form having a large specific area and should be distributed evenly throughout the drying chamber. Often, the food is cut or ground; others are blanched (a treatment that stops biological activity) or precooked. Sulfur dioxide may be used to inhibit browning reactions. Usually, fruit juices are concentrated before freeze-drying is attempted.

Prefreezing Large industrial freeze-drying plants contain a freezing room that is separated from the dryer itself. The freezing process leads to ice crystals. Two different methods to freeze can be distinguished: on drums or belts (contact cooling) and on trays by cold air circulation (convective cooling). As for pharmaceutical products, the freezing process has an influence on the quality of the product. Again,

Figure 11.11 Minifast 10: a small-scale freeze-dryer for pharmaceuticals having a shelf area of 1.1 m^2. (Courtesy of IMA Edwards Freeze Drying Solutions, Dongen, The Netherlands.)

the two important independent variables are the freezing rate and the minimum temperature.

The refrigeration rate has a great influence on the structure, the consistency, the color, and the aroma retention. A slow rate promotes the formation of large ice crystals. Large crystals lead to large pores promoting sublimation; however, large crystals can damage the cells. A slow rate leads to relatively large solid particles between the large ice crystals. Thus, aroma retention is promoted. The solution remaining after the ice crystallization is concentrated, and as the freezing proceeds, concentration gradients develop between the intracellular and extracellular solutions. These gradients may affect the quality adversely, and to avoid this, quick freezing is recommended. Whereas the freezing rate is expressed in $K{\cdot}min^{-1}$ when pharmaceutical products are freeze-dried, the freezing rate is expressed in $cm{\cdot}h^{-1}$ when foodstuffs are freeze-dried. Usually, when freeze-drying foodstuffs, freezing

Figure 11.12 Industrial lyophilizer having 20 m^2 of shelf area before installation. From left to right: the drying chamber–cylindrical ice condenser–refrigeration system. The front of the drying chamber will be sealed into the wall of the sterile area. (Courtesy of IMA Edwards Freeze Drying Solutions, Dongen, The Netherlands.)

Figure 11.13 Lyomax freeze-dryer for pharmaceuticals having a shelf area of 30 m^2. (Courtesy of IMA Edwards Freeze Drying Solutions, Dongen, The Netherlands.)

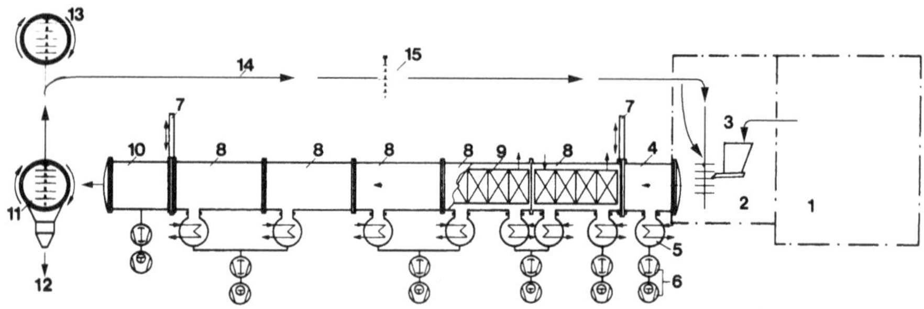

Figure 11.14 Flow diagram of a continuous five-tunnel freeze-drying plant. 1, Freezing room; 2, loading room; 3, tray-loading station; 4, entry lock; 5, condenser; 6, pumping system; 7, gate valve; 8, freeze-drying tunnel; 9, heating-plate system with suspended carriers; 10, exit lock; 11, unloading station; 12, product discharge; 13, washing equipment; 14, rail system for carrier transport; 15, suspended carriers.

rates are in the range 0.3 to 3 $cm{\cdot}h^{-1}$. Typically, when freeze-drying foodstuffs, the minimum cooling medium temperatures are somewhat higher than the minimum cooling medium temperatures when pharmaceutical products are freeze-dried: −40 to −50°C.

Primary Drying Figure 11.14 depicts a continuous five-tunnel freeze-dryer. Batch dryers also exist. The prefrozen materials are loaded on trays and subsequently enter the dryer via an entry lock. On entering the dryer, the trays are connected to an overhead rail to enable transport through the dryer and contact heating. Heating can be effected by means of oil, water, or vacuum steam. Vacuum steam is particularly suitable because it gives a uniform service-side temperature (temperature variations between 45 and 120°C are possible). The chamber pressure is in the range 0.1 to 1 mbar. The remarks concerning the shelf temperature and the chamber pressure made for the freeze-drying of pharmaceutical products apply here also. About 50 to 80 kg of fresh products can be dried per square meter and per 24 h on utilizing a special tray design. The dried material leaves the dryer via an exit lock. The trays (combined to a carrier) are washed and recycled.

Figure 11.15 shows the front view of a tunnel section, and Figure 11.16 shows a lock to a tunnel with a vacuum-tight gate valve in the background.

Secondary Drying Bound water is removed in this stage. Neither the temperature of the heating system nor the pressure in the drying chamber is different from the temperature and pressure at the primary drying. The secondary drying stage is entered after a certain residence time. Normally, postdrying is carried out until the residual moisture is in the range 2 to 3% by weight.

Packing The freeze-dried products are packed in a manner to prevent moisture pickup.

Figure 11.15 Front view of a tunnel section.

Scale Food-processing equipment is typically built with shelf areas over 200 m^2. Continuous belt lyophilizers have been developed that accept prefrozen material, usually in the form of granules or frozen droplets. The material enters the drying chamber through a vacuum lock and falls onto the highest of a series of moving belts, running in opposite directions. It is possible to obtain a desired temperature profile

Figure 11.16 Tunnel freeze-dryer lock with a gate valve in the background.

by means of different heating loops. Material falls from the end of each belt onto the next lower belt and finally, into an exit vacuum lock. High specific evaporation rates are achieved (e.g., 5 $kg{\cdot}m^{-2}{\cdot}h^{-1}$). Drying times between 6 and 9 min can be registered for granules having a size between 0.5 and 2 mm and a water content of 60% by weight. An example is the continuous freeze-drying of coffee extract. The milled and agglomerated product is subsequently packed in small packages to be used in hotel rooms, for example. Depending on the type of feed, the capacities of the continuous tunnel dryers range from 5 t of intake per 24 h to 20 t per 24 h. The optimum freezing and drying conditions for the continuous tunnel dryer are assessed empirically by means of pilot-plant drying (dryer size, e.g., 0.6 to 1 m^2).

11.3 VACUUM PUMPS

Liquid Ring Pumps

In most cases, the liquid ring consists of water. A typical water ring pump is depicted in Figure 11.17. The vane-impeller is arranged eccentrically in the properly adjusted housing, and the moving impeller gives impetus to the rotating liquid ring. Due to the eccentricity of the impeller to the housing, a crescent-shaped cavity is formed between the impeller hub and the liquid ring as the impeller rotates. The impeller vanes divide this cavity into many segments having different volumes. The segments increase in size in the region of the intake port when the impeller rotates and gas or vapor is sucked in.

In the region of the opposite discharge port, the pumped medium in the segments (which are here becoming smaller) is compressed and ejected. The flow of the pumped gas or vapor is indicated by means of arrows on the right side of Figure 11.7. Water ring pumps are rotary machines resembling piston pumps. The water ring performs the function of the piston. The arrows on the left side of Figure 11.17 indicate the flow of a liquid medium lubricating the bearings and cooling the electromotor. The lubrication and cooling are necessities because Lederle-Hermetic's water ring pumps are hermetically closed. For these functions it is possible to use the water of the water ring pump. Alternatively, if it is necessary to avoid contamination of

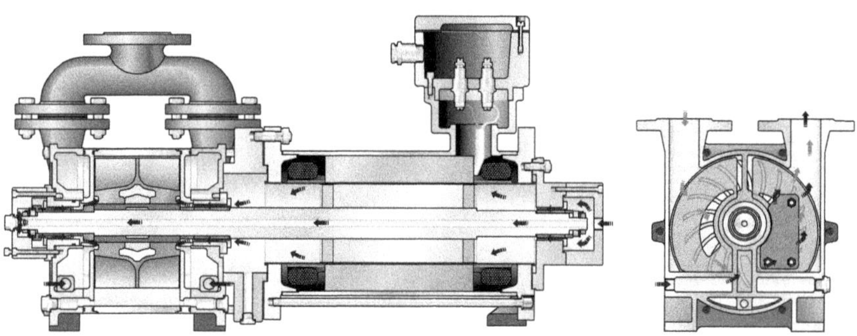

Figure 11.17 Water ring pump. (Courtesy of Hermetic-Pumpen GmbH, Gundelfingen, Germany.)

the bearings by impurities from the process, the liquid circulation to lubricate the bearings and to cool the motor is separated from the liquid system of the water ring.

Heat is generated in liquid ring pumps by the compression; in addition, heating may be caused by condensation of vapors and absorption of gases, and by the cooling of the principal gas at elevated temperatures. The heat can be carried away by a continuous supply and purge of ring liquid. Alternatively, it is possible to recycle indirectly cooled ring liquid. It is also possible to combine these two methods.

Typical Process Variables The liquid ring pumps have capacities in the range 1 to 27,000 $m^3 \cdot h^{-1}$. The minimum operating pressure range is 1013 to approximately 33 mbar. Using water at 15°C as the ring fluid, the lowest possible inlet pressure that is practical as well as possible is approximately 33 mbar. The reason is that the saturated vapor pressure of water at 15°C is 17 mbar, and to avoid boiling in the pump housing, a minimum inlet pressure of 33 mbar should be maintained. If other ring liquids (e.g., oil) are used, inlet pressure is limited to between about 10 and 30 mbar, due to outgassing. Liquid ring vacuum pumps are often used in combination with gas ejector pumps, and it is then possible to obtain inlet pressures in the range between 33 and approximately 5 mbar. The gas ejector is, in this combination, placed directly at the intake of the vacuum pump. Atmospheric air, nitrogen, or the process gas itself may be used to drive the gas ejector.

Typical Applications In a pilot plant, the vacuum in a vacuum tray dryer is created by means of a combination of an air-driven ejector and a water ring pump. For a typical operation, the absolute pressure in the drying chamber is 20 mbar.

Oil-Sealed Pumps

There are two basic designs of oil-sealed rotary pumps. Figure 11.18 shows the two-vane pump in which gas is trapped between the vanes and the stator before it

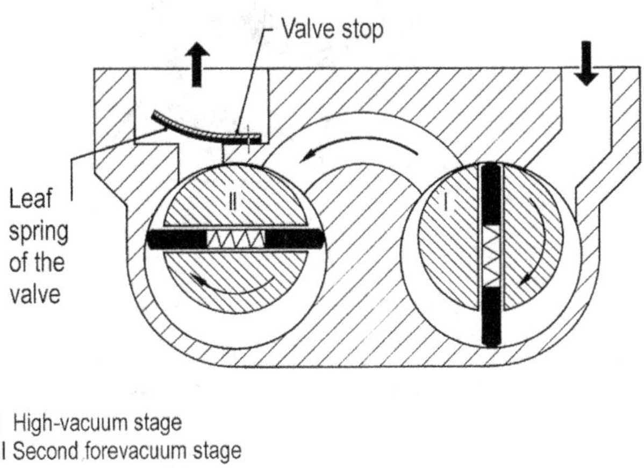

Figure 11.18 Two-stage two-vane pump. (Courtesy of Oerlikon Leybold Vacuum GmbH, Cologne, Germany.)

is swept out through the outlet valve. The main path of leakage between the inlet and the outlet is, in this design, the rotor-to-stator gap at the point between these two ports. Generally, a long path seal, which is provided by a groove in the stator and which has the same diameter and is on the same center as the rotor, is used to reduce this leakage to a minimum. Thus, a long path clearance is obtained, which is sealed with oil between the inlet and the outlet, and which reduces gas carryover between the two ports to a minimum.

Figure 11.19 illustrates the rotary piston pump in which a single vane is slotted into the stator by the use of a hinge pin. The single vane is also part of a sleeve that fits around the rotor. Furthermore, it is hollow and acts as an inlet valve, closing off the pumping chamber from the inlet when the rotor is at top center.

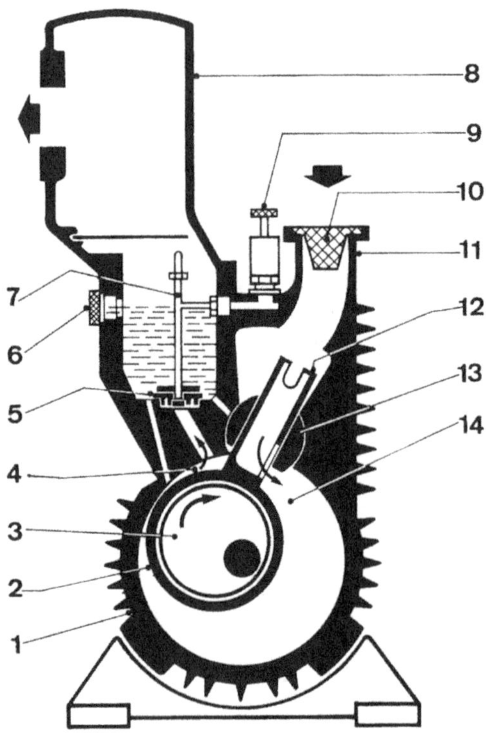

1 Casing
2 Cylindrical piston
3 Eccentric
4 Compression chamber
5 Oil-sealed pressure valve
6 Oil-level sight glass
7 Gas ballast channel
8 Exhaust pot
9 Gas ballast valve
10 Dirt trap
11 Intake port
12 Slide valve
13 Hinge bar
14 Pumping chamber (air is flowing in)

Figure 11.19 Rotary piston pump. (Courtesy of Oerlikon Leybold Vacuum GmbH, Cologne, Germany.)

Smaller pumps tend to be of the two-vane design, whereas larger pumps are more frequently of the rotary piston design. In two-stage pumps it is usual to feed outgassed oil from the outlet stage to seal and lubricate the inlet stage; thus, it is ensured that the best ultimate result can be obtained.

Typical Process Variables The capacities of available oil-sealed vacuum pumps range from 1 to 1500 $m^3 \cdot h^{-1}$. The operating pressure range of a single-stage pump is 1000 to $5 \cdot 10^{-2}$ mbar, whereas the operating pressure range of a two-stage pump is 1000 to 10^{-3} mbar.

Gas Ballast This is a feature useful to avoid condensation of vapor in the compressed gas in the vacuum pump. Atmospheric air or, if required, a dry or inert gas is admitted to the pump during the compression stage to increase the proportion of noncondensable gas in the pump by the time compression has reached a level where the outlet valve lifts (approximately 1200 mbar). The result is that the partial pressure of the vapor being pumped does not, when the outlet valve lifts, exceed its saturated vapor pressure at pump temperature, so that condensation does not occur. Further assistance in preventing the condensation of vapors within the pump is provided by the temperature rise, due to the extra work in compressing the gas introduced as gas ballast. When gas ballast is used, the pressure attainable for a single-stage pump is about 0.5 mbar, whereas it is about 10^{-2} mbar for a two-stage pump.

Typical Applications Two-vane pumps are typically applied in large pharmaceutical freeze-drying plants [9]. The bulk of the water vapor leaving the drying chamber sublimes on a cold surface having a temperature of −70°C, for example. The gases then enter a two-stage vane pump that delivers them to a Roots pump (dry vacuum pump). This combination can compress large gas flows at pressures lower than 30 mbar, and it is suitable for low pressures.

Dry Vacuum Pumps

Roots Pump

A typical Roots pump is illustrated in Figure 11.20. It consists of two figure-of-eight-shaped rotors that are synchronized by external gears and rotate in opposite directions within the stator. The clearance between the rotors and between the rotors and the stator wall is generally between 0.05 and 0.25 mm. Because the rotors run dry, back leakage occurs through these clearances at a rate dependent on the pressure difference across the pump and the gas being pumped.

Loss of clearance, which may result in damage, can be caused by overheating. Overheating is a potential source of trouble, as oil to absorb the heat of compression is absent. One possibility is to limit the compression ratio by means of a relief valve, through which gas is recycled from the outlet to the inlet. Larger pumps are equipped with a cooler in the outlet line. Cooled exhaust gases are recycled to the pump chamber. Different from oil-sealed pumps, the Roots pump does not have internal compression, as the swept volume remains constant.

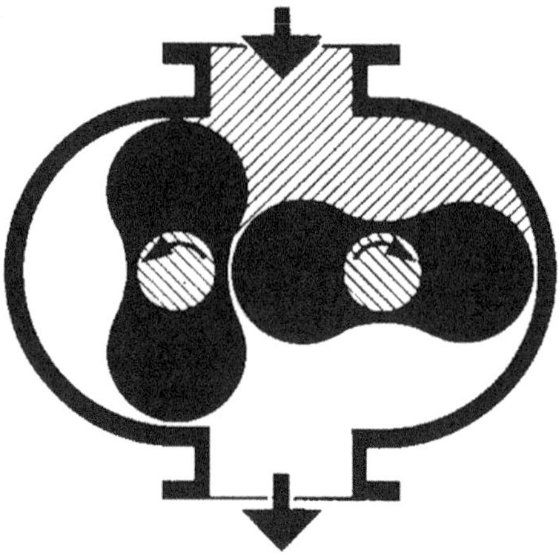

Figure 11.20 Roots pump. (Courtesy of Oerlikon Leybold Vacuum GmbH, Cologne, Germany.)

Typical Process Variables The highest compression ratios, 40 to 50, are attained when the outlet pressures are in the range 1 to 2 mbar. The compression ratio is approximately 5 when the outlet pressure is atmospheric. The consequence is that multistage Roots pumps are required when a vacuum of 10^{-2} mbar, for example, must be maintained and the gases are delivered at atmospheric pressure. The capacities of Roots pumps are in the range 75 to 100,000 $m^3{\cdot}h^{-1}$. The capacities of multistage Roots pumps are in the range 25 to 1000 $m^3{\cdot}h^{-1}$. Gas ballast or purging is often used on a number of the stages to ensure that vapors being pumped do not condense.

Typical Applications Roots pumps are typically applied in large food freeze-drying plants [10]. The bulk of the water vapor leaving the drying chamber sublimes on a cold surface having a temperature of, for example, −50°C. The gases then enter a multistage Roots pump delivering them to an air ejector followed by a water ring pump. This combination is adequate for large capacities. A further typical application of a two-stage Roots pump is the evacuation of the locks of large freeze-drying plants [10].

Claw Pump

A typical claw pump is illustrated in Figure 11.21. It consists of two claw-shaped rotors that are synchronized by external gears and rotate in opposite directions within the stator. The clearance between the rotors and the stator wall is normally 0.1 to 0.2 mm. Because the rotors run dry, back leakage occurs through these clearances. However, back flow is almost eliminated by the shape of the rotors and the inlet and outlet ports. They are designed to act as a valve. The result is that at high pressure, the

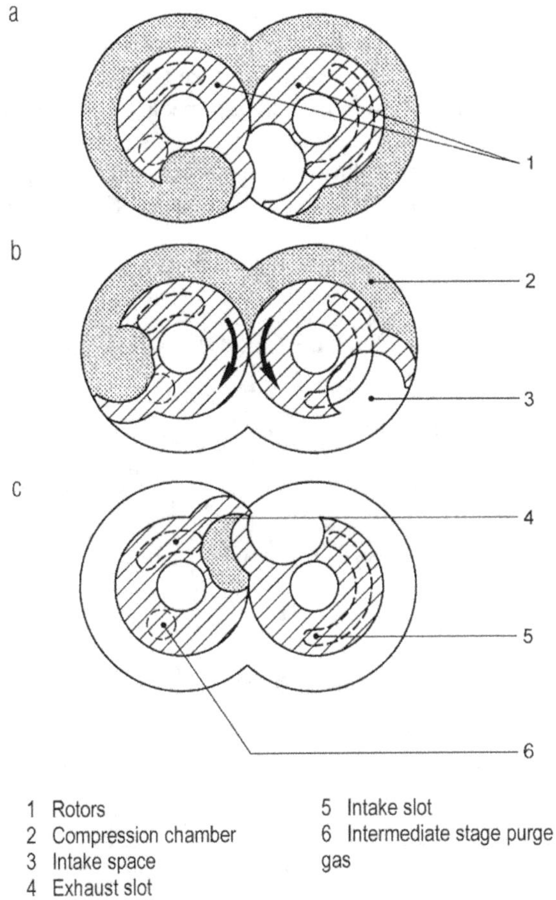

Figure 11.21 Claw pump. (Courtesy of Oerlikon Leybold Vacuum GmbH, Cologne, Germany.)

compression ratio is much higher than can be achieved with a Roots pump. Usually, claw pumps are provided as multistage units.

Typical Process Variables The capacities of multistage claw pumps are in the range 25 to 500 $m^3 \cdot h^{-1}$. Their capacity can be extended to 1200 $m^3 \cdot h^{-1}$ when equipped with an inlet Roots stage. Claw pumps can maintain pressures as low as 10^{-4} mbar while delivering gases at atmospheric pressure. For gas ballast, see the section on the Roots pump.

Typical Applications The application of three-stage claw pumps having a capacity of 250 $m^3 \cdot h^{-1}$ for filter–dryers in a pharmaceutical plant is typical. The minimum pressure in the drying chamber is 1 mbar. The combination raises the pressure to atmospheric pressure. It is possible to condense vapors from the gas stream at atmospheric pressure in a condenser.

Screw Pump

The mechanism of a screw pump consists of two intermeshing screw rotors enclosed in a close-fitting stator. The clearances between the two rotors and between the rotors and the stator wall are very small. There are specially shaped ports in the stator for the inlet and the outlet. The rotors are normally coated to improve the resistance to chemical attack and also to reduce the clearances within the pumps to a minimum. Figure 11.22 shows three different constructions. The upper configuration is

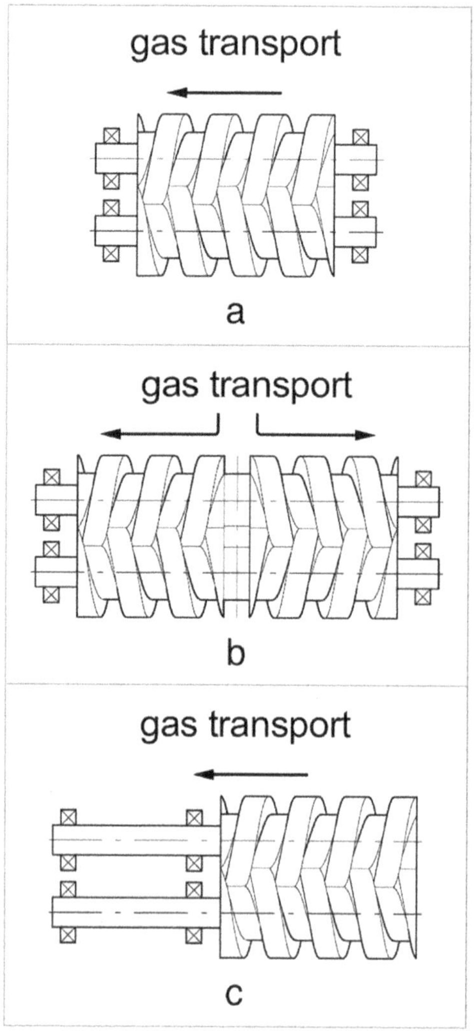

Figure 11.22 Possible screw pump configurations: (a) standard concept; (b) advanced double-flow arrangement; (c) cantilevered design. (Courtesy of Oerlikon Leybold Vacuum GmbH, Cologne, Germany.)

the conventional design and it implies the presence of a bearing and a shaft seal on the vacuum side of the pump. This has proven to be a weak point. The middle construction avoids this disadvantage but is rather complex, as it implies flow in two opposite directions. Finally, the lower configuration is a cantilevered design that is simpler than the two-flow design and also does not have the disadvantage noted. Moreover, disassembly and assembly is simple. The screw pump is particularly suitable for applications where slugs of liquid from the process vessel are likely to reach the pump.

Typical Process Variables The capacities of screw pumps are in the range 25 to 2700 $m^3 \cdot h^{-1}$. Screw pumps can maintain pressures as low as 10^{-2} mbar. The pressure can, with a special screw design, even be a decade lower. When the machine is cold, the clearances between the two rotors and the rotors and the stator wall are in the range 0.1 to 0.3 mm. When the machine is running, they can be even smaller than 0.1 mm. For gas ballast, see the section on the Roots pump.

Typical Applications Screw pumps are used in the freeze-drying of coffee concentrate. They are used in combination with Roots pumps whereby the Roots pump creates the lowest pressure, and the screw pump compresses the gas to atmospheric pressure.

REFERENCES

[1] Stein, W.A. (1976). Drying batchwise with different vacuum contact dryers. *Verfahrenstechnik, 10*, 769–774 (in German).

[2] Stein, W.A. (1977). Drying batchwise with different vacuum contact dryers. *Verfahrenstechnik, 11*, 108–111 (in German).

[3] Entrop, W., Jensen, R. (1975). Vacuum or pressure mixing. *Process Engineering, 56*, 43.

[4] Lücke, R. (1976). Local heat transfer coefficients in a dryer with plough-type agitating elements. *Verfahrenstechnik, 10*, 774–777 (in German).

[5] Fischer, J.J. (1963). Drying in vacuum tumblers. *Industrial Engineering Chemistry, 55*, 19–24.

[6] Thijssen, H.A.C., Rulkens, W.H. (1969). Effect of freezing rate on rate of sublimation and flavour retention in freeze-drying. Presented at the Institut International du Froid, Transactions Committee X Meeting, Lausanne, Switzerland, June 5.

[7] Willemer, H. (1977). The condition of aqueous solutions during freezing demonstrated by electrical resistance measurements and low-temperature freeze-drying. *Transactions, Joint Meeting of the I.I.R.—Commissions C1 and C2,* Karlsruhe, Germany.

[8] Rossi, G., Snowman, J.W. (1987). Industrial freeze-drying. *Technology Magazine of the BOC Group*, 3–13.

[9] Heldner, M. (1997). Pharmaceutical freeze-drying plants. *Vakuum in Forschung und Praxis, 9*, 281–288 (in German).

[10] Barfuss, H. (1993). Freeze-drying in laboratory and industrial applications. *Vakuum in der Praxis, 5*, 85–88. (in German).

12

STEAM DRYING

Steam drying is a drying method by which the drying occurs by direct contact between the product to be dried and superheated steam. Usually, water is evaporated. The superheated steam cools down and the heat transferred is used to evaporate water from the product. The superheated steam becomes saturated and is recycled and reheated in a closed loop while the amount of steam raised by water evaporation is bled from the loop. Thus, the evaporated water becomes available as (slightly superheated) steam that can be used as a utility for a multiple-effect evaporation plant, for example.

Steam drying is convective drying in which the permanent gases normally used at convective drying (i.e., air and nitrogen) are replaced by a vapor (i.e., steam). Thus, in principle, the same convective dryers as those used for air and nitrogen can be used. The dryer types used until now for steam drying are the fluid-bed dryer, the flash dryer, and in one known case, the spray dryer (see Section 12.5). A fluid-bed dryer is discussed in Section 12.1, and a flash dryer gets attention in Section 12.2. The advantages and disadvantages of continuous steam drying are discussed in Sections 12.3 and 12.4, respectively. Section 12.5 contains additional remarks concerning continuous steam drying.

Drying with superheated steam is an emerging separation method with great potential advantages concerning energy saving, emission reduction, fire and explosion prevention, and product quality. The energy saving is often crucial for the applicability of the method, as the investment is high.

Drying in the Process Industry, First Edition. C.M. van 't Land.
© 2012 John Wiley & Sons, Inc. Published 2012 by John Wiley & Sons, Inc.

Most industrial steam dryers operate at a pressure higher than the atmospheric pressure as the usability of the steam raised increases with its pressure. Most industrial dryers also function continuously, as large water flows must be evaporated.

It is also possible to evaporate solvents by means of this separation method. If the solvent is immiscible with water, one deals with steam distillation. A typical example of a batch steam dryer, which is also suitable for the evaporation of solvents, is discussed in Section 12.6.

On taking a look at the continuous applications realized so far, it strikes that the products dried are often of a biological nature. Examples are sugar beet pulp, DDGS (distillers dried grain with solubles) from the production of bioethanol, and wood chips. The possible reason is that the products have to be transported into and out of the dryer continuously while there is an overpressure in the dryer. With abrasive materials, the organs for these functions would erode quickly.

12.1 SUGAR BEET PULP DRYER

Figure 12.1 shows a continuous fluid-bed steam dryer for sugar beet pulp. The sugar beets arrive at the sugar plant and are, after cleaning, cut into pieces. Sugar is leached out of the pieces by means of warm water. The wet particles remaining after this process step are called sugar beet pulp. Their size is in the range of several millimeters to several centimeters. Conventionally, it is, after pressing for water removal, dried

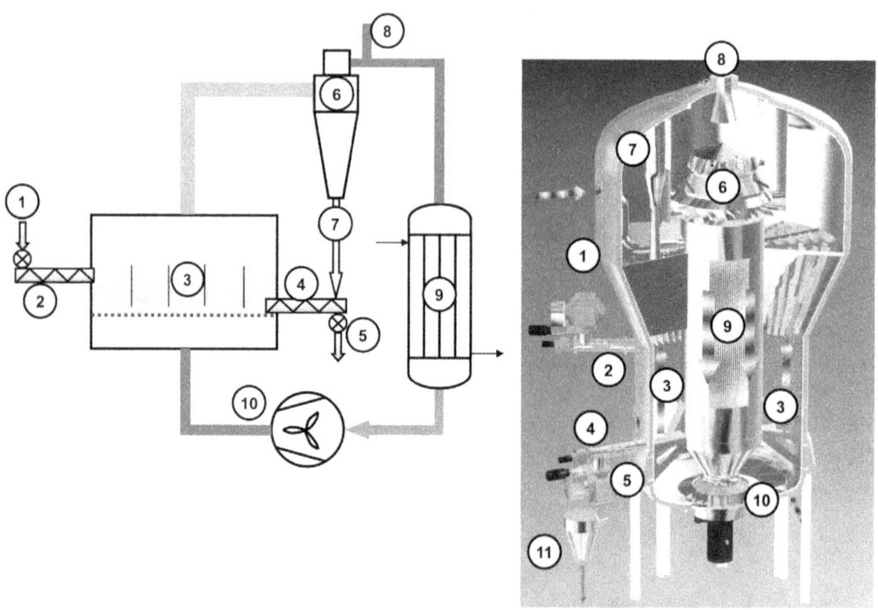

Figure 12.1 Exploded view of fluid-bed steam dryer. (Courtesy of Braunschweigische Maschinenbauanstalt AG, Braunschweig, Germany.)

convectively in rotary dryers using hot air temperatures of up to 800°C. The material entering the dryer has a moisture content of approximately 4 kg of water per kilogram of dry matter. The dried product contains, on a wet basis, approximately 10% by weight of water and can be used for the production of feed. An aspect of this practice is that the spent drying gases possess a penetrant odor, and this is one reason for the introduction of steam drying in this field.

Slightly superheated steam with a pressure of 25 bar absolute, for example, condenses in the shell of a heat exchanger (9) that is part of the steam dryer. The condensation temperature of that steam is 224°C. The heat exchanger is a shell-and-tube exchanger of the single-pass type at both the process side and the medium side. Steam also condenses in heating panels in the steam fluid-bed dryer. The heat transferred to steam flowing through the tubes and past the heating panels is used for the drying step.

Pressed sugar beet pulp is fed to the dryer by means of a feed lock (1) and a feed screw conveyor (2). The pulp is fluidized by circulating steam (3) and flows, through an annular space surrounding the heat exchanger, and supported by a perforated distribution plate, in plug flow to the product discharge system. The latter comprises a discharge screw conveyor (4), a discharge lock (5), and an expansion cyclone (11) with a separate rotary discharge lock.

The pressure inside the dryer is 3.5 bar absolute, for example. Steam having that pressure condenses at 139°C. By the same token, water boils at 139°C when the pressure is 3.5 bar absolute. Thus, the temperature of the pulp containing free water is 139°C. Steam is circulated in the steam dryer by means of a fan (10). Steam entering the tubes at the top of the heat exchanger has a temperature of 139°C and, on leaving this piece of equipment at the bottom, a temperature of 169°C, for example. On flowing through the fluid bed, this steam cools down to 139°C again, because that is the temperature of the water on the pulp and thermal equilibrium between the vapor and the liquid is attained. The sensible heat that is thus made available to the product, is converted into latent heat for evaporating water from the product. The latter step is the drying step.

Saturated steam leaving the fluid bed enters the superheater via a vaned ring, a cylinder, and a second vaned ring (6). The first vaned ring imparts a rotating movement to the steam. By centrifugal action, any entrained dust particles are concentrated at the cylinder wall. The rotating separated dust reaches a steam-driven ejector (7), enters this ejector, and is forced directly into the last cell of the annular space. The product is extracted out of this last cell by the discharge screw conveyor (4). Steam entering the superheater is almost dust-free.

There is a makeup of steam in the fluid-bed dryer. Hence, there must also be a bleed from the steam loop. This steam bleed leaves the unit at the center of the rotary separator through the top-end vapor pipe (8). It can, for example, be used to evaporate water in a multiple-effect evaporation plant. Per kilogram of steam of 25 bar absolute, approximately 1 kg of steam of 3.5 bar absolute is raised.

The feed lock (1) and the discharge lock (5) are critical items. They serve to prevent steam leaking to the surroundings while feeding wet pulp and, simultaneously, extracting dried pulp. The locks are carefully machined and controlled pieces of equipment. The requirements would be less stringent if the pressure in the dryer would be lower. However, lowering the pressure in the dryer would mean compromising on the usefulness of the steam generated, as its condensation temperature decreases on lowering the pressure.

The steam flow circulating through the dryer is a compromise. On one hand, the flow must be chosen so that the pulp is properly fluidized; however, too much entrainment of the pulp should be avoided. On the other hand, a large flow enables a high tube velocity promoting heat transfer from the heat exchanger tube walls to the circulating steam.

Plug flow of the pulp is ensured by designing 16 cells between the feed point and the exit. Large pulp particles pass from cell to cell via an underflow, whereas small particles are, in the first instance, entrained. Subsequently, they fall down into the next cells because the gas velocity decreases due to the expanding freeboard. The residence time of the pulp is approximately 8 min.

The product is not fluidized in the last cell. As far as the product is concerned, the high temperature of the steam and the product in the next-to-last cell means a low moisture content, and vice versa. Thus, the temperature of the steam leaving the next-to-last cell can be used for process control. This temperature is used to adjust the high-pressure steam flow. The dryer pressure is kept constant by adjusting the quantity of low-pressure steam passed on to a plant using this steam.

BMA sells four different dryer sizes with diameters of the upper cylindrical part varying from 8.0 to 12.0 m. The water evaporation capacities are in the range 25 to 56 $t \cdot h^{-1}$ when the steam pressure is 25 bar absolute. For a steam pressure of 15 bar absolute, the capacities are in the range 17 to 40 $t \cdot h^{-1}$. The fan powers required are in the range 0.6 to 1.35 MW. The heat exchanger areas must be large because the heat transfer coefficient in the tubes is rather low. The fluid-bed steam dryer at the Puttershoek plant of Suiker Unie in The Netherlands has a water evaporation capacity of 42.4 $t \cdot h^{-1}$ and a heat transfer area of 6169 m^2. The steam pressure is 23 bar absolute and the steam temperature is 230°C. The condensation temperature of 23 bar absolute steam is 218°C. The pressure inside the dryer is 3.5 bar absolute. The area is provided by 4563 tubes having a length of 9.67 m, an internal diameter of 40 mm, and an outside diameter of 44 mm. This dryer is no longer in operation because the plant has been closed. The dryer was bought from Niro in 1994. The diameter of the upper cylindrical part is 10 m.

The inside and outside tube diameters are 40 and 44 mm. Thus, the wall thickness is 2 mm. Mechanical strength considerations would allow the wall thickness to decrease if the tube diameters would be smaller. It would thus be advantageous to decrease the tube diameters to 29 and 32 mm, for example. When installing the same heat transfer area, that would save 25% on steel. However, smaller tubes suffer from plugging more readily. Plugging could be caused by small particles entrained by the circulating steam and sticking to the hot tube wall.

Niro, and later, BMA, installed 14 of these dryers between 1990 and 2004.

The diameter of the upper part of the dryer shown in Figure 12.1 is larger than the diameter of the lower part of the dryer. One reason is that the velocity of the steam leaving the bed is then relatively low in the upper part, which leads to relatively little entrainment of solid particles. BMA recently introduced a cylindrical steam dryer, which is cheaper to build. A dryer having an evaporative capacity of 80 $t{\cdot}h^{-1}$ and a diameter of 10 m was started up successfully in the United States in 2009. The dried product is DDGS (distillers dried grains with solubles). DDGS is a coproduct of the distillery industries. It is the dried residue remaining after the starch fraction of corn is fermented with selected yeasts and enzymes to produce ethanol with carbon dioxide as a by-product.

12.2 GEA EXERGY BARR–ROSIN DRYER

Figure 12.2 gives a schematic presentation of the GEA Exergy Barr–Rosin dryer. The dryer is a flash dryer. Several characteristics of the dryer are similar to the characteristics of the steam fluid-bed dryer described in Section 12.1. The drying of municipal sludge is a typical application, furthermore, sawdust and wood fiber and sugar beet pulp are dried in this dryer.

Slightly superheated steam having a pressure in the range 10 to 20 bar absolute condenses in the shell of heat exchangers that are part of the steam dryer. The heat exchangers are shell-and-tube exchangers of the single-pass type at both the medium

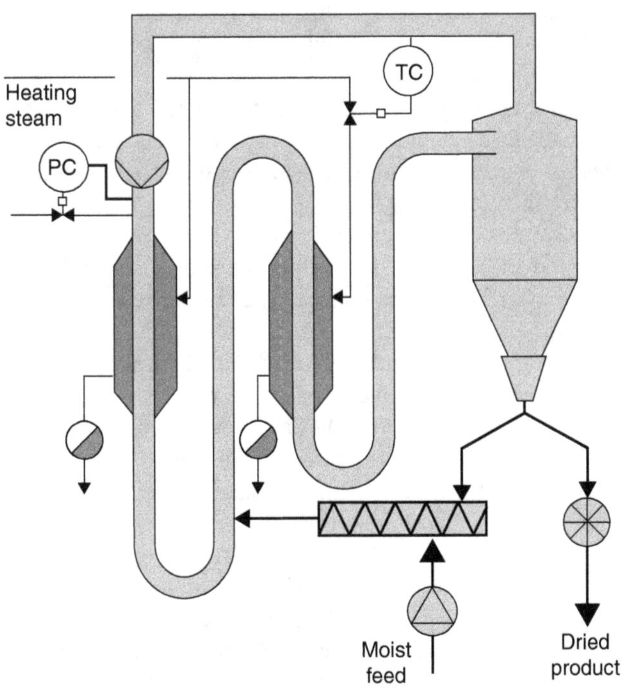

Figure 12.2 Process flow diagram of flash steam dryer. (Courtesy of GEA Barr-Rosin Ltd, Maidenhead, UK.)

and process sides. In Figure 12.2 the heat exchangers are drawn as jacketed pipes; however, the process flow passes through a number of parallel tubes. The condensation temperature of the primary steam is in the range 179 to 211°C. The heat transferred to the steam flowing through the tubes is used for the drying step.

The wet solids are fed to the steam dryer by means of a pressure-tight rotary valve or a plug screw. Sticky, pasty, and sludgelike materials require backmixing of dried product to improve the handling characteristics of the feed. The feed is entrained by the circulating steam, having a pressure in the range 1 to 5 bar absolute. Steam having a pressure of 2 bar absolute condenses at 120°C. By the same token, water boils at 120°C when the pressure is 2 bar absolute. Thus, the temperature of the product containing free water is 120°C in that case. Heat is transferred from the tube walls to the circulating steam, and this heat is passed on to the wet feed, where the sensible heat is converted into latent heat. The latent heat is used to evaporate water from the feed, and the steam thus raised combines with the circulating steam flow. To reduce the product moisture to the desired level, this process can be repeated by passing the two-phase flow through additional heat exchangers.

The dried product enters a cyclone, together with the low-pressure steam. The cyclone separates solid particles larger than several micrometers from the gas, and the dried product is discharged from the dryer by means of a pressure-tight rotary valve. The steam is recycled by a centrifugal fan to the inlet of the first heat exchanger and the excess steam is removed continuously from the system. It can, for example, be used to evaporate water in a multiple-effect evaporation plant. Per kilogram of high-pressure steam, approximately 1 kg of low-pressure steam is raised.

The residence time of the solid particles is in the range 5 to 60 s. The gas velocities in the dryer and in the heat exchanger tubes are in the range 20 to 40 $m \cdot s^{-1}$. The heat exchanger tubes have a diameter of 2 or 3 in. See the short discussion of the heat exchanger tube size in Section 12.1.

The high temperature of the steam leaving the cyclone means a low moisture content of the dried product, and vice versa. Thus, the temperature of the steam leaving the cyclone can be used for process control. The temperature is used to throttle the high-pressure steam. Any pressure loss of this steam is accompanied by a lowering of its condensation temperature, and vice versa. The dryer pressure is kept constant by passing on low-pressure steam to a different plant.

GEA Barr–Rosin has installed up to 30 flash steam dryers. Kudra and Mujumdar [1] discuss two parallel steam dryers of this type for the drying of peat in Sweden. Each dryer produces 20 $t \cdot h^{-1}$. The total heat-exchanging area is 5500 m^2. Thus, for large evaporation capacities, large areas are required. The electricity consumption of the compressors and blowers is 10 MW.

Figure 12.2 shows, schematically, a flash steam dryer in which backmixing is practised. For products with better handling characteristics, it is also possible to do without this option. Furthermore, in Figure 12.2, low-pressure steam with entrained product is passed through one of the two heat exchangers. This could, for certain products, lead to incrustations in the tubes. The low-pressure steam passing through the other heat exchanger contains hardly any entrained dust, as it has been cleaned by the cyclone. For this heat exchanger, the risk of incrustation in the tubes is smaller. It is possible to avoid passing low-pressure steam containing entrained wet product through heat exchanger tubes by connecting the heat exchangers in series before the feeding point. The circulating low-pressure steam then carries the product between the feeding point and the cyclone only. In that case, however, the product residence time can be only several seconds.

12.3 ADVANTAGES OF CONTINUOUS STEAM DRYING

1. The steam raised by the steam drying step can be used for industrial purposes. Use of the latter steam is important for the success of the operation, as the investment for a steam dryer is high.
2. Steam drying is a safe operation, as the presence of oxygen is excluded, so fires and dust explosions cannot occur.
3. It is environmentally acceptable as the drying does not require large amounts of air that are recycled to the atmosphere.
4. The process step can favor the product quality, as the drying occurs in an environment of water and steam only (e.g., case hardening is less likely to occur). However, whether the relatively high temperature at the drying step is acceptable for the product must be checked.
5. Dust and odor emissions from the equipment do not occur, as the operation is contained. The elimination of an odor problem was the motivation for the installation of a fluid-bed steam dryer for sugar beet pulp at the Puttershoek plant of Suiker Unie in The Netherlands. However, it turned out that after installation of the fluid-bed steam dryer, another odor from a different part of the plant was still present. The latter odor had been masked by the odor from the conventional pulp drying step. An additional incinerator had to be installed to cope with the latter odor.

12.4 DISADVANTAGES OF CONTINUOUS STEAM DRYING

1. The investment is high, mainly because of the necessity to install a large heat-exchanging area. The overall heat transfer coefficient is much smaller than the heat transfer coefficient in an evaporator. This is caused by the small coefficient for the transfer of the heat from a metal wall to superheated steam.
2. The feed and discharge locks are technically complicated because of the need to prevent steam from escaping through these locks.
3. The interdependencies in a plant increase as the drying step not only produces a dry product, but also steam that is used for, for example, a concentration step in the plant.
4. The steam raised contains fine particles and gaseous components from the drying step. Hence, the condensate obtained from this steam will also contain solid particles and, probably, dissolved material.

5. The product temperature is rather high. In a steam atmosphere, the evaporation of free water requires a temperature of at least 100°C at atmospheric pressure.
6. All parts that come into contact with steam, even if it is superheated, must stay above the condensation temperature of the steam at the prevailing pressure. Thus, good insulation is required. Good insulation is also required to protect the operators.

12.5 ADDITIONAL REMARKS CONCERNING CONTINUOUS STEAM DRYING

Continuous steam dryers have been built as fluid-bed dryers and flash dryers, and, in one case, as a spray dryer [2]. A typical fluid-bed dryer is described in Section 12.1, and a typical flash dryer is described in Section 12.2. The incentive to install a steam spray dryer for detergents was the need to avoid odor emission. The application of a fluid-bed dryer requires the fluidizability of the product to be dried. This means particle sizes in the range of several millimeters to several centimeters. The residence time of the product to be dried is independent of the low-pressure steam velocity. For the fluid-bed dryer of Section 12.1 it is about 8 min. The application of a flash dryer requires the possibility to entrain the product by means of circulating low-pressure steam. This means particle sizes of maximum 1 mm. Furthermore, the product residence time is now coupled to the low-pressure steam residence time. It is in the range 5 to 60 s.

In a fluid-bed dryer, direct contact between the product to be dried and the heat exchanger tubes is avoided because the circulating steam is, in principle, freed from solid particles before it enters these tubes. In a flash dryer, if direct contact between the product to be dried and the heat exchanger tubes must be avoided, the residence time of the product can be several seconds only. See Section 12.2.

In Sections 12.1 and 12.2, the heat source is steam having a pressure in the range 10 to 25 bar absolute. It is also possible to supply the heat by the combustion of natural gas. The hot gases then flow through the shell of a shell-and-tube heat exchanger while the circulating low-pressure steam passes through the tubes. Electric heating is a further possibility.

12.6 EIRICH EVACTHERM DRYER

The Eirich Evactherm dryer operates batchwise. The system pressure is normally in the range 800 to 1000 mbar; however, system pressures as low as 50 mbar are possible. Typical applications are the drying of galvanic sludges, zinc-containing sludges from incinerator plants, and special ceramic products such as ferrites or varistor masses. The dryer is also used to recover organic solvents from paint sludges and spray cans. Because of the relatively low pressure, the drying system does not aim at raising steam for industrial purposes. The steam and the organic solvents are condensed and processed as need be.

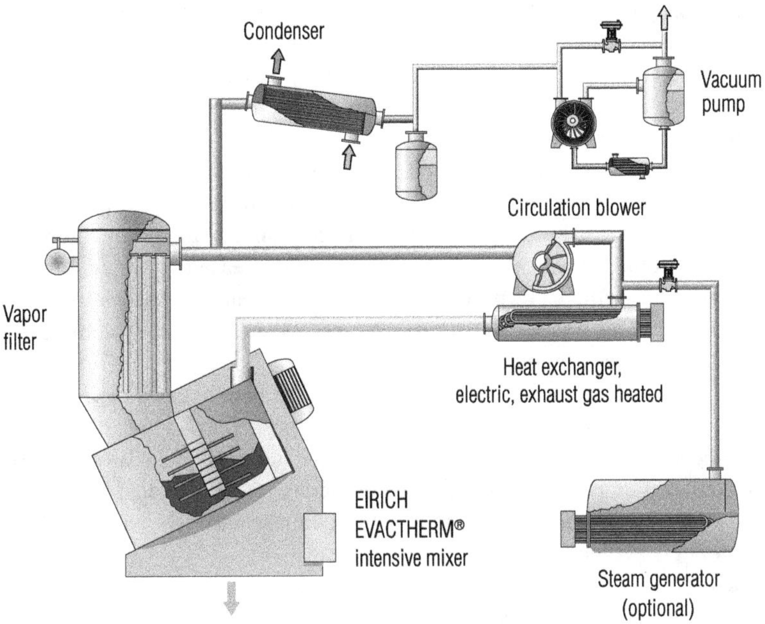

Figure 12.3 Batchwise steam drying. (Courtesy of Maschinenfabrik Gustav Eirich GmbH & Co. KG, Hardheim, Germany.)

Figure 12.3 exhibits the Eirich Evactherm drying plant. The tilted mixing bowl is the heart of the drying plant and is available in sizes up to 7 m^3. It rotates while an excentric stirrer rotates as well, usually in the opposite direction. The bowl is enveloped by a stationary and vacuum-tight casing. The rotational speed of the stirrer with mixing arms is greater than the rotational speed of the bowl. The bowl speed is fixed, whereas the stirrer speed is variable. One stationary scraper prevents the rotation of the wet product as a whole and supports the discharge through the central outlet. Moreover, the scraper prevents the occurrence of incrustations on both the bowl bottom and the bowl wall. Due to the batchwise operation, the material properties change during the drying process. The feed usually has low to medium viscosity. During drying, the viscosity increases strongly. The processed material passes through a plastic phase, after which a free-flowing material is obtained. The stirrer speed is adapted to the various drying phases. Superheated steam having a pressure and a temperature of 1030 mbar and up to 450°C, for example, is introduced into the wet product by means of an extended pipe. First, water evaporation is considered. The maximum product temperature can be adjusted and controlled by pressure adjustment. The boiling temperature of water at 500 mbar is 80°C, so the temperature of the wet product is, in the first drying phase, a maximum of 80°C. The added steam cools down to 120°C, for example, and sensible heat is transferred to the wet product. This heat is converted into latent heat at the evaporation of water from the product. Steam that is still slightly supersaturated leaves the mixer, passes through a filter to retain entrained solid particles and through a fan, and is reheated in an indirect heat exchanger. In relatively small plants, the heat exchanger is an electric one. In large plants, the heat is obtained by the combustion of natural gas. The excess steam is condensed in a surface condensor under pressure control. The condensate flows into a receiver and the progress of the drying step can be followed by checking the amount of condensate.

If the final moisture content desired is very low, the product temperature must be increased to reach the product moisture specification. The final product is then usually a powder. If it is desired to obtain dustfree granules, the product temperature can, prior to product removal from the dryer, be decreased by rapid lowering of the pressure, after which the product is cooled by means of evaporative cooling. For example, 46°C is reached at a pressure of 100 mbar. The system vacuum is maintained by a water ring pump.

The feeds of the Evactherm are usually cold. In the first instance their temperature must be increased to the temperature of the first drying stage. It is possible to heat the feed in a short time by using saturated steam. The steam condenses on the product. The drying can start as soon as the drying temperature is reached. It is also possible to heat the feed in a longer time by means of hot air or nitrogen. Now condensation on the product is avoided.

It is also possible to practise steam drying at atmospheric pressure in the Evactherm. Vacuum casing is not required in that case.

If organic solvents are to be removed, it is useful to distinguish between solvents that are miscible with water and solvents that are immiscible with water. The removal of miscible solvents is single-stage distillation, whereas the removal of immiscible solvents is steam distillation. The two-phase system boils at the latter operation when the sum of the vapor pressures of the two liquids equals the system pressure. This phenomenon defines the temperature at which the drying step occurs.

Example 12.1 Toluene is removed from the mixing bowl by means of steam distillation. The steam enters at a temperature of 350°C. The bowl pressure is 400 mbar. In the first instance, toluene is removed from the bowl. For this stage it is requested to calculate the product temperature in the mixer, the composition of the vapor leaving the mixer, and the amount of water left behind in the mixer per kilogram of superheated steam. In the second instance, condensed water is removed from the bowl. For this stage it is requested to find the product temperature in the bowl and to calculate the amount of water removed per kilogram of steam.

Water data See Section 4.13.

Toluene data

Specific heat liquid toluene: 1.7 $kJ \cdot kg^{-1} \cdot K^{-1}$
Heat of evaporation at 61.5°C: 358 $kJ \cdot kg^{-1}$

The product temperature is 61.5°C. At this temperature, the toluene vapor pressure is 185 mbar and the water vapor pressure is 215 mbar. The sum of the two vapor pressures is 400 mbar.

	Vapor Pressure at 61.5°C (mbar)	Pressure (%)	kmol per 100 kmol	$kg{\cdot}kmol^{-1}$	Mass (kg)	Mass (%)
H_2O	215	53.75	53.75	18	967.5	18.5
Toluene	+ 185	+ 46.25	+ 46.25	92	+ 4255.0	+ 81.5
	400	100.00	100.00		5222.5	100.0

The vapor leaving the mixing bowl consists of 18.5% by weight of water and 81.5% by weight of toluene. The amount of water left behind in the mixer results from an enthalpy balance (reference: 0°C). The enthalpy of 1 kg of steam entering the bowl is

$$2500 + 1.9 \cdot 350 = 3165 \text{ kJ}$$

x kilograms of vapor leaves the mixer per kilogram of steam entering the bowl.

Thus, the enthalpy balance is

$$0.185{\cdot}x(2500 + 1.9{\cdot}61.5) + 0.815{\cdot}x(358 + 1.7{\cdot}61.5) + (1 - 0.185{\cdot}x)4.2{\cdot}61.5 = 3165 \text{ kJ}$$

$$x = 3.57 \text{ kg}{\cdot}\text{kg}^{-1}$$

This amount of gas contains $0.185{\cdot}x = 0.66$ kg of steam. $1 - 0.66 = 0.34$ kg per kilogram of steam entering the bowl condenses in the bowl. The removal of water condensed in the bowl starts after the toluene removal.

In the second phase, the temperature rises to 76°C, this temperature is the water saturation temperature at 400 mbar. The amount of water evaporated also results from an enthalpy balance (reference: 0°C). x kilograms of steam leaves the mixer per kilogram of steam entering the bowl.

$$x(2500 + 1.9{\cdot}76) = 3165 \text{ kJ}$$

$$x = 1.2 \text{ kg}{\cdot}\text{kg}^{-1}$$

In Example 12.1, 61.5 and 76°C are calculated and known as the temperatures of the two drying stages: the stage of toluene removal and, subsequently, water removal. In practice, these two temperatures are possibly 20 K higher because thermal equilibrium is not reached completely.

Example 12.1 shows that it is possible to remove organic solvents at relatively low temperatures by means of steam distillation. Toluene boils at 110°C at atmospheric pressure. Toluene can be removed from the mixing bowl at 67°C when the pressure is 400 mbar. This is a well-known advantage of steam distillation. Even high-boiling organic liquids such as nitrobenzene, which has a boiling point of 211°C at atmospheric pressure, can be distilled at 99°C at atmospheric pressure with superheated steam. The condensate contains 15.8% by weight of nitrobenzene in that case. The cause of this phenomenon is the great difference in molecular weight between, on one hand, water, and, on the other hand, the immiscible organic liquid.

REFERENCES

[1] Kudra, T., Mujumdar, A.S. (2002). *Advanced Drying Technologies*, Marcel Dekker, New York, p. 103.

[2] van Deventer, H.C. (2004). *Industrial Superheated Steam Drying*, TNO, Apeldoorn, The Netherlands, p. 16.

13

RADIATION DRYING

Radiation drying is drying by means of electromagnetic waves. A distinction can be made between dielectric drying and infrared drying. Dielectric drying is treated in Section 13.1, and infrared drying is dealt with in Section 13.2.

Dielectric drying results when heat is generated by the application of certain electromagnetic fields, the energy source being an electric network. Radio-frequency (RF) drying and microwave drying are the two types of dielectric drying. Radio waves have frequencies in the range 10^6 to 10^8 Hz, and microwaves have frequencies in the range 10^8 to 10^{10} Hz. Characteristic wavelengths of RF waves and microwaves are 10 and 0.1 m, respectively. The relatively short wavelength of the microwaves (as compared to radio waves) is the reason for calling them microwaves.

Infrared (IR) drying is accomplished by a further type of electromagnetic radiation. The wavelengths used are in the range 1 to 6 μm. IR radiation can be generated in one of two ways: electrically, by passing an electric current through a resistance, or by a burning gas that heats a ceramic plate, which then emits radiation.

The wavelengths and frequencies of electromagnetic waves are coupled by means of the formula $v = f{\cdot}\lambda$, where v is the propagation velocity in a vacuum (i.e., $2.9978{\cdot}10^8$ $\text{m}{\cdot}\text{s}^{-1}$), f is the frequency in hertz (i.e., s^{-1}), and λ is the wavelength in meters. Figure 13.1 gives a qualitative idea of the penetration depths of the three different types of radiation drying. Drying is one of the successful areas of RF, microwave, and IR processing. The tempering of frozen food is also a typical successful area of microwave processing.

Drying in the Process Industry, First Edition. C.M. van 't Land.
© 2012 John Wiley & Sons, Inc. Published 2012 by John Wiley & Sons, Inc.

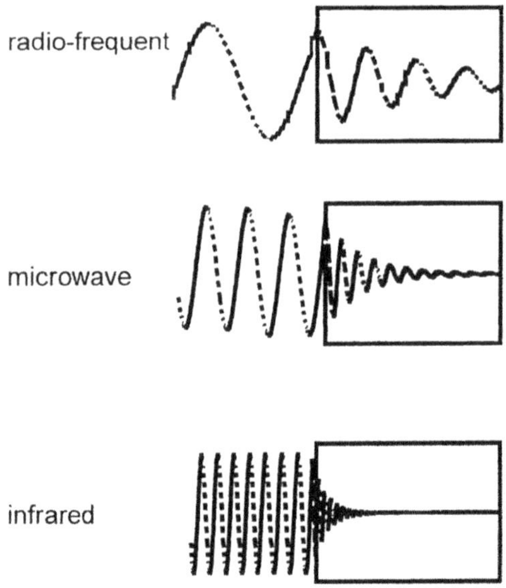

Figure 13.1 Radiation penetration depth.

13.1 DIELECTRIC DRYING

Dielectric drying offers advantages when the material being dried is not particulate (e.g., textiles) or is particulate with a large particle size (centimeters or decimeters); in other words, dielectric drying is successful when the surface/volume ratio is relatively small. Heat input by convection, conduction, or IR radiation is rather inefficient in these cases. Dielectric heating has the unique ability to generate heat within the product.

Dielectric drying uses an electric network as the energy source. The use of this energy source may increase in the future as fossil fuels become scarce, and their use implies the emission of carbon dioxide. Thus, it could be interesting to stimulate dielectric drying.

Dielectric equipment generates the same type of electromagnetic waves as are used by telecommunication systems. Therefore, in both the United States and Europe, a number of frequencies have been reserved for dielectric heating. The frequencies used most frequently are, for RF, $1.3{\cdot}10^7$ and $2.7{\cdot}10^7$ Hz, and, for microwaves, $9.15{\cdot}10^8$ and $2.45{\cdot}10^9$ Hz.

The quantum energy of electromagnetic radiation is $h{\cdot}f$, where h is Planck's constant (i.e., $6.60{\cdot}10^{-34}$ J·s). For microwaves that have a frequency of $2.45{\cdot}10^9$ Hz, the product equals $1.617{\cdot}10^{-24}$ J. The energy of the C—C bond in the ethane molecule, for example, is $6.08{\cdot}10^{-19}$ J. Thus, this bond energy is $3.76{\cdot}10^5$ times greater than the quantum energy of this microwave type. It follows that as far as the potential for ionization is concerned, microwaves are harmless for humans and animals. Gamma-rays do have this potential, as their frequencies are much higher than microwave frequencies. The quantum energies of radio waves are still smaller than the quantum energies of microwaves, as their frequencies are smaller than microwave frequencies. However, microwaves have the potential of heating human and animal tissues. Hence, regulations exist concerning the microwave field strength outside microwave equipment.

For dielectric heating to be utilized it must be possible to introduce the heat into the material being dried (i.e., the material must be susceptible). This means that the material must have an electrical conductivity intermediate between that of conductors and insulators (i.e., it should be a semiconductor). Semiconductors are also called dielectric materials. The coupling of the electromagnetic field into the process material can proceed by two mechanisms: (1) a dipole orientation and (2) an ionic orientation (i.e., resistance heating). A molecule possesses a dipole when the center of the positive charge does not coincide with the center of the negative charge. A well-known example of this is the water molecule. Molecules that have a dipole will be oriented by the electromagnetic field, and this movement will, through friction with adjacent molecules, generate heat. The ionic orientation comprises the influence of the alternating electric field on freely movable ions. At microwave drying, the dipole orientation is generally more important than the ionic orientation. At RF drying, the ionic orientation is usually more important. For example, at the drying of tobacco having a moisture content of 13.2% by weight (wet basis), the ionic mechanism is important at $2.712 \cdot 10^7$ Hz, whereas the dipole orientation dominates at $2.45 \cdot 10^9$ Hz [1]. Insulators (e.g., glass and Teflon) are transparent to radio waves and microwaves; conductors (e.g., stainless steel) reflect them, so they can be used to guide microwaves. Figure 13.2 shows a large microwave dryer, and Figure 13.3 illustrates an RF dryer.

Figure 13.2 A 1-MW microwave drying system with $3 \cdot 10^6$ Btu·h^{-1} auxiliary hot air system. (Courtesy of Microdry Inc., Crestwood, KY; http://www.microdry.com.)

Figure 13.3 A 50-kW RF dryer for bakery products. (Courtesy of Strayfield Limited, Reading, UK.)

Theory

Power Equation The electromagnetic fields used are RF and microwave. Equation (13.1) is valid for both fields:

$$v = f \cdot \lambda \tag{13.1}$$

where

v = propagation velocity, m·s^{-1}
f = frequency, s^{-1}
λ = wavelength, m

The propagation velocity in a vacuum is $2.9978 \cdot 10^8$ m·s^{-1}. This is the speed at which the light of the sun travels to the Earth through space. On entering a medium (e.g, a glass), the frequency remains constant, however, both v and λ vary.

The formula expressing the heat generation at both RF drying and microwave drying is [2]:

$$P = 2\pi \cdot f \cdot \varepsilon_0 \cdot \varepsilon_r' \cdot \tan\phi \cdot E^2 \tag{13.2}$$

where

P = power generation, W·m^{-3}
ε_0 = dielectric permittivity of free space (i.e., $8.854 \cdot 10^{-12}$ F·m^{-1})

ε_r' = relative dielectric constant of the material (i.e., the ratio at a specified frequency of the capacity of a capacitor, with the material as dielectric, to the capacity of the same capacitor in vacuo)

$\tan\phi$ = loss tangent (i.e., the tangent of the phase angle between the field in the material and the applied field)

E = field strength, $V{\cdot}m^{-1}$

The product $\varepsilon_r'{\cdot}\tan\phi$ is termed the *loss factor*. The symbol ε_r'' is generally used for the loss factor. A rule of thumb is that materials exhibiting, at a given set of temperature and frequency, a loss factor greater than 0.02 can be considered receptive to RF heating. Microwave heating can be effective at even smaller values of the loss factor. Table 13.1 gives the dielectric loss factors of some common materials. Concerning equation (13.2), one conclusion can be drawn now on taking a look at this table. The loss factors for a typical microwave frequency are usually greater than the loss factors for a typical RF frequency. Equation (13.2) states that the specific power input in $W{\cdot}m^{-3}$ is proportional to the frequency. Thus, for equal field strengths, the specific microwave power input is usually between 100 and 1000 times greater than the specific RF power input. Indeed, microwaves are more powerful than RF waves.

Often, RF heating is operated with high field strengths to compensate for the low frequencies. A practical limit is 10^5 $V{\cdot}m^{-1}$ for most materials. Clean relatively dry air has an electrical breakdown strength of $3{\cdot}10^6$ $V{\cdot}m^{-1}$.

Table 13.1 illustrates a number of additional aspects. It is well known that water, due to its dipole, is receptive to both microwaves and RF waves. Ice is barely receptive, the reason being that its molecules are fixed in a crystal lattice and can neither rotate nor translate. Microwaves act on the rotation of molecules and on the translation of ions. Both methanol and ethanol are receptive to RF waves and microwaves. Because of the presence of the hydroxyl group in the molecule, both molecules have a dipole. However, on going from methanol to ethanol, the alkane character of the molecule becomes more important and the effect of the dipole decreases. Polyethylene is very symmetric and the polymer molecules do not possess a dipole. Finally, paper possesses a small loss factor that is large enough to allow postdrying of paper by means of RF waves. The most important constituent of paper is cellulose.

Table 13.1 Dielectric Loss Factors of Various Materials at Two Frequencies

Compound	Temperature (°C)	10^7 Hz	$3{\cdot}10^9$ Hz
Ice	−12	0.067	0.0029
Water	25	0.36	12.0
Methanol	25	0.81	15.3
Ethanol	25	0.80	1.63
Polyethylene	25	<0.00045	0.0007
Paper	22	0.25[a]	0.5

Source: [3] for paper and [4] for the other compounds.

[a] At $2.712{\cdot}10^7$ Hz.

Table 13.2 Dielectric Data of Wet α-Lactose Monohydrate at $2.45 \cdot 10^9$ Hz

Water Content, Wet Basis (% by weight)	ε'_r	tan ϕ	Loss Factor, Powder	Loss Factor, Densified Powder[a]
0	1.61	0.0048	0.0077	0.014
3	1.72	0.026	0.045	0.083
7.5	2.40	0.069	0.17	0.25
15	3.68	0.135	0.50	0.648

Source: [5].
[a] Bulk density 1000 $kg \cdot m^{-3}$.

Uniform Drying In practice, mixtures are dealt with. Table 13.2 contains data for a mixture of water and α-lactose monohydrate. The latter material is an important constituent of pharmaceuticals. The data were measured at $2.45 \cdot 10^9$ Hz, a typical microwave frequency. This frequency is generally adopted for domestic microwave ovens and for many industrial microwave dryers. The fourth column contains data of powder, and the fifth column contains data of densified powder. The loss factors of the latter material are higher than the loss factors of the former material. This is due to the higher water concentration in the latter material. The table shows that wet material is more receptive to microwaves than relatively dry material. This is often the case, although not always. The phenomenon is an advantage of dielectric drying because energy is no longer absorbed when the material is dry. Or, put differently, microwaves are more active in wet areas than in dry areas. Thus, dielectric drying has the potential of producing a uniformly dried product. The phenomenon also means that the process is controlled by itself.

Comparing RF and Microwaves Table 13.3 contains data on the maximum specific power input and the heating rate. The table is based on equation (13.2). A material having a specific mass of 2000 $kg \cdot m^{-3}$ and a specific heat of 2000 $J \cdot kg^{-1} \cdot K^{-1}$ is taken. A field strength of 10^5 $V \cdot m^{-1}$ is assumed. The table contains data on the specific power input in $W \cdot m^{-3}$ and the heating rate in $K \cdot s^{-1}$ for a typical RF frequency and a typical microwave frequency and two loss factors. An implicit assumption is that the field strength remains constant when the radiation penetrates the material. That is usually not the case, as on penetrating the material, the radiation starts to heat the material, and hence the field loses strength. Keeping this in mind,

Table 13.3 Maximum Specific Power and Heating Rate

	Specific Power ($W \cdot m^{-3}$), Loss Factor		Heating Rate ($K \cdot s^{-1}$), Loss Factor	
Frequency (Hz)	0.1	1.0	0.1	1.0
10^7	$5.56 \cdot 10^5$	$5.56 \cdot 10^6$	0.14	1.4
$3 \cdot 10^9$	$1.67 \cdot 10^8$	$1.67 \cdot 10^9$	41.8	418

it is possible to draw two conclusions qualitatively. The loss factor is very important for the specific power input, and microwaves are more powerful than RF.

Penetration Depth At penetration depth x, the following differential equation is valid:

$$-\frac{dE}{dx} = a \cdot E(x)$$

Integration yields

$$E(x) = E_0 \cdot e^{-ax} \tag{13.3}$$

where a is a material constant. This is *Lambert–Beer's law*. The law expresses that the field strength decreases exponentially with the path traveled. This means that the heat development in W·m^{-3} also decreases.

The following equations are approximately valid [6]:

$$\delta = \begin{cases} \dfrac{\lambda\sqrt{\varepsilon_r'}}{2\pi \cdot \varepsilon_r' \cdot \tan\phi} & \text{if } \tan\phi \ll 1 \quad (13.4) \\[2ex] \dfrac{\lambda}{2\pi\sqrt{2\varepsilon_r' \cdot \tan\phi}} & \text{if } \tan\phi \gg 1 \quad (13.5) \end{cases}$$

where δ is the path traveled when the field strength is reduced to $1/e$ of its orginal value. The field strength is then approximately 37% of its original value (i.e., the field strength of the dielectric waves when they enter the material). δ is called the penetration depth. $\tan\phi \ll 1$ for the data in Table 13.2. Calculations for the noncompacted powder result in:

Water Content (% by weight)	Penetration Depth (m)
0	3.19
3	0.57
7.5	0.18
15	0.07

Breakdown Breakdown is the occurrence of an electrical discharge between two metal parts in an electric field. It is caused by ionization of the gas molecules between the two electrodes. It can be compared to the corona between the two electrodes of a fluorescent lamp. The probability of an electric discharge at dielectric drying can be minimized, however, it cannot be completely excluded.

For dry air, it is useful to distinguish among three ranges:

1. Pressures higher than 50 mbar. The molecular concentration is high and thus the molecules cannot, due to collisions, readily get the velocity required for ionization. At 50 mbar, field strengths up to 10^5 V·m^{-1} can be used, and at 1000 mbar up to $3{\cdot}10^6$ V·m^{-1}.
2. Pressures in the range 1 to 50 mbar. In this range, ionization is possible. At 1 mbar, field strengths up to 10^4 V·m^{-1} can be used.
3. Pressures lower than 1 mbar. The molecular concentration is low. Although the molecules can get a high velocity, ionization is less probable due to the few collisions. At 1 mbar, a field strength of 10^5 V·m^{-1} can be used.

Efficiency Dielectric drying accomplishes drying by means of electrical energy. If electrical energy (i.e., electricity) is raised by the combustion of fossil fuel, the efficiency is approximately 40%. That is, 40% of the chemical energy present in the fossil fuel is converted into electricity. The efficiency of the conversion of the electric energy from the grid into microwaves having a frequency of $9.15{\cdot}10^8$ Hz is approximately 90%, and for microwaves having a frequency of $2.450{\cdot}10^9$ Hz, approximately 70%, whereas the efficiency is 65 to 70% for radio waves. The reason is that heating up of the equipment also takes place, and this heat is usually carried away by cooling water or cooling air. The efficiency of the acceptance of the electromagnetic radiation by the dielectric material is in the range 80 to 90%. First, not all radiation is accepted by the material, and second, part of the energy accepted by the material is lost due to heat radiation. This means that the overall efficiency is 25 to 30%. This also means that dielectric drying is not an economic option for bulk materials, but it is a useful technique for special applications.

A total of 2 to 2.5 kWh is used per kilogram of evaporated water in the Belgian textile industry (RF). The drying of latex utilizes 2.3 kWh per kilogram of evaporated water [7]. The evaporation of 1 kg of water at 0°C requires 2500 kJ, which is 0.696 kWh.

RF Drying

Dielectric drying by means of RF employs a generator to convert the main frequency of 50 or 60 Hz to the operating frequency. The first step is to transform the electrical energy from the grid to electrical energy having high tension. The frequency is not changed. The next step is the rectification to dc energy having high tension. Then an oscillator circuit produces electrical energy having both a high tension and a high frequency. Dielectric tubes containing the oscillator circuit have a lifetime of 5000 to 10,000 h. However, the manufacturer's guarantee is usually much less. The applicator is that part of the installation in which the product is heated, which can take one of three different forms: (1) two flat metal plates, (2) a stray field electrode system, and (3) a staggered through-field electrode system.

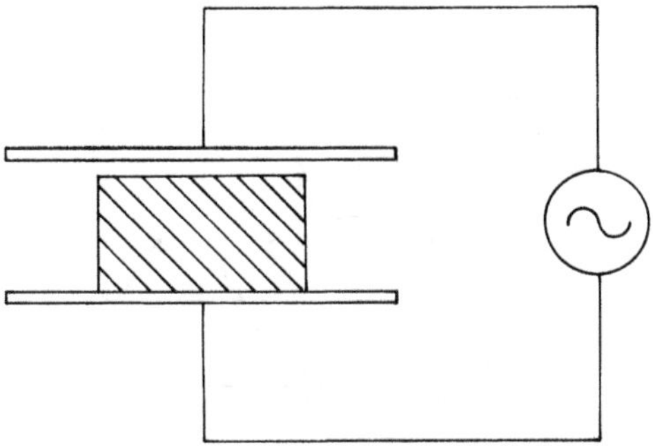

Figure 13.4 RF applicator built as two flat metal plates.

The first option is exhibited in Figure 13.4. It can be used for large objects. The clearance between the top of the object and the upper condensor plate should be as small as possible. This design is used for the drying of hanks of yarn in the textile industry, for example. Figure 13.5 outlines the second option. Here, a horizontal nonuniform field passes through (usually) thin webs (up to 10 mm). The postdrying of paper to obtain a uniform low moisture content is, for example, the drying of a thin web. Electrodes that are arranged both above and below the material can be seen in Figure 13.6. Relatively thick sheet material can be processed (third option).

RF drying usually exhibits a self-limiting or leveling effect: A wet load causes an increase in electrode voltage, but a dry load will cause a power reduction.

Stalam is an Italian manufacturer of RF dryers functioning batchwise or in continuous mode. Eighty percent of their production is used for textile drying and 15% is used for food drying. Their largest single dryer has a capacity of 150 kW. Stalam uses the frequency $2.712 \cdot 10^7$ Hz. RF heating has become an accepted and reliable method of drying bakery products. The RF dryer has proven to be effective to control the product moisture content accurately and uniformly when installed after the baking oven. Furthermore, the capacity increases (see Figs. 13.7 and 13.8). Figure 13.8 illustrates a case in which the capacity of an existing conveyor dryer could be extended from 500 to 700 $kg \cdot h^{-1}$.

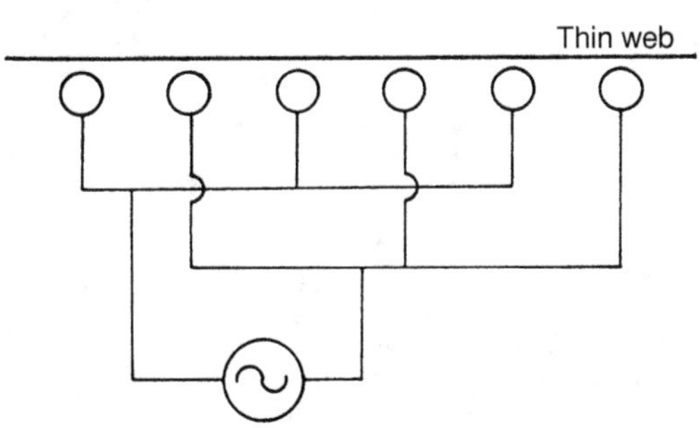

Figure 13.5 Thin web being treated with RF waves.

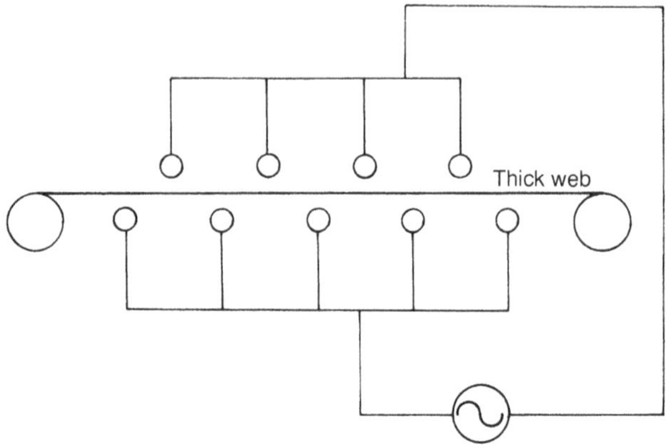

Figure 13.6 Through-field RF electrode system for thick webs.

Strayfield is an English manufacturer of RF dryers functioning batchwise or in continuous mode. Their dryers are also used in both the textile industry and the food industry. Figure 13.3 shows a stainless steel version of a continuous 50-kW dryer for bakery products. They also use the frequency $2.712 \cdot 10^7$ Hz.

Microwave Drying

Microwave drying utilizes unconventional electric equipment. After generation by means of a magnetron or klystron (lifetime 5000 to 10,000 h), the microwave energy

Figure 13.7 RF dryer for bakery products. (Courtesy of STALAM S.p.A., Nove, Italy.)

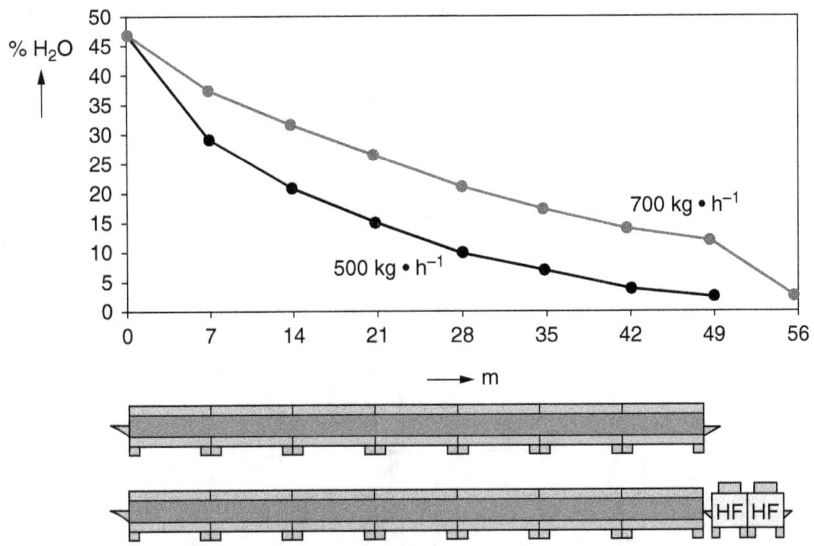

Figure 13.8 Typical capacity extension of a baking oven by the addition of an RF postbaking dryer. (Courtesy of STALAM S.p.A., Nove, Italy.)

is transported to the applicator. At the frequency $2.45 \cdot 10^9$ Hz, the wavelength is 12 cm. Usually, waveguides are employed for the transport (however, coaxial cable is also possible for low power). A waveguide is a hollow rectangular conduit made of metal; the dimensions match the nature of the transported microwaves. Waveguides can also be used as applicators, as in the passing of filamentary materials through the field in a waveguide. Traveling-wave applicators are also used (see Fig. 13.9).

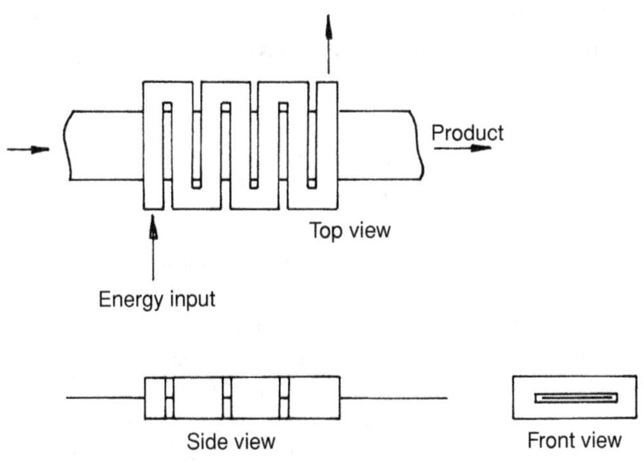

Figure 13.9 Traveling-wave applicator.

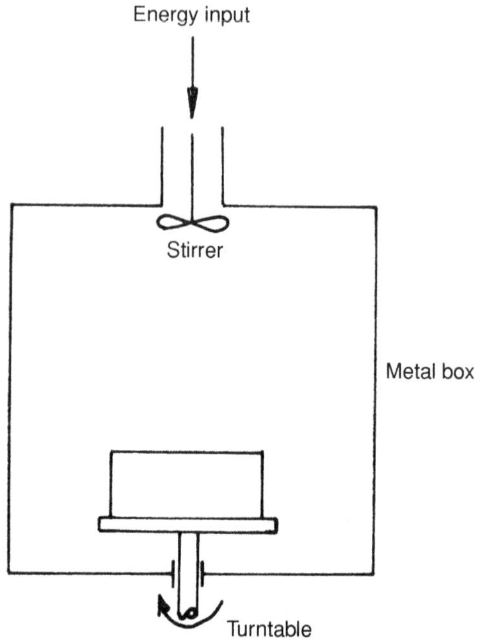

Figure 13.10 Cavity applicator.

A thin sheet of material is passed through the slots. Figure 13.10 exhibits a cavity applicator, a type used widely, such as in home microwave ovens. Basically, this applicator consists of a metal box with continuous and batchwise operations possible. It is important that the microwave field in the load be uniform. Steps to ensure this are:

1. Moving or rotating the load by means of conveyors or turntables.
2. Installing mode stirrers to avoid the occurrence of stationary ventral segments and nodes.
3. Using multiple ports.
4. Using multiple sources with slightly different frequencies.
5. Selecting a set of cavity dimensions.

A decrease in the evaporation load usually does not automatically lead to a decrease of the tubes' power output. In this aspect, microwave drying does not resemble RF drying. Steps must be taken to avoid overheating the tubes in case the processed material changes its susceptibility. First, the generator and the drying cavity are often physically separated to protect the generator from reflection. Second, provisions are designed between the generator and the drying cavity to absorb reflected power. For a continuous dryer, the leakages at the product inlet and outlet are controlled by reject filter traps using quarter-wavelength choke elements or dissipative traps through which an absorbing liquid flows. The latter provision can be a combination of an upper and a lower plastic pipe through which water flows. In terms of safety, it is more important to contain microwaves than RF. The reason is the frequency difference. Containing microwaves is more difficult than containing RF because the wavelength of the former is smaller.

The German company Püschner in Schwanewede indicates that the largest module for the generation of microwaves having a frequency of $9.15 \cdot 10^8$ Hz is a 100-kW module. Their largest power module for the generation of microwaves having a frequency of $2.45 \cdot 10^9$ Hz is a 20 to 30-kW module. The reason for this difference is the control of the heat transfer. The generation of microwaves is accompanied by heat development. A magnetron is in principle a vacuum diode containing a cylindrical anode and a central cathode. A small wavelength (and a high frequency) require a small distance between the two electrodes, and vice versa. Thus, the dimensions of a magnetron generating microwaves having a frequency of $9.15 \cdot 10^8$ Hz are larger than the dimensions of a magnetron generating microwaves having a frequency of $2.45 \cdot 10^9$ Hz. A relatively large magnetron has a relatively large area for heat transfer. At $2.45 \cdot 10^9$ Hz, a capacity of 75 kW, for example, can be obtained by having a number of elements in parallel.

Microwave drying is found less often than RF drying. The two reasons are that microwave drying is, as to the investment, more expensive than RF drying and that microwave drying requires more precautions than RF drying. The small wavelength of the microwaves requires special provisions to contain them. However, it is attractive that reasonable microwave powers can be combined with modest field strengths. That minimizes the risk of a breakdown.

In microwave drying, the center of a spherical object is heated more strongly than the periphery. This is called the *lens effect*. An example is the heating of a potato.

Applications of Microwave Drying The first application is a shortcut pasta microwave hot-air drying system sold by Microdry Incorporated at Crestwood in Kentucky. A typical macaroni dryer produces 4000 lb of product per hour. The drying system consists of three parts:

1. A conventional hot-air predryer. The dwelling time is 35 min here and the moisture content is reduced from 30 to 18% by weight.
2. The microwave–warm air stage. A further reduction to 13 to 13.5% occurs in approximately 12 min. The microwave system operates at approximately 30 kW. The air temperature is 180°F.
3. An equalizing stage to arrest the drying and to allow temperature gradients to equalize and cool the product. This stage lasts approximately 1 h.

Conventional convective drying takes 8 h.

The second application is the drying of sugar lumps [8]. Conventionally, sugar lumps are dried in infrared dryers. Crystalline sugar is mixed with approximately

1% by weight of water. Sugar lumps are then made by means of a rotating press. The energy consumption at microwave drying is 20 to 40% less than at infrared drying, whereas the two product qualities are comparable. Microwave drying occupies less floor space than infrared drying. In 1995, 62,500 t of sugar lumps were produced in The Netherlands. A typical microwave dryer for sugar lumps has an hourly capacity of 1600 kg.

The third application is the small-scale fluid-bed drying of a wet porous ceramic powder [9]. The study was carried out to find a method to immobilize dissolved radioactive waste on a fine powder. The fluid bed had a diameter of 0.3 m. At each test, the initial charge was 10 kg of powder having particle sizes in the range 20 to 50 μm. The charge was fluidized by means of ambient air having a typical velocity under the distribution plate of 0.034 $m{\cdot}s^{-1}$. Next, water was sprayed onto the powder until the moisture content was, for example, 0.15 kg of water per kilogram of powder. Microwave drying was started when the test moisture content was reached. The microwave frequency was $2.45{\cdot}10^9$ Hz and the microwave power was continuously adjustable from 0 to 6 kW. The fluidization was continued during the drying. A bed temperature of 70°C, for example, was adjusted by means of the microwave power input. The drying times were in the range 1 to 2 h. A typical energy consumption figure is 6000 $kJ{\cdot}kg^{-1}$. To simulate the liquid radioactive waste, work was also carried out with aqueous solutions of sodium nitrate. The presence of the dissolved salt decreases the drying rate. The experiments were successful. Lump formation and incrustation did not occur. Lumps and incrustations did form in rotating contact dryers.

Drying with warm air only is not feasible because the particle size is small. The powder is elutriated at relative high air velocities. Making the powder coarser is not an option either, as the specific area in $m^2{\cdot}kg^{-1}$ is inversely proportional to the particle size.

The fourth application is the combined contact drying and microwave drying under vacuum of a pharmaceutical product manufactured by Laboratoires Glaxo in France [10,11]. The product is a granule and the evaporated liquid is an alcohol. The applied dryer type is depicted in Figure 13.11. Prior to the installation of an industrial 1200-L batch dryer made by GEA Pharma Systems (formerly, Machines Collette) in Belgium, experiments were carried out in three different dryers having volumes of 75, 300, and 600 L. Table 13.4 is a survey of the experiments carried out. The dryer has the form of a closed vessel in which a stirrer rotates intermittently. In the 300-L dryer, the stirrer completes a full rotation once per 5 min, and this takes 6 s. It was assessed that continuous rotation causes bad product compressibility. The product is dried by heat transfer from a heating jacket and by microwaves. The absolute pressure in the dryer is 5 mbar on both the 75 and 300-L scales. On going from the 75- to 300-L scale, it is possible to keep the drying time of approximately 40 min constant by the simultaneous action of contact drying and microwaves. This is achieved by keeping the specific microwave power input in W/kg of material processed constant. The drying time increases from 30 min on the 75-L scale to 135 min on the 300-L scale when contact drying is the only drying mode. The cause is the decrease in the heating area per kilogram of material processed on scaling up. The quality of the

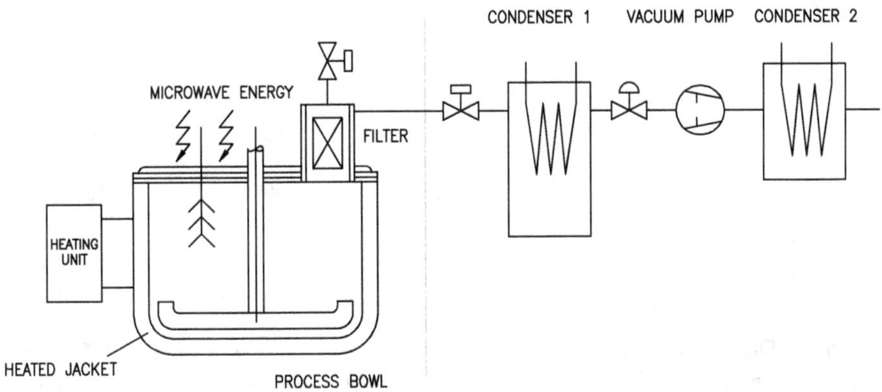

Figure 13.11 Combined contact and microwave drying. (Courtesy of GEA Pharma Systems nv, Wommelgem, Belgium.)

product dried by means of microwaves and a warm water jacket was equal to the quality of the product dried by a warm water jacket only.

Runaway In many instances, the product stops absorbing microwave energy when the product gets dry. However, this is not always the case. Potato starch still absorbs microwave energy when the product is dry [5]. The cause is that the product contains 8% by weight of bound water when the product is dry (see Table 13.5). This table contains data of drying experiments carried out at an absolute pressure of 200 mbar and a temperature of 60°C. On applying microwaves ($2.45 \cdot 10^9$ Hz) on this product at 0% water by weight, the temperature rises to 100°C and higher. The loss factors of powder and compacted powder remain approximately constant when the moisture content varies from 0 to 15% by weight.

Product Damage Drying wood and clay by means of microwaves can cause product damage. This is the case when the steam raised within the material cannot escape and there is a pressure buildup. This can cause the cracking of the material. This does not occur readily in the drying of textile, as it has an open structure.

Table 13.4 Microwave Dying Tests at Laboratoires Glaxo

Dryer volume (L)	75	75	300	300	300	600	600
Product[a]	p	p	p	p	a	p	a
Drying method[b]	c	c/mw	c	c/mw	c/mw	c/mw	c/mw
Batch size (kg)	30	15	150	150	150	200–300	
Drying time (min)	60	30	135	30	40–45	35–50	

[a] p, placebo; a, active material.
[b] c, contact drying; mw, microwave drying.

Table 13.5 Microwave Drying of Wet Potato Starch

H_2O (% by weight)	ε'_r	$\tan\phi$	$\varepsilon'_r \cdot \tan\phi$ Powder	$\varepsilon'_r \cdot \tan\phi$ Densified Powder[a]
0	2.86	0.3	0.85	1.30
3	3.42	0.34	1.15	1.54
7.5	3.13	0.28	0.89	1.50
15	3.58	0.29	1.05	1.83

[a] Bulk density 1000 $kg \cdot m.^{-3}$

Product Discoloration A case has been reported of the discoloration of a pharmaceutical product in batchwise combined contact drying and microwave drying. In the indirect drying of the product without microwaves, the phenomenon did not occur. The measured temperature and pressure conditions were equal at the two tests. The probable explanation is that the microwaves have caused hot spots in the material.

Process Control of Batchwise Microwave Vacuum Drying Both the field strength in the dryer and the product temperature are measured. The energy supply must be reduced or stopped if one of these two variables increases. An increase in the field strength indicates the end of the drying process if the dry product has a low loss factor. An increase in the product temperature indicates the end of the drying process if the dry product has a high loss factor. Furthermore, it is possible that the stirring power is a function of the product's moisture content. Finally, it is possible to check the progress of the drying step by measuring the amount of evaporated and condensed moisture.

Process Safety in Batchwise Microwave Vacuum Drying

1. Purge with N_2.
2. Pressure not lower than 50 mbar. *Background*: At pressures below 50 mbar, the probability of a corona (breakdown) increases. Note that the pharmaceutical application discussed does not satisfy this requirement.
3. Pressures not higher than 100 mbar. *Background*: The pressure at a deflagration can rise to approximately eight times the initial pressure. In case a deflagration occurs, the pressure remains below the atmospheric pressure.
4. The field strength should be chosen carefully.

13.2 INFRARED DRYING

The drying technique is very suitable for the drying of surface coatings (paints), webs (e.g., paper and textile), and objects (e.g., ceramic tiles). Unlike dielectric drying, the penetration depth of IR is not substantial (e.g., several millimeters). It can

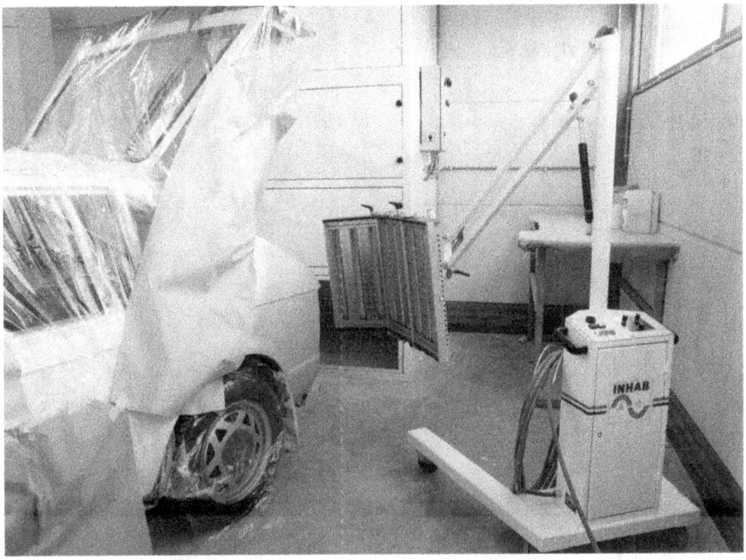

Figure 13.12 IR spot heater.

offer advantages compared to convective and conductive drying. The advantages may concern both the manufacturing cost price and the way the drying operation affects the quality of the product.

Generally, the objectives of IR radiation are drying, hardening, or accelerating a chemical reaction. The discussion is restricted to drying. Generally, the wavelength of IR radiation is used for classification: (1) short, 0.76 to 2.0 μm, (2) medium, 2.0 to 4.0 μm, and (3) long, 4.0 to 10.0 μm. The first commercial application in 1930 concerned the drying of cars (see Fig. 13.12).

IR radiation can be generated by means of electricity or by burning gas. Electric energy is expensive compared with energy generated directly by combusting gas. However, using electricity may, in the future, become an interesting option, as fossil fuels will become scarce. In addition, electric drying is clean because there are no exhaust gases. The amount of air to be supplied then only serves to entrain the vapor; hence, the dust problem is reduced. Gas systems are often fitted to existing heating and drying systems as postdryers or predryers to increase their capacity. Gas radiant heating is not capital intensive.

Theory

The IR radiation received by a body is absorbed, reflected, or transmitted. The part of the absorbed energy is the only effective part for drying. The factors influencing the percentage of the energy useful for the drying are (1) the properties and thickness of the coating, (2) the properties and thickness of the substrate, and (3) the geometrical configuration.

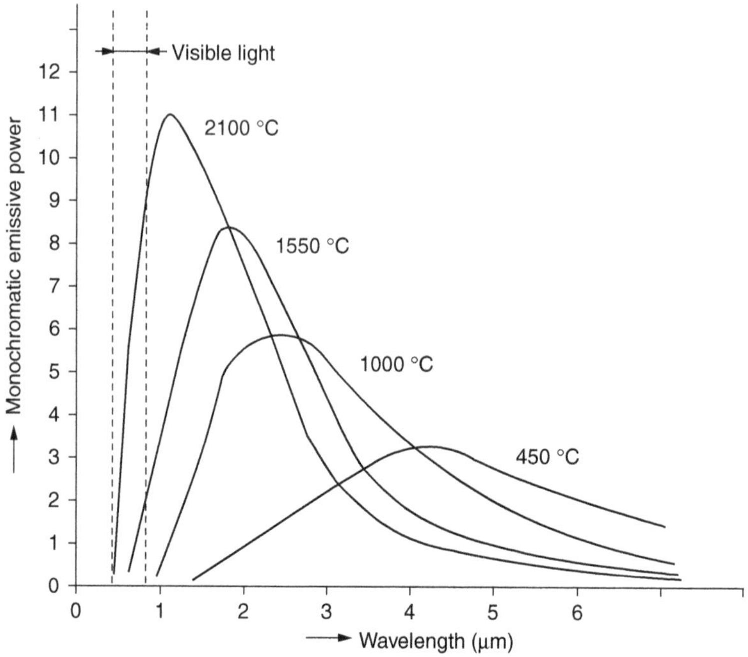

Figure 13.13 Emission spectra of a black body.

A distinction is made between black bodies, white bodies, gray bodies, and transparent bodies. A *black body* absorbs all radiation received (no reflection, no transmission). A black body emits a characteristic radiation. The spectral distribution of the energy is dependent on the temperature only (see Fig. 13.13). The amount of energy radiated increases when the temperature increases (areas under the curved lines). *Stefan–Boltzmann's law* expresses this for a black body quantitatively:

$$Q = \sigma \cdot T^4 \qquad \mathrm{W \cdot m^{-2}} \tag{13.6}$$

where σ is Stefan–Boltzmann's constant: $5.675 \cdot 10^{-8}$ $\mathrm{W \cdot m^{-2} \cdot K^{-4}}$. The wavelength at which the maximum energy prevails decreases when the temperature increases. This fact is known because bodies (e.g., metals) start to glow and even emit visible light when they become hot. For example, the wavelength at which the maximum energy density of the sun's light occurs is approximately 0.5 μm. Visible light ranges from 0.4 to 0.7 μm. *Wien's law* expresses this quantitatively:

$$\lambda_{\max} \cdot T = c_{\mathrm{w}} \tag{13.7}$$

where c_{w} is Wien's constant: 2897 μm·K.

A *white body* reflects all radiation received (no absorption, no transmission). Such a body cannot be heated by means of infrared radiation.

A *gray body* reflects part of the energy received and absorbs part of it, whereas the fraction absorbed is independent from the wavelength. Energy is not transmitted. Many bodies preferentially absorb radiation having a certain wavelength. For example, water absorbs IR radiation of wavelengths 2.7 and 6.0 μm preferentially. A further example of preferential absorption: The fact that a tomato is red means that light having the wavelength corresponding with red is reflected, and other wavelengths are more or less absorbed. It stands to reason that it is important to find out what wavelengths are absorbed by the articles to be dried. Thus, the emission spectrum of the source and the absorption characteristics of the receiving body must be matched.

Transparent bodies transmit all infrared radiation (no reflection, no absorption). Gases can often be considered transparent. However, water vapor is an exception because it absorbs IR radiation strongly. This points to the necessity to entrain the water vapor after evaporation.

There is a difference in the drying characteristics of surface coatings on metal and surface coatings on wood. Both are usually dried by means of medium-wavelength IR radiation (2.6 to 4.0 μm). The surface coatings tend to absorb IR radiation, which has wavelengths of approximately 3 μm, preferentially. Paints on metal do so, and the radiation, which has other wavelengths, is reflected on the metal's surface. Paints on wood behave similarly; however, the radiation passing through the surface coating is absorbed by the wood. The wood is thus heated, and in turn heats the paint by conduction from below. The net effect is that the paints on wood are dried homogeneously, whereas surface coatings on metals may exhibit dry spots that hamper solvent release. The energy exchange between two black bodies can be calculated by means of equation (13.6).

Example 13.1

Starting points

Temperature first body: 800°C
Temperature second body: 75°C

The emitting and receiving boundaries are infinitely large parallel planes. The gas between the two black bodies is transparent.

Calculation

$$Q = \sigma(T_1^4 - T_2^4) = 5.675 \cdot 10^{-8}[(800 + 273)^4 - (75 + 273)^4]$$
$$= 74,393\ \text{W}\cdot\text{m}^{-2}$$

It will be difficult to get the value calculated in Example 13.1 in actual practice. Ideal black bodies do not exist, and there will be so-called convective losses (gas between the two bodies that is heated and escapes). Stefan–Boltzmann's law is modified for gray bodies into

$$Q = \varepsilon \cdot \sigma \cdot T^4 \qquad \mathrm{W \cdot m^{-2}} \tag{13.8}$$

ε is the emission coefficient. Most dull and rough surfaces have ε values of about 0.6 to 0.8, but bright metal surfaces have low emissivities in the range 0.01 to 0.1. *Kirchhoff" s law* states that at each wavelength the emission coefficient of a gray body equals the absorption coefficient ($\alpha = \varepsilon$). It can be proved that the heat transfer between two parallel planes (gray bodies, no transmission) can be expressed as follows:

$$Q = \frac{\sigma\left(T_1^4 - T_2^4\right)}{1/\varepsilon_1 + 1/\varepsilon_2 - 1} \qquad \mathrm{W \cdot m^{-2}} \tag{13.9}$$

The gas between the two planes must be transparent and the radiation that is not absorbed is reflected diffusely (i.e., in all directions). The two planes must be very large (to prevent reflected radiation from escaping).

A gray body can be convex and surrounded completely by a second gray body. Then the following expression applies for the smaller body:

$$Q = \frac{\sigma\left(T_1^4 - T_2^4\right)}{1/\varepsilon_1 + (A_1/A_2)(1/\varepsilon_2 - 1)} \qquad \mathrm{W \cdot m^{-2}} \tag{13.10}$$

If $A_2 >> A_1$, and if ε_2 is close to 1,

$$Q = \varepsilon_1 \cdot \sigma\left(T_1^4 - T_2^4\right) \qquad \mathrm{W \cdot m^{-2}} \tag{13.11}$$

Approximately 90% of the energy consumed by an electric infrared generator is converted into radiation. The percentage for gas combustion is smaller.

Theory today does not mean too much for the design of infrared radiation drying equipment. Carrying out small-scale experiments to obtain design data is common. These tests produce drying times, energy consumption data, and an insight into product quality. Temperature–time curves are obtained (see Fig. 13.14).

An important aspect is the time delay on starting and stopping: (1) 0.76 to 2.0 μm: several seconds, (2) 2.0 to 4.0 μm: approximately one min, and (3) 4.0 to 10.0 μm: approximately 15 min. A long delay affects process control. Furthermore, a shutdown can mean that the radiation continues for some time. Often, it is necessary to provide facilities that isolate the source from the object (swing-out arm, barrier).

Equipment

Electrical Equipment Infrared radiation is obtained by passing an electric current through a resistance. The resistance heats up and thereby attains a high temperature. Lamps and tubes are known examples. Figure 13.15 shows a cross section of a typical tube emitter. Tungsten and Cr/Ni are being used as construction materials

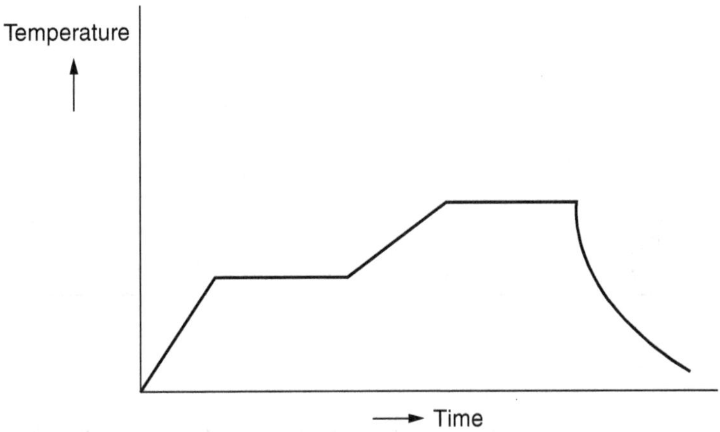

Figure 13.14 Temperature–time curve at IR heating.

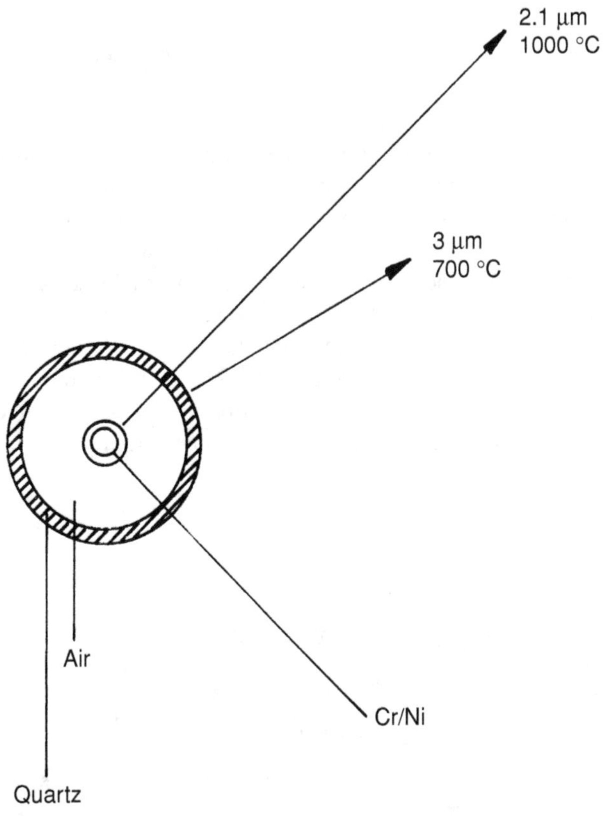

Figure 13.15 Cross section of a typical tube IR emitter.

Table 13.6 Primary Characteristics of Electrical IR Equipment

Indication	Lamp	Lamp	Tube	Tube	Tube
Material resistance	W	W	Cr/Ni	Cr/Ni	Cr/Ni
Temperature resistance (°C)	1900	2100	1100		
Wavelength resistance (μm)	1.3	1.2	2.1		
Wall material	Quartz	Quartz	Quartz	[a]	[a]
Wall (°C) temperature	170	400	700	400/800	650
Wall wavelength (μm)			3	2.7–4.3	3.1
Intermediate material	Vacuum	Gas	Air	[b]	[b]

[a] Stainless steel.
[b] Ceramic powder.

for resistances. The stainless steel emitters are preferred for drying purposes and are in use for textile and for paper. Tungsten is oxidized by air at approximately 2000°C (see Table 13.6). Electrical oven systems usually have no size limit and are tailored to a client's requirements.

Gas-Fired Equipment A distinction is made between surface combustion and volume combustion.

Surface Combustion Surface combustion is the most common method of IR heating. A perforated ceramic plate is commonly employed, with the gas–air mixture flowing through the small channels and igniting (see Fig. 13.16). Gradually, as the ceramic plate (thickness, e.g., 12 mm) heats up, the flames with an approximate length of 2.5 mm travel back into the small channels. This causes the surface to glow (900 to 950°C) and to emit radiation having, according to equation (13.7), its maximum power density at a wavelength of 2.3 μm. The plate's design (porosity) guarantees a temperature of the other side of the plate of 200 to 250°C. This temperature is too low to ignite the gas–air mixture; thus, the combustion process is safeguarded. Typical power supplies due to the combustion of, for example, natural gas are in the range

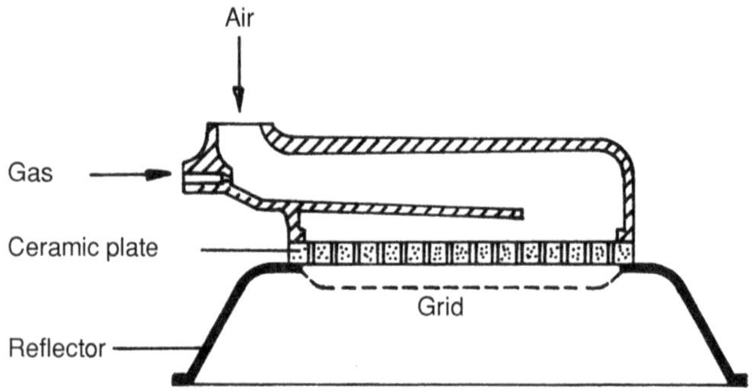

Figure 13.16 Gas-fired IR heater.

50–100 kW·m^{-2}. A fraction of the power is actually supplied as radiation power. The fraction depends on the application. Note that on using gas-fired equipment, the product comes into contact with exhaust gases.

According to equation (13.6), the radiation power emitted is proportional to T^4. On increasing the temperature from, 950°C to 1450°C, for example, the power emitted would increase with a factor of 4. However, on increasing the natural gas flow and the airflow to the burner to increase the surface temperature, the flames would leave the ceramic plate. A higher surface temperature can be attained with volume combustion.

Process control is not perfect; a turndown to about 60% of the nominal capacity is the limit. Process control is therefore often executed via the on–off mode for groups of burners. The catalytic burner is employed for leather drying. The gas reacts with oxygen without the emission of visible light, and the surface temperature is approximately 400°C. Propane and butane combustion are more complete than methane combustion. Gas oven systems usually have no size limit and are tailored to a client's requirements. Gas heaters fitted as a predryer or postdryer to an existing drying system are often standardized.

Volume Combustion The maximum temperature of the emitting surface can be raised to 1450°C by using volume combustion. Gas-fired equipment using volume combustion is marketed by GoGas Goch in Dortmund, Germany (see Fig. 13.17). The burner is divided into two regions. Region A mainly has to insulate the hot from the cold part of the burner and to mix the air and gas intimately. Region C houses

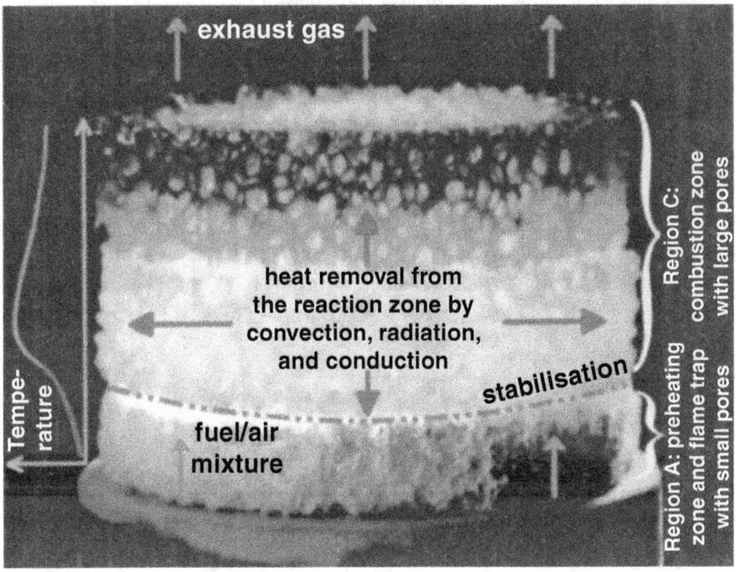

Figure 13.17 Heat transport mechanisms at volume combustion. (Courtesy of GoGas Goch GmbH & Co. KG, Dortmund, Germany.)

the combustion per se in a ceramic foam. The combustion takes place in many small cells. With a surface temperature of 1450°C, the maximum power density of the emitted radiation is now at a wavelength of 1.7 μm. The maximum power supply due to the combustion of natural gas is 1000 $kW{\cdot}m^{-2}$. A fraction of this power is actually supplied as radiation power. The fraction depends on the application. The turn-down ratio of this burner type is 6. It is thus much higher than the turn-down ratio of the previous burner type. A further aspect is that combustion in this burner type is more complete than the combustion in the previous burner type. The CO and NO_x values in the flue gas from this burner type are considerably lower than in the flue gas from the previous burner type.

Testing

Successful tests provide the basis for an infrared drying system's design. Static or moving drying trials are carried out at the equipment manufacturers' pilot plants. It is also customary to carry out site trials using a portable machine. A theoretical analysis can provide a feel for the direction to look for, however, tests are a prerequisite for a good decision.

Process Safety

Historically, there has been some concern in industry regarding the potential fire hazard of radiant heaters. It was stated that IR radiation equipment emitting in the wavelength range of 2 to 4 μm remains hot for approximately 1 min in the event of sudden stoppage. The material being processed may catch fire. Conventional heat shields, mechanical heater rotation, and retraction techniques have been used to decrease the risk, but these have not been fail-safe.

It is possible to buy surface burners that can be turned off and touched within 3 to 4 s. Rapid modulation within the operating range to match the thermal load is possible.

Another aspect is the possibility that an evaporated solvent may catch fire. For a typical range of solvents used in paint stoving applications, 60 m^3 of fresh air at 16°C is required per liter of solvent evaporated [12]. The power supplies to the heaters should be cut off in the event of a reduction or loss of air supply.

REFERENCES

[1] Metaxas, A.C., Meredith, R.J. (1993). *Industrial Microwave Heating*, Peter Peregrinus, London, pp. 73, 78.

[2] Reference [1], pp. 70–73.

[3] Metaxas, A.C. (1996). *Foundations of Electroheat: A Unified Approach*, Wiley, Chichester, UK, p. 459.

[4] von Hippel, A.R. (1954). *Dielectric Materials and Applications*, MIT Press, Cambridge, MA, pp. 291–370.

[5] Vromans, H. (1994). Microwave drying of pharmaceutical excipients; comparison with conventional conductive drying. *European Journal of Pharmacy and Biopharmacy*, *40*, 333–336.

[6] Reference [1], pp. 79, 80.

[7] van Loock, W.M.A. (1987). Some applications of electromagnetic energy in Belgium. *Transactions of the Symposium on Dielectric Drying*, ETECE, Arnhem, The Netherlands (in Dutch).

[8] Derckx, H.A.J.M., Torringa, H.M. (1997). Equipment for drying sugar cubes. *Dutch Patent 1006216* (in Dutch).

[9] Sizgek, E., Sizgek, G.D. (2002). Drying characteristics of porous microspheres in a microwave heated fluidised bed. *Chemical Engineering Technology*, *25*, 287–292.

[10] Robin, P., Lucisano, L.J., Pearlswig, D.M. (1994). Rationale for selection of a single-pot manufacturing process using microwave/vacuum drying. *Pharmaceutical Technology*, *18*(5), 28–36.

[11] Pearlswig, D.M., Robin, P., Lucisano, L.J. (1994). Simulation modeling applied to the development of a single-pot process using microwave/vacuum drying. *Pharmaceutical Technology*, *18*(6), 44–60.

[12] British National Committee for Electroheat (1985). *Infra-red Process Heating*, BNCE, London.

14

PRODUCT QUALITY AND SAFEGUARDING DRYING

14.1 PRODUCT QUALITY

The drying operation may affect the quality of the product processed. The effect can be positive, neutral, or negative. A positive effect is noticed at the manufacture of fluid cracking catalysts as described in Section 8.1. Spray drying of the feed slurry results in the production of strong spherical particles. A neutral effect is the impact the drying step has on the bulk density of spray-dried products as reported in Section 8.5. The dusty character of flash-dried vacuum-pan salt, as reported in Section 3.6, can be characterized as a negative effect. The decrease in the weight-average particle size with a factor of 3 to 4 in a vigorously agitated contact dryer, as stated in Section 10.3, is also a negative effect.

Usually, at an elevated temperature, the product is in contact with oxygen and moisture, and this may result in oxidation or hydrolysis. Common analytical techniques such as DTA and DSC can be used to check the impact of the drying step.

Following are three case studies concerning the product quality at the drying step.

Case 1: Caking of Product 1

A plant producing a water-soluble organic product was started up. The crystalline material was shipped in 50-lb bags. Customer complaints concerning caking of the material in the bags were received.

The crystals produced were separated from the aqueous mother liquor by means of a pusher centrifuge. The wet crystals were dried concurrently in a convective drum dryer. The product leaving the dryer at a temperature of 50°C was transported

Drying in the Process Industry, First Edition. C.M. van 't Land.
© 2012 John Wiley & Sons, Inc. Published 2012 by John Wiley & Sons, Inc.

pneumatically to a silo, using ambient air. In the summer, normal day temperatures were 30°C and normal relative humidities were 80%.

The dry product absorbs water from the atmosphere at 25°C when the relative humidity is 65%. This relative humidity corresponds with 14 g of water per kilogram of dry air. At 40°C, water is absorbed from the atmosphere when the relative humidity is 80%. That relative humidity corresponds to 40 g of water per kilogram of dry air. The conclusion was drawn that at slightly elevated temperatures, this product has a great affinity for water. It was recommended that the dry product be transported to the silo by means of air of 5°C. This air can be raised relatively cheaply by means of a chilled-water unit. Cold air of 5°C and having a relative humidity of 100% contains 6 g of water per kilogram of dry air. Ambient air having a temperature of 30°C and a relative humidity of 80% contains 22 g of water per kilogram of dry air. Thus, the water content of the transport air is reduced by a factor of almost 4. Moreover, the transport air cools the product effectively, thereby decreasing the product's tendency to absorb water from ambient air.

The plant modification was introduced successfully. Complaints concerning caking were no longer received.

Case 2: Caking of Product 2 [1]

Initially, complaints concerning caking of a chemical were not received. However, complaints were received when the product was shipped overseas in large bags.

The crystals produced were separated from the aqueous mother liquor by means of a decanter centrifuge. The wet crystals were dried in a flash dryer. The moisture content of the dried product was 0.2% by weight. The dried product still contains 6% by weight of water of crystallization. X-ray diffraction patterns of the product were taken in a heated camera in which the sample's temperature was increased linearly with time. It appeared that 59 and 106°C are two crystal structure transformation temperatures. The structure obtained above 106°C is thermodynamically stable. On cooling this structure to ambient temperature, it remains intact. It was also found that this crystalline structure contains slightly less water of crystallization than the plant product.

The conversion to the stable modification, albeit at a low rate, also occurs at ambient temperature thereby liberating free water. This free water caused the caking problems.

The solution was the installation of a plug-flow fluid-bed conditioner–cooler. Conditioning for 30 min at 140°C appeared to produce a salable product. Too high a conditioning temperature gave a colored product. Conditioning was followed by cooling. The fluid-bed conditioner–cooler was installed between the dryer's cyclone and the product hopper.

Case 3: Dielectric Drying of Grindstones [2]

Drying grindstones having a mass of 200 kg with warm air took a minimum of 4 weeks. The drying was followed by a thermal treatment. A total of 20 to 40% of the grindstones had to be rejected after this treatment because of moisture remaining after the conventional drying step. Microwave drying lasted 30 h, and in the central part of the grindstone, the residual moisture content was maximum 0.4% by weight.

14.2 SAFEGUARDING DRYING

A classification of powders that may give rise to a hazard is [3]:

1. Materials defined as deflagrating or detonating explosives according to tests drawn up by the United Nations.
2. Compounds which, on heating, even in the absence of oxygen, exhibit exothermic decomposition with rapid evolution of large volumes of gas (e.g., peroxides and blowing agents). A dangerous buildup of pressure without fire may occur if the amount of gas is large enough.
3. Combustible powders which, when heated in air, can oxidize exothermally or decompose.

The fire hazard and dust explosion hazard presented by the third category of materials is the main concern in this section. Drying operations should be safeguarded to prevent personnel injury, to prevent plant damage, and to remain a reliable manufacturer.

A fire is a chemical reaction with oxygen in which light, heat, and flame are produced. Fires can occur whenever a combustible material is being dried with air. If a combustible solvent is being evaporated rather than water, the likelihood of a fire increases. In that case, at convective drying, nitrogen is usually taken as the drying gas, whereas at contact drying, nitrogen is then used as the stripping gas or drying is carried out under vacuum. Since many dryers are fired directly, a further possibility is the uncontrolled combustion of the heating medium (e.g., gas or oil), leading to the spreading of flames and ignition of the material being dried.

An explosion is a fast release of energy and, for drying, the relevant type of explosion is the dust explosion. The occurrence of a vapor explosion can be excluded if the drying operation is inertized or carried out under vacuum. A dust explosion occurs when an air–explosible dust mixture is ignited. It is a fire that propagates rapidly and, in a confined space, causes pressure rise. The dust concentration must be between the minimum and maximum explosible limits.

A dust explosion usually manifests itself as a deflagration. A deflagration can, under defined circumstances, become a detonation. However, the conversion into a detonation requires path lengths exceeding drying equipment sizes. An exception to the latter is that conversion may occur in a line with a large L/D ratio [4]. Both explosion types are described. Deflagrations are distinguished from detonations because it is, when a deflagration occurs, possible to alleviate the explosion effects by measures such as venting and suppression. Such alleviation is usually not possible for detonations because their flame velocity is too high.

Deflagration The word *deflagration* is derived from the Latin *deflagrare*, meaning "to burn down." Burning, even of solids, requires the reacting components to be in the gaseous state. A deflagration is a fire in which a flame front travels rapidly, but at subsonic speed, through a gas. Propagation of the explosion is by means of heat conduction. Typically, on confinement with air, a deflagration involving an organic material causes pressure to rise from the initial value to up to a maximum of 10 times the initial pressure. The main reason is that when these materials react completely with the oxygen available in the confined space, the adiabatic temperature rise is a maximum of approximately 2450 K, and this causes the pressure to rise with a factor of 10. The explanation is that the calorific values in $kJ \cdot kg^{-1}$ of many organic materials (e.g., sugar powder and coffee creamer) are approximately equal. Furthermore, the number of gaseous molecules after the combustion is larger than the number of gaseous molecules before the combustion. This effect adds to the pressure increase due to the thermal effect. However, a dust explosion of aluminum powder causes the pressure to rise with a factor of maximum 13.

In the case of dust explosions, the flame travels through a suspension of the dust in a gas and not through a pure gas. Although this is true, prior to combustion, the solid particle must decompose into gaseous components.

The pressure exerted by a deflagration in a confined space can cause considerable damage. A peak pressure of 8 bar absolute can be obtained if the initial pressure is 1 bar absolute, whereas drying equipment has a comparatively low strength, typically yielding at 1.2 bar absolute. Besides the peak pressure, the rate of pressure rise is also important. The maximum rate of pressure rise is a measure of the severity of an explosion.

Detonation In the gas phase, a detonation is defined as a deflagration having a flame velocity greater than the speed of sound. The propagation of the explosion is by means of a pressure wave. Typically, confined gas-phase detonations cause much higher pressures than confined deflagrations; peak pressures of 20 to 40 bar absolute (initial pressure 1 bar absolute) are mentioned in the literature. An example of a detonatable gas mixture is hydrogen and oxygen. The rates of pressure rise for confined gas-phase detonations far exceed those experienced for deflagrations.

The process safety during drying can be evaluated by measuring product safety characteristics of the material since, under normal circumstances, a chemical reaction other than burning does not occur.

Principally, two different methods for dealing with the risks of fires and explosions can be distinguished: prevention and cure. Safeguarding by means of prevention relies on process conditions excluding the possibility of undesired events occurring. Curing comprises minimizing the effect of a fire or an explosion. All things being equal, prevention is to be preferred to curing; however, sometimes a balance must be struck between cost and efficiency [5].

Finally, it must be noted that dust explosions and fires can give rise to a series of events; for example, a dust explosion may result in a fire, which, in turn, may cause another dust explosion, and so on. Next, we treat fires, deal with dust explosions, and make some additional comments regarding a few important dryers.

Fires

Many of the materials that are dried are flammable, and at the elevated temperatures prevailing and in the presence of oxygen, they may be ignited or they take fire spontaneously (self-ignition). There are three distinct types of fires for the materials being processed: (1) ignition of dust clouds, (2) ignition of dust layers or deposits, and (3) ignition of bulk materials.

Because of the high rate at which the flame front propagates during the combustion of a dust cloud, the process is known as a *dust explosion*. This topic is dealt with next. Burning of dust layers and burning of bulk dust are both generally considered to be fires.

VDI [6] has discussed methods to test the flammability of dust layers. The dust layers are ignited by a heat source. Measurement of the smolder temperature is a typical example. A circular dust layer having a diameter of 100 mm and a thickness of 5 mm is deposited on a hot metal plate. At a constant plate temperature, the sample is observed for a period of 2 h for a flame or smoldering phenomenon to appear. The plate temperature is varied in steps of 10 K. The lowest temperature causing a flame or smoldering phenomenon is the smolder temperature. The ignition of a layer leads to a small fire, which can ignite suspended dust. A further possibility is the transfer of smoldering or glowing material to other parts of the process, in which suspended dust occurs.

Example 14.1 A biological powder exhibits a smolder temperature of 290°C. The particles are smaller than 200 μm.

On drying a combustible material, it is recommended to subtract 75 K from the smolder temperature to obtain the maximum temperature of a hot surface in the drying system. However, if two-thirds of the minimum dust cloud ignition temperature is lower than the obtained temperature, the former value is selected. The minimum dust cloud ignition temperature is discussed in a later section.

Other powder layer tests simulate the conditions in dryers, such as tray and conveyor dryers, in which hot air circulates above a layer of material, and also simulate the condition of deposits on the internal surfaces of all types of dryer.

Methods to check self-ignition of bulk powder are mentioned as well [6]. The process that causes the ignition of a dust layer which is surrounded by a heat source and air, after self-heating of the dust, is called *self-ignition*. The degree of oxidizing self-heating depends on the sample size and duration of the temperature exposure. Increased sample volume will result in self-ignition at lower ambient temperatures. The reason is that the length of the heat diffusion path from the sample center to the sample periphery increases and that the sample specific outer area in $m^2 \cdot m^{-3}$ for heat transfer decreases on scaling up. Bulk powder tests simulate the conditions in hoppers, silos, or bags, and at the bottom of some dryers where material collects in bulk.

A quick test for the determination of the self-ignition temperature of small dust samples is the Grewer oven [6]. It is an oven with six wire mesh baskets each having a volume of 8 cm^3. The sample and a reference sample (graphite) are heated at a rate of 1 to 2 $K \cdot min^{-1}$ in an airstream of 100 to 120 $L \cdot h^{-1}$ up to 350°C. The surrounding temperature is noted at the moment the temperature of the sample, which is heated by the airstream, exceeds the temperature of the reference.

Example 14.2 In the Grewer oven, an organic chemical exhibits a self-ignition temperature of 180°C. At this temperature the sample starts to burn and the sample temperature rises to 530°C.

Hot storage tests produce data that are suitable for practical applications if the test conditions and the size of the wire mesh basket reflect actual process conditions (e.g., size of product buildup, amount stored, and retention time) [6]. To extrapolate for larger quantities, tests with varying sample amounts are needed. Cylindrical wire mesh baskets having a diameter equal to its height of various sizes (e.g., 100, 400, and 1000 cm^3) are filled with the dust to be tested and kept at a constant temperature in an oven while the oven also contains air. An air purge of 100 $L \cdot h^{-1}$ may also be used. The temperature of the sample center is recorded and a temperature rise above the storage temperature is classified as self-heating. A sample temperature exceeding 400°C means self-ignition. The highest storage temperature not causing self-ignition is the self-ignition temperature. The test is continued for 24 h, for example.

Example 14.3 The organic chemical tested in the Grewer oven exhibits a self-ignition temperature of 130°C in a 400-cm^3 hot storage test. At 140°C, and after approximately 6 h, the sample starts to burn and the sample temperature rises to 650°C. At 130°C, the sample temperature does not rise.

The examples given prove that increasing the sample size from 8 cm^3 to 400 cm^3 lowers the self-ignition temperature. On varying the sample size used at the hot storage tests, it is possible to obtain a series of self-ignition temperatures. On plotting $1/T$ as a function of $^{10}\log(V/A)$, a straight line is obtained [6]. V and A are the sample volume and sample area, respectively. On storing, for example, many cubic meters of a combustible material in contact with air, the self-ignition temperature may come down to values as low as 30°C, and this is sometimes the cause of a spontaneous fire in a storage.

In general, the results of the tests can give guidance on the selection of safe operating conditions, but the flammability of any gases evolved if the material decomposes must also be taken into consideration. Internal agitators can potentially provide ignition sources by mechanical friction that heats a bearing. A further potential ignition source is the discharge of powder charged electrostatically. Normal firefighting methods should be adhered to for drying operations (e.g., provision of sprinkler systems).

Alternatives for measurement of the smolder and self-ignition temperatures can be found in Abbott [3]. It concerns the powder layer test, the aerated powder test, and bulk powder tests.

Dust Explosions

Dust from a combustible solid, when suspended in air, may, under certain circumstances, cause a dust explosion. Frequently, dust explosions have occurred in the past during grain milling, grain storage, and coal-mining operations, and have also resulted from drying operations in the chemical industry. These explosions have often resulted in severe personal injury and in material damage. Dust explosions in grain warehouses causing 11 casualties occurred at Blayle in France in 1997. This accident gave rise to the formulation of the Atex Guidelines. If an explosible dust can occur anywhere in a drying system, the possibility of an explosion cannot be excluded unless preventive measures are adopted.

For example, explosible dust clouds can form in the baghouse of a spray dryer (dust collection is hazardous because it serves to collect all the fines in one place) or in a batch fluid-bed dryer at the end of the drying cycle (dry material has a greater propensity for explosion than wet material). Fuel, oxygen, and heat are all required for a dust explosion, and it is often not easy to exclude a potential source of heat in a specific drying system. The ignition sources for fires can also trigger an explosion of a dust cloud.

Dust and vapor explosions are similar in that they are rapidly advancing fires which, on confinement, cause pressure increase. However, there are two principal differences:

1. The ignition sensitivity of vapor–air mixtures is normally much greater than that of dust–air suspensions.
2. Dust–air suspensions are usually heterogeneous since dust tends to settle, whereas vapor–air mixtures are homogeneous and do not segregate readily.

Primary and secondary dust explosions are known. Primary dust explosions can lead to a dispersion of secondary dust in the air, and this resuspended dust may cause a secondary explosion. The possibility of a chain reaction is an inherent risk when dealing with combustible dusts.

Dust Explosion Possibilities It is important to determine whether or not a specific dust is explosible. A flammable material can, in principle, cause a dust explosion. However, there are exceptions; that is, materials exist that can burn but are nonexplosible. The modified Hartmann apparatus is used widely [6] (see Fig. 14.1). The ignition source is either an electric spark or a heated electric coil. The spark energy is approximately 10 J. The dust to be tested is examined either as received or as a sample consisting of particles smaller than 63 μm. Particle sizes smaller than 63 μm, (e.g., 30 or 40 μm) are susceptible to a dust explosion. The specific area in $m^2{\cdot}kg^{-1}$ is large and the particles do not settle readily. A pretreatment comprises, for example, drying under vacuum for an hour at 50°C. Alternatively, drying occurs at 75°C at atmospheric pressure. Dust is dispersed from the base by means of air at 8 bar absolute to give in the 1.2-L volume dust concentrations up to 1000 $g{\cdot}m^{-3}$. If a flame extends at least half the length of the tube, the dust is considered "positive."

The severity of any dust explosion depends on (1) the chemical composition of the material, (2) the surface nature of the material, (3) the dust concentration, (4) the particle-size distribution (smaller particles cause stronger explosions because the interfacial area is inversely proportional to the particle size and a small particle is

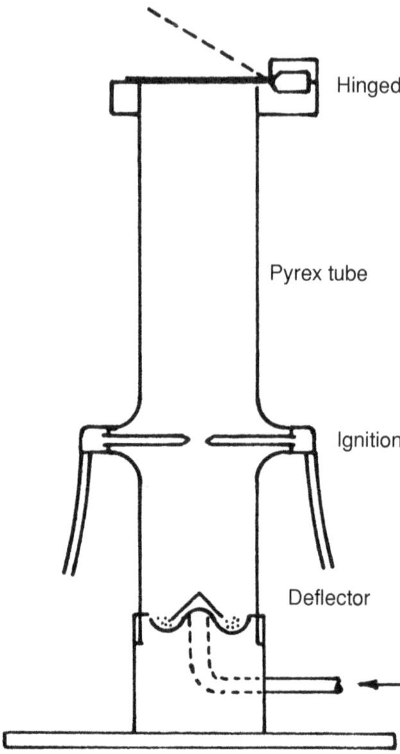

Figure 14.1 Modified Hartmann apparatus.

also a small heat sink), (5) the moisture content (water phlegmatizes), (6) the strength of the ignition source (weak sources can have a moderate effect, and vice versa [7]), (7) the oxygen concentration, (8) the degree of turbulence, and (9) the temperature and pressure. Note that this list is not exhaustive but covers the principal factors.

Hybrid mixtures are suspensions of dust in air–solvent vapor mixtures. A synergistic effect may exist between the vapor and the dust.

Explosible dusts have a minimum and a maximum explosible concentration. A dust explosion cannot occur when either the dust concentration is below the minimum explosible concentration or when the dust concentration exceeds the maximum explosible dust concentration. The explanation is that the progress of a deflagration is hindered because of phlegmatization by the air (at low concentration) or by the dust (at high concentration). It is usual to establish minimum explosible concentrations in a 20-L spherical apparatus (Fig. 14.2), using a pyrotechnic igniter supplying 10 kJ to ensure ignition. Usually, minimum explosible concentrations are in the range 10 to 60 $g{\cdot}m^{-3}$. Usually, for a given dusty material, the minimum explosible limit decreases if the particle size decreases. At the minimum explosible concentration of a typical dust, the visibility is in the range 1 to 2 m. Sometimes, it is possible to safeguard a system by ensuring that the concentration of the suspension is always

Figure 14.2 Twenty-liter spherical explosion apparatus. (Courtesy of Adolf Kühner AG, Birsfelden, Switzerland.)

below the minimum explosible concentration. The maximum explosible concentration is not well defined and has been measured for only a few dusts, as it has little practical importance.

The minimum oxygen concentration required to support a dust explosion varies from 3 to 15% by volume, depending on the chemical nature of the dust, its particle size, its moisture content, and its temperature. It can be determined in a modified Hartmann apparatus or in a 20-L spherical apparatus. Usually, the minimum oxygen concentration decreases sharply with an increase in temperature. It is possible to be protected against dust explosions by keeping the oxygen in the system below this concentration; however, usually because of economic considerations, relief venting and suppression are more common.

Susceptibility of a Dust Cloud to Ignition The susceptibility of a dust cloud to ignition is determined by the measurements of its minimum dust cloud ignition temperature and minimum ignition energy in air. The latter property is important because of the possibility of electrostatic discharges.

The minimum dust cloud ignition temperature is an important parameter that is frequently required for the selection of safe operating temperatures for dispersion dryers, such as spray dryers, fluid-bed dryers, and flash dryers. Minimum ignition temperatures are measured by passing a dust–air suspension through a laboratory furnace and determining the temperature at which ignition takes place. Well-known laboratory ovens are the Godbert-Greenwald oven and the BAM oven [6]. BAM stands for Bundesanstalt für Materialprüfung, a German institute. Typically, values above 400°C are found, but some materials can be considerably below this level (e.g., sulfur dust has a value of 190°C). The value obtained from the test is normally higher than the "real" figure since the air–dust mixture entering the furnace is cold, whereas in practice the suspension is usually preheated. It is recommended two-thirds of the measured minimum dust cloud ignition temperature in °C be taken as the maximum temperature of a hot surface in the drying system. However, if the smolder temperature minus 75 K is lower than two-thirds of the minimum dust cloud ignition temperature, it is recommended to select the former. The smolder temperature was discussed in detail earlier.

Example 14.4 For a material, the self-ignition temperatures according to the 400-cm^3 hot storage test and the Grewer test (8 mL) were given earlier as 130 and 180°C, respectively. The minimum dust cloud ignition temperature of this material is 440°C. Basically, a dust explosion is a fire proceeding at a high rate. So, in a way, the self-ignition temperature of this material increases from 130 to 440°C in going, via 8 cm^3 (180°C), from 400 cm^3 to a single particle.

Warning The minimum dust cloud ignition temperature has sometimes been taken erroneously as the self-ignition temperature of a material. The subsequent selection of too high an air temperature for a fluid-bed granulator has caused a fire (see also [8]).

Normally, the minimum ignition energies of dust suspensions are much higher than those for vapor–air mixtures. The values for hydrogen and methane are 0.03 and 0.3 mJ, respectively. The minimum ignition energy of ethylene oxide is approximately 5 mJ for the conditions found in the ethoxylation reactor. For dusts, the minimum ignition energies are in the range 10 to 500 mJ [9]. However, it has been shown that the values of dusts can approach those of vapors if the ignition spark's discharge time is long enough. Adjusting this reduces the "shock-wave" effect that pushes the particles away from the spark [10]. Usually, the minimum ignition energy decreases when the particle size decreases.

Dust explosions can be initiated by electrostatic discharges. During powder handling, electrostatic charges can build up. Discharge of the accumulated charge in the form of a spark may initiate a dust explosion. The minimum ignition energy and the volume resistivity of the dusty material characterize the electrostatic hazard.

It is possible to measure minimum ignition energies in a Hartmann apparatus that gives ignition energies ranging from 1 mJ to 2 J. Many values can be found in the literature. A value of less than 100 mJ indicates that, in principle, it would be possible to ignite dust by means of an electrostatic discharge from personnel.

Example 14.5 The smolder temperature of the biological powder mentioned in Example 14.1 (particle size smaller than 200 μm) is 290°C. For a sample of this material consisting of particles smaller than 125 μm, the minimum ignition energy is greater than 1000 mJ whereas the volume resistivity is $1.6 \cdot 10^{10}$ Ω·m. The latter figure means that it is possible to charge the powder electrostatically by means of such operations as pneumatic transport, milling, screening, or pouring. However, a dust explosion becomes improbable because of the high minimum ignition energy. For the possibility of the ignition of a dust explosion by static electricity, the minimum ignition energy would have to be lower than approximately 50 mJ.

Example 14.6 Wood particles of approximately 50 μm can be explosive. The minimum ignition energy is less than 25 mJ and the minimum explosible concentration is 35 $g \cdot m^{-3}$. Wood containing at least 15% by weight of water (wet basis) is incombustible.

Explosion Effects P_{max}, the maximum pressure, and the maximum rate of pressure rise, $(dp/dt)_{max}$, are discussed in this part. The highest pressure occurring in a confined space that is due to a dust explosion is termed P_{max}. If the equipment exceeds certain minimum sizes, P_{max} is independent of the equipment size; however, $(dp/dt)_{max}$ is strongly dependent on the equipment size. Both variables can be determined conveniently in a 20-L spherical apparatus. A typical record is given in Figure 14.3. The maximum rate of pressure rise so determined can be used to calculate

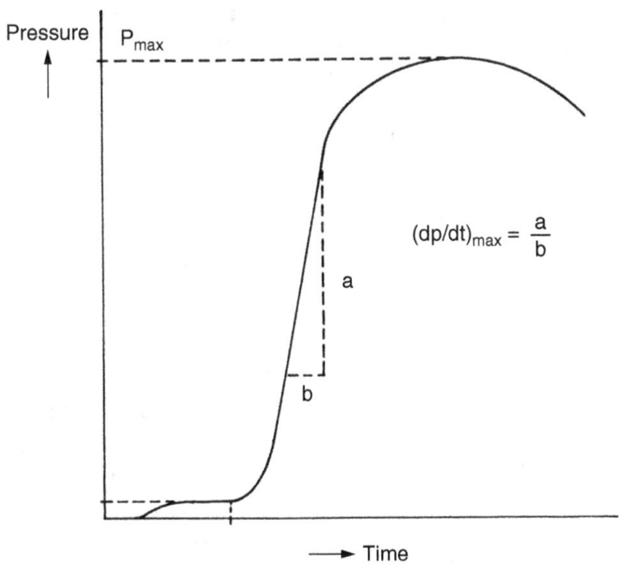

Figure 14.3 Pressure–time record of a dust explosion test.

the deflagration index K_{st} by means of the "cube-root law":

$$\left(\frac{dp}{dt}\right)_{max} \cdot V^{1/3} = K_{st} \tag{14.1}$$

This law holds for vessels where the length-to-diameter ratio is small and for a single-point source of ignition. In industrial practice, explosions (usually due to static electricity or friction) approximate a single-point source. The law is applicable to small- and large-scale equipment and can therefore be used for scaling up. K_{st} is a constant for a given dust. Equation (14.1) states that the maximum rates of pressure rise measured in large pieces of equipment are smaller than those established in small apparatus. This can be explained by assuming that the linear burning velocity is constant and that, therefore, the combustion in smaller equipment is complete in a shorter period of time than the combustion in larger equipment. A typical value for the initial linear burning velocity of a dust explosion is 3 $m{\cdot}s^{-1}$. The K_{st} value so determined is used for the sizing of relief vents; the higher the rate of pressure rise, the greater the vent area required. Furthermore, the cube root law shows that the size of the relief vent can be relatively small for large-volume equipment. The applicability of this law is one reason for permitting the amount of vent area per cubic meter to decrease on scaling up.

Standard practice is to investigate dry dusts using only materials of particle size less than 63 μm. The dust is dried carefully at 50°C, for example, under vacuum or at 75°C at atmospheric pressure. The dust concentration is varied over a wide range until there is no further increase in either the explosion pressure or the rate of pressure rise. The maximum explosion pressure and, more significantly, the maximum rate of pressure rise increase with decrease in particle size. A pyrotechnic igniter with a total energy of 10 kJ is used as the ignition source [6]. The tests are usually carried out at atmospheric pressure; however, it is possible to adjust other pressures. The stainless steel 20-L sphere is provided with a water jacket in order to be able to work at elevated temperatures.

Three classes are used for the classification of K_{st} values:

Class	K_{st} (bar·m·s^{-1})
1	$0 < K_{st} \le 200$
2	$200 < K_{st} \le 300$
3	$K_{st} > 300$

The most serious explosion effects might, theoretically, be expected with a stoichiometric concentration. However, it has been found experimentally that for both P_{max} and $(dp/dt)_{max}$, the highest values are measured with dust concentrations several times that of the stoichiometric mixture. Generally, P_{max} does not exceed 10 bar absolute and is proportionally smaller if the initial pressure is lower. A typical dust concentration at which both P_{max} and $(dp/dt)_{max}$ can be determined is 750 $g{\cdot}m^{-3}$.

Safeguarding Against Dust Explosions As mentioned earlier, two fundamentally different categories of safeguarding exist: one that is based on prevention and the other on curing. Since methods based on prevention are preferable, this group is covered first.

Prevention

INERTING Inerting is maintaining the oxygen concentration below the value required to cause ignition using nitrogen, carbon dioxide, or other inert gas. The first gas is used often. It is the most effective means of preventing dust explosions. It is essential to monitor the oxygen content of the gas. The main reason that this method is not applied more often is the cost. The inert gas is recycled and the drying system is under slight overpressure to prevent oxygen ingress. The overpressure causes leakages and nitrogen makeup is required. For many dusts, a reduction of the oxygen concentration in the air to approximately 50% of the normal value will be sufficient to prevent explosions [11].

Self-inertization is a special type of inertization. It can be used if water is to be evaporated (see Fig. 14.4). The heat required for the drying operation is supplied by combusting a natural gas flow in an airflow which is slightly larger than the airflow that is required stoichiometrically. Thus, the oxygen level in the combustion gases is low (e.g., 4% by volume). The hot flow of combustion gases is mixed with a cold recycle gas flow, which also has a low oxygen level. The mixed gas is used for drying. The gas flow leaving the dryer is split up into two gas flows. The smaller of the two flows is purged, whereas the larger of the two flows passes through a condenser. After cooling to condense water in the condenser, the latter gas flow is recycled.

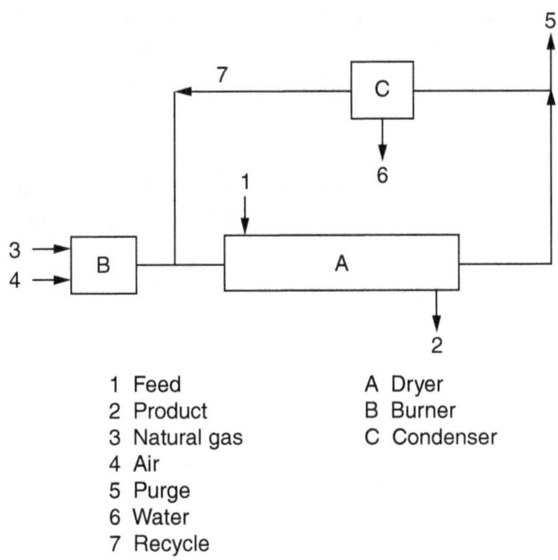

Figure 14.4 Simplified self-inertization outline.

The temperature of the drying gas can be adjusted by varying the ratio recycle gas flow/combustion gas flow. Also in this case, the system is under slight overpressure to prevent air ingress.

ELIMINATING IGNITION SOURCES Another important preventive aspect is the use of the proper class of electrical hardware and wiring to avoid electrical sparks. The electrical equipment should be dusttight or nonsparking and should not produce surface temperatures capable of igniting the dust. Also, electrostatic discharges must be controlled. Efficient grounding of equipment is of paramount importance, and the installation of bonding strips across joints is essential. Antistatic materials can be used for conveyor belts, clothing, footwear, and flooring. Steel fibers or carbon fibers interwoven into the fabric of the filter bags of fluid-bed dryers, for example, are helpful. The entrance of stray metal on combined milling–drying must be avoided, and metal detectors are often installed to check the feed. It is advisable to keep the drying air temperature below the minimum ignition temperature of a deposit (margin at least 20 K). The maximum temperature of hot surfaces has already been discussed. Direct firing entails the risk of incandescent particles entering the dryer with the drying air.

Example 14.7 An inorganic, incombustible product was dried in a direct-fired fluid-bed dryer. The hot air was raised by combusting milled coal. One day, a dust explosion occurred in the baghouse. Due to incomplete combustion, coal particles had accumulated in the baghouse and a dust cloud had been ignited, probably by incandescent particles carried over from the burner.

Example 14.8 On starting up a batch fluid-bed dryer in which an organic combustible chemical had to be dried, blue sparks were noticed in the vicinity of the filter bags. The cause was electrostatic charging of the powder, followed by discharges. The remedy was the installation of filter bags with interwoven conducting fibers. The fibers were grounded.

GOOD HOUSEKEEPING The avoidance of unnecessary dust suspensions and accumulations will considerably reduce the possibility of a dust explosion.

Example 14.9 A convective dryer is inside a building of height 5 m. The floor is covered with dust. The bulk density of the dust is 500 kg·m^{-3} and the lower explosible limit is 20 g·m^{-3}. On resuspending the dust in the air, the critical dust layer thickness for an explosion to occur in this building is

$$\frac{5 \cdot 1 \cdot 1 \cdot 20 \cdot 10^{-3}}{500 \cdot 1 \cdot 1} = 2 \cdot 10^{-4} \text{ m } (200\ \mu\text{m})$$

This example points to the need to avoid the accumulation of dust on floors, as the resuspension in the air of even thin dust layers gives rise to a hazardous situation.

Curing

CONTAINMENT For containment of a dust explosion, the plant components must be built or strengthened to be able to withstand the full force of an explosion. It is possible to distinguish between *pressure resistant* and *pressure-shock resistant*. The former possibility means that the system will not be damaged, whereas the latter possibility means that possibly, depending on the character of the explosion, some parts of the system will have to be replaced after the dust explosion. Containment can be applied to small pieces of equipment, such as mills. Normally, drying equipment can withstand overpressures of only 0.2 bar gauge, for example.

PLANT SEPARATION To prevent the spread of an explosion, processing plants ought to be adequately separated. Rotary valves and screw conveyors may act as "chokes."

RELIEF VENTING The principle of explosion relief venting is that, at a predetermined pressure rise, an aperture opens to vent the explosion products safely (see Fig. 14.5). For a short period after the vent opens, the pressure may continue to rise, so sufficient area should be provided to ensure that the pressure peak does not damage the vessel. Venting is used widely as a safeguarding method. It does not prevent a dust explosion from occurring but protects the equipment against damage. This method can be used only if the emission of material is allowable and a safe discharge area for the products can be found. It must be ascertained that the relief vents are not hindered from opening by rust or ice. It will usually not be possible to emit the burning dust and vapor into the plant and ducts are often used. However, a careful design is required because the flow through the duct entails a pressure loss.

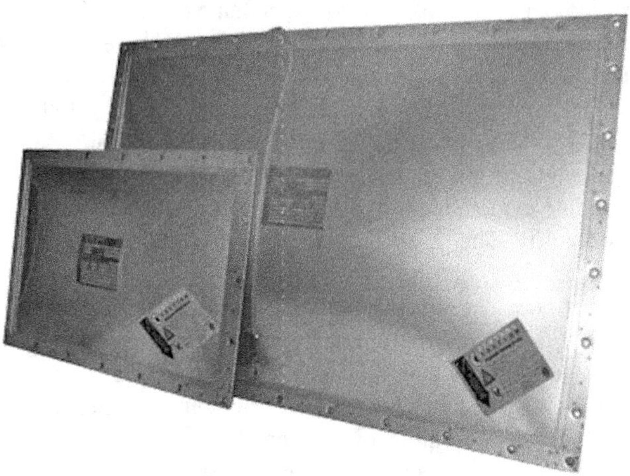

Figure 14.5 Domed explosion vents. (Courtesy of Rembe GmbH Safety + Control, Brilon, Germany.)

VDI [12] describes the sizing of relief vents using data of experiments in a 20-L sphere, that is, the maximum pressure and K_{st}. For the applicability of the methods, for K_{st} values smaller than or equal to 300 bar·m·s^{-1}, the maximum pressure should be between 5 and 10 bar absolute. For the same reason, for K_{st} values greater than 300 bar·m·s^{-1} but smaller than or equal to 800 bar·m·s^{-1}, the maximum pressure should be between 5 and 12 bar absolute. The background of these boundary conditions is that the dust explosions must be deflagrations. The results can be used for equipment volumes of up to 10,000 m^3. There is also a boundary condition concerning the length/diameter ratio of the piece of equipment. It is required to select P_{stat}, the pressure at which the vent opens (e.g., 0.1 bar gauge) and P_{red}, the maximum explosion pressure in the vented piece of equipment (e.g., 0.2 bar gauge).

The NFPA [13] also describes the sizing of relief vents using data of experiments in an approximately spherical calibrated test vessel of at least 20 L capacity, that is, P_{max}, the maximum pressure, and K_{st}. The following limitations are applicable to the equations and graphs:

1. 5 bar absolute $\leq P_{max} \leq$ 12 bar absolute
2. 10 bar·m·s^{-1} $\leq K_{st} \leq$ 800 bar·m·s^{-1}
3. 0.1 m^3 $\leq V \leq$ 10,000 m^3
4. $P_{stat} \leq$ 0.75 bar gauge

V is the enclosure volume. There are also boundary conditions concerning the length/diameter ratio of the piece of equipment. It is required to select P_{stat} (e.g., 0.1 bar gauge) and P_{red} in the vented piece of equipment (e.g., 0.2 bar gauge).

The German company Rembe at Brilon markets a discharge device called the ECO-Q-Rohr. The design is based on the invention of the Davy safety lamp in 1815 for use in coal mines. This lamp had a wick and an oil vessel originally burning a heavy vegetable oil. Davy had discovered that a flame enclosed inside a copper mesh of a certain fineness cannot ignite methane, the main component of flammable gases in mines. The minimum explosible concentration of methane in air is between 4 and 5% by volume. The screen acts as a flame arrestor. Rembe's relief discharges into a closed tube having the same diameter as the relief. A typical tube has a diameter of 0.55 m and a length of 0.9 m. The walls consist of several layers of metal gauze that in case of a dust explosion, cool emitted material. Thus, discharge into a plant building is a possibility. The dusts should belong to dust classes 1 or 2 and have a maximum K_{st} value of 250 bar·m·s^{-1}.

Different venting devices may be used for pressure venting (e.g., rupture disks or explosion doors). The usual rupture disks are made of polyethylene or aluminum having a specific mass of less than 0.5 kg·m^{-2}. These devices do not obstruct venting. When using explosion doors, which close the vent area after the explosion, the cooling of the hot gases of combustion may create a vacuum in the vessel. Vacuum breakers have to be present to prevent deformation of the vessel [12].

SUPPRESSION Suppression is a safeguarding method that is activated when the start of a dust explosion is detected. An extinguishing agent is distributed to dissipate the

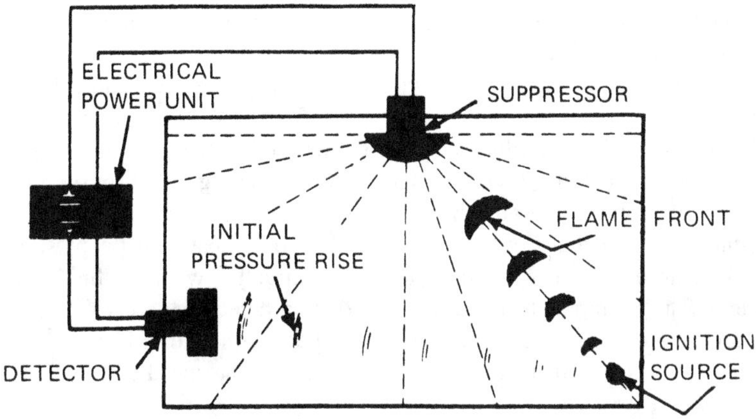

Figure 14.6 Simplified suppressor outline. (Courtesy of Kidde Fenwal Inc., Ashland, MA.)

explosion. When an explosion, which is initiated at atmospheric pressure, is suppressed, the rise of pressure is normally limited to 0.2 bar gauge. The operation of this protection is discontinuous; the safeguarding does not function during normal operation but starts working upon a signal. Suppression requires more than one stage to achieve safety (e.g., pressure detection, conveying a signal, and the subsequent reaction). Its functioning is dependent on equipment checking and maintenance. Proper operation is not continuously apparent. An outline of this system is shown in Figure 14.6.

The delopment of the suppresssion system was inspired by experiences during World War II. At that time the Royal Air Force was suffering severe losses of combat aircraft. More than 50% of the losses caused by gunfire were the consequences of fuel-tank explosions; structural damage from bullets had relatively little effect on the planes.

The RAF started to investigate whether tank explosions could be stopped after they had started, and their investigations were sucessful. However, with the arrival of surface-to-air and air-to-air missiles, the protection system became obsolete. A hit by a missile meant destruction of the aircraft regardless of the fuel-tank protection [14]. Then the applicability of this safeguarding method was checked for the process industry. The first field installations appeared during the 1950s: coal dust, starch, plastics, and pharmaceuticals. Whether this protection method is fast enough might give cause to wonder.

A sample calculation illustrates the feasibility of this approach. A dust explosion starts at a 2-m distance from a detector. The signal of the pressure rise travels at the speed of sound to the detector (342 m·s^{-1}), and therefore the detection time is approximately 6 ms. For all practical purposes, the electrical actuation is instantaneous. A typical dispersal speed of the suppressing agent is 60 m·s^{-1}. The suppressor is located at a 2-m distance from the developing fireball. The suppressing agent requires about 34 ms to arrive. So there is a 40-ms elapse between detection and dissipation. Now the question is: How far has the dust explosion advanced in 40 ms? A typical flame speed is 3 m·s^{-1}. Remember that a dust explosion starts out with a relatively low linear flame speed. If it is allowed to progress, the flame speed would accelerate sharply. However, we are dealing with the first stage, and consequently, the dust explosion has advanced 12 cm in 40 ms.

This sample calculation illustrates that the protection method can be quite adequate. However, detection should be right at the beginning and it must be possible for the flow of suppressant to reach the ignition source unhindered.

An important advantage of this protection method is that a dust explosion is not followed by an emission, which is particularly attractive when dealing with toxic substances. The example elucidates that a radiation detection hardly offers a bonus over a pressure detection. Basically, both the speed of sound (342 m·s^{-1}) and the speed of light (2.9978·10^8 m·s^{-1} in a vacuum) exceed the speed of flame (3 m·s^{-1}) considerably.

Well-known suppressants are monammonium phosphate (a powder) and sodium bicarbonate (also a powder). The former is used in the chemical industry and the latter is used in the food industry.

During normal operation of the plant, the suppressant is contained in pressurized sealed containers with direct access to the interior of the drying system. When a developing explosion is detected the seal is broken by means of a fast-acting electromechanical actuator, a gas generator (like the ones used for the activation of airbags), or an electrically fired detonator. Subsequently, the contents of the containers are rapidly dispersed throughout the volume of the vessel. Kidde Fire Protection in Thame, England uses containers with 4 or 16 kg of extinguishing powder; the containers are pressurized with dry nitrogen having a pressure of 60 bar gauge. The discharge times of these containers are 30 and 140 ms, respectively.

A limitation of this technique is the path the suppressant can be expected to travel, with possibly a maximum value of 4 m [15]. Successful testing on a 250-m^3 scale has been reported [16]. As a general summary, Kidde Fire Protection stated that their explosion-suppression systems are effective against dust explosions provided that the explosibility rate constant K_{st} is smaller than 500 bar·m·s^{-1}. For more violent materials and for weak components that are located within a building such as larger spray dryer complexes, the combination of explosion venting and explosion suppression is an alternative.

Suppression has found application at, for example, the flash drying of starch [17]. Figure 14.7 exhibits an installed suppression system, Figure 14.8 depicts a membrane pressure detector, and Figure 14.9 shows a high-rate discharge explosion suppressor with a capacity of 5.4 L.

Additional Remarks for a Few Important Dryers

A distinction is made between convective dryers and contact dryers. Both types can be equipped with safeguarding systems. Some systems can be provided with additional options. For example, the pressure detector that initiates a suppression can, at the same time, (1) activate an electromechanically operated bursting disk, (2) introduce a suppressant into an area of the plant that is away from the place where the

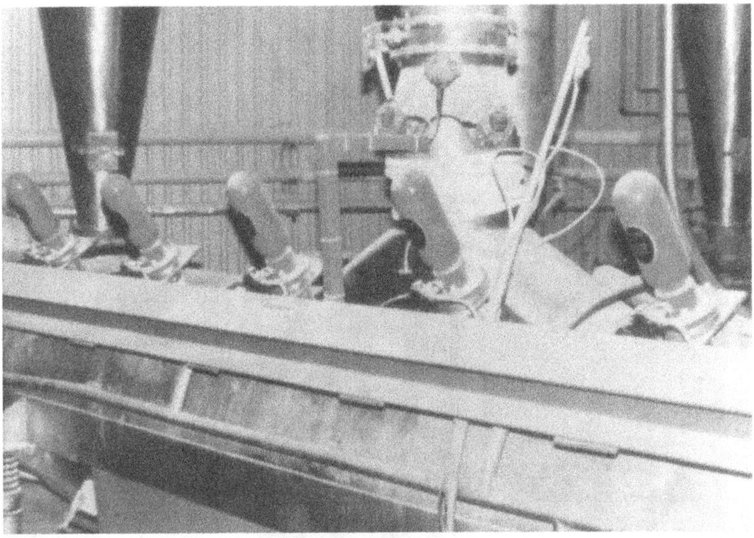

Figure 14.7 Suppressor installation. (Courtesy of Kidde Products Limited, Colnbrook, UK.)

explosion started to prevent a second explosion or fire (advance inerting), (3) isolate one area of a plant from another by means of explosion isolation valves, (4) flood hazardous areas (deluging), or (5) shut down the drying plant.

The aforementioned provisions require activation to be effective. An option that prevents the passage of flames from one part of a drying system into a different one

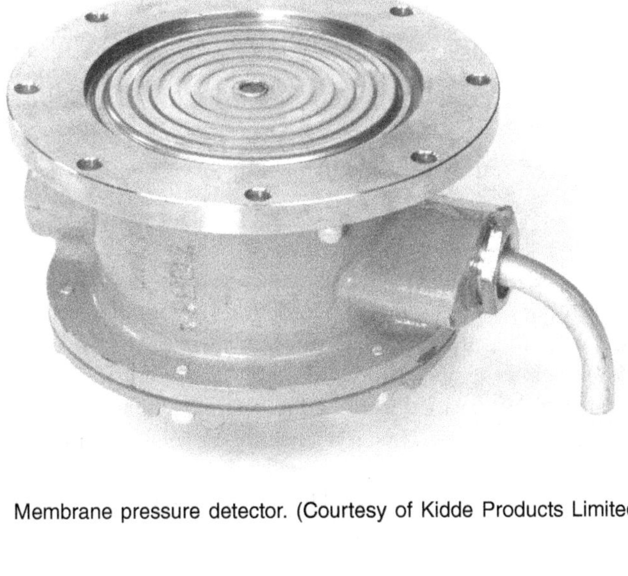

Figure 14.8 Membrane pressure detector. (Courtesy of Kidde Products Limited, Colnbrook, UK.)

Figure 14.9 High-rate discharge explosion suppressor. (Courtesy of Kidde Products Limited, Colnbrook, UK.)

and that does not need activation is the flame arrestor, which is usually mounted in a line.

Convective Dryers Dealing with solvents almost automatically calls for inerting. An inert gas recycle with solvent condensation is necessary. Venting is a widely used safeguarding method if the product is a water-wetted flammable material. The choice of a safe air inlet temperature is an important issue. If drying air having a certain temperature flows past a deposit, a fire occurs if the material is combustible and a certain deposit size is surpassed. Generally, dependent on the size, the self-ignition temperature of combustible deposits is in the range 100 to 200°C. The inlet air temperature should be below the self-ignition temperature of a characteristic deposit size. The inlet air temperature should also be below the ignition temperature of a dust cloud. However, the ignition temperatures for dust explosions are generally much higher than the self-ignition temperatures of deposits. The latter temperatures are, for example, 400°C or higher.

Fluid-Bed Dryers The ignition of a deposit on the distributor plate, which would have a temperature equal to the inlet temperature, is a potential fire hazard. Special attention must be paid to the hazards that may be caused by static electricity. Due to the way that fluid-bed dryers operate, the product can become electrostatically charged. It is essential to ground all metal parts if a combustible powder is processed. For such products it is also important to equip batch fluid-bed dryers with conductive filter bags.

The most convenient location for installing vents in the drying chamber of continuous dryers is in the roof. For fluid-bed dryers operating batchwise, a single vent on the side of the chamber will be satisfactory.

For both continuous fluid-bed dryers and fluid-bed dryers operating batchwise, it can be stated that suppression is suitable only if the air velocity is low enough to prevent the suppressant from being carried away when it is released.

Rotary Dryers These dryers are generally used for relatively coarse material. This dryer type is less suitable for flammable materials. Because the shell is rotating, it is difficult to locate parts of a protection system on it; therefore, relief or suppression must be arranged on or from the heads. Internal flights or baffles may impede free access of explosion products to the vents. The use of suppressants is limited to relatively short drums, as it must be possible for them to reach all parts of the drum rapidly.

Flash Dryers Dust deposits are most likely to occur at the top bend of the vertical drying leg and at the bottom if the feed particles are too large. Suppression is effective to safeguard the drying tube if the suppressant is injected close to the feed point. The suppressant should remain in the dryer and not be exhausted by the airflow. The airflow should be shut off by means of fast-acting valves once the suppressant is distributed along the length of the drying tube.

Spray Dryers Incrustations are often formed near the spray device on the inner surface of the dryer. Self-ignition or smoldering of these incrustations may be caused by prolonged thermal exposure. Their crumbling off may cause a fire or a dust explosion.

The roof of the drying chamber is often the most convenient place to install a vent.

Suppression is, in most cases, not attractive for large spray dryers. The moving airstream tends to remove part of the suppressant as it is released. To fill large volumes rapidly, a very large amount of suppressant would be needed.

The analysis of carbon monoxide in the exhaust airflow has found an application at the spray drying of concentrated milk to produce milk powder. The technique used is infrared spectroscopy. The carbon monoxide concentration in the spent drying gases increases when a deposit starts to decompose. In practice, the difference between the levels in the fresh air and the exhaust air is determined. The dryer must be inspected if the difference is 0.5 ppm by volume. The dryer operation must be stopped in a normal way if the difference is 3 ppm. Extinguishing with water must be considered if the difference is 6 ppm because a fire may occur. The Dutch company Hobré at Purmerend is one supplier of this instrument.

Example 14.10 The air-inlet temperature of a spray dryer producing milk powder from concentrated milk was 200°C. The concentrated milk was atomized by means of a set of single-fluid nozzles. The powder produced was received in a continuous fluid-bed dryer at the bottom of the spray dryer. One day, a fire was ignited in which all the material in the dryer combination burned. The cause was the formation of an incrustation at one of the nozzles. The self-ignition temperature of the chunk was below 200°C and thus the incrustation took fire. The burning material crumbled off and fell in the fluid bed, which caught fire as well. The spray dryer was equipped with an instrument measuring the concentration of CO in both the air inlet flow and the exiting gas. The instrument had indeed indicated that the CO level in the exiting gas was too high before the onset of the fire. However, the signal had not been incorporated in a control loop, and the operators had not noticed the reading.

Combined Milling and Drying Altenburger Maschinen Jäckering in Hamm, Germany, a manufacturer of combined milling–drying systems, have, until today, not experienced a dust explosion in the mill of a combined milling–drying system. If the dry product, together with oxygen in the drying air, can give rise to a dust explosion, the mill is built pressure-shock resistant. Dust explosions have occurred in the filter which is provided with a rupture disk if a dust explosion could occur.

Baghouses Many convective dryers are equipped with baghouses. Frequently, dust can accumulate in this equipment and the potential for a dust explosion or a fire can be considerable. The contents can be ignited by means of a flame front entering from the dryer through the line. An example of a dust explosion in a baghouse was given earlier. Relief venting is a common protection method.

A series of incidents occurred in the wood industry, probably initiated by glowing dust particles. The standard practice is now to locate the filter at a safe distance and to apply relief venting.

The location at a safe distance (e.g., in the yard), is not always possible in the chemical industry. However, it is important to realize that the yielding of the vent relief emits burning material which should not be allowed to create a harmful situation.

Contact Dryers Generally speaking, contact dryers can be safeguarded more easily than convective dryers. Inerting is not too difficult because the strip gas flow is much smaller than the gas flow on convective drying. Contact dryers can often be built strong enough to be able to contain an explosion.

However, there are aspects that must be observed closely. The product acquires the wall temperature, and this must not cause ignition. Often, contact dryers are equipped with an agitator, and this feature entails the risk of hot spots (bearings), frictional heat, or sparks that are due to tramp metal or a slight shaft misalignment. Furthermore, the dust concentration is often within the explosible limits. Pfaudler tumblers or the like are vulnerable to the generation of static electricity. It must be realized that dust explosions can also occur in vacuum dryers, but rarely at pressures below 50 mbar.

Relief venting and suppression can be practiced; however, there are specific ins and outs. To give an example: Both relief venting and suppression are not adequate for plate dryers because of their relatively complicated construction. Inerting is the answer in this instance.

REFERENCES

[1] Bühler, W., Liedy, W. (1989). Characterization of product qualities and its application in drying process development. *Chemical Engineering and Processing*, *26*, 27–34.

[2] Segerer, H. (1998). Producing technical ceramics by microwave drying. *American Ceramic Society Bulletin*, *77*, 64.

[3] Abbott, J. (1990). *Prevention of Fires and Explosions in Dryers*, The Institution of Chemical Engineers, Rugby, UK, p. 4–25.

[4] Bartknecht, W. (1981). *Explosions Course Prevention Protection*, Springer-Verlag, Berlin, p. 69.

[5] Gerritsen, H.G., van 't Land, C.M. (1985). Intrinsic continuous process safeguarding. *I&EC Process Design & Development*, *24*, 893–896.

[6] VDI (1990). *Test Methods for the Determination of the Safety Characteristics of Dusts* (*VDI 2263, Part 1*), VDI, Düsseldorf, Germany, pp. 2–23.

[7] Hay, D.M., Napier, D.H. (1977). Minimum ignition energy of dust suspensions. *IChemE Symposium Series No. 49.*

[8] VDI (2004). *Dust Fires and Dust Explosions; Hazards, Assessment, Protective Measures; Explosion Protection in Fluid-Bed Dryers; Hints and Examples of Operation* (*VDI 2263, Part 5.1*), VDI, Düsseldorf, Germany, p. 31.

[9] National Fire Protection Association (2007). *NFPA 68 Standard on Explosion Protection by Deflagration Venting*, NFPA, Quincy, MA, p. 53.

[10] Reference [4], p. 85.

[11] Field, P. (1982). *Dust Explosions*, Elsevier, Amsterdam, The Netherlands, p. 91.

[12] VDI (2000). *Pressure Venting of Dust Explosions* (*VDI 3673, Part 1*), VDI, Düsseldorf, Germany, pp. 1–40.

[13] Reference [9], pp. 1–82.

[14] Martin, A.J. (1978). Explosions: the pause that suppresses. *Waste Age*, *9*, 36–41.

[15] Reference [11], p. 162.

[16] Moore, P.E. (1986). Towards large volume explosion suppression systems. Are they too expensive and can they cope? *The Chemical Engineer, 430*, 43–45.

[17] Meinhold, T.F., Carney, R.D. (1967). Over 300 kinds of starch processed in one dryer. *Chemical Processing*, *30*, 20, 21.

15

CONTINUOUS MOISTURE-MEASUREMENT METHODS, DRYER PROCESS CONTROL, AND ENERGY RECOVERY

Four different aspects are treated in this chapter: (1) continuous moisture-measurement methods for solids, (2) continuous moisture-measurement methods for gases, (3) dryer process control, and (4) energy recovery from exiting flows.

Measuring the moisture content of the feed or the product can be useful for process control; the same reasoning can be applied to the measurement of the gas's moisture content. However, in many instances, practical dryer process control methods are based on temperature measurements. The large energy consumption of dryers per se and the energy prices sometimes call for an energy recovery. Practical recovery methods are restricted to recovery from gases.

15.1 CONTINUOUS MOISTURE-MEASUREMENT METHODS FOR SOLIDS

A distinction can be made between absolute and inferential methods. In absolute methods, water is extracted from the material by oven drying, desiccation, distillation, titration, or by reaction (e.g., Karl Fischer method). They do not need calibration. A physical property of the material under consideration, which varies with the

Drying in the Process Industry, First Edition. C.M. van 't Land.
© 2012 John Wiley & Sons, Inc. Published 2012 by John Wiley & Sons, Inc.

quantity of water present, is measured in inferential methods. They do need calibration. The degree of binding of the water molecules to the host material affects the results of most inferential methods of moisture measurement. Hence, the calibration results for one material do not apply to a different one.

Absolute methods are not discussed in this section because invariably they are carried out batchwise and only continuous methods are dealt with. The continuous solids moisture-measurement methods discussed are (1) electrical capacitance, (2) infrared reflection, and (3) microwaves.

When selecting a method, a feasibility study is carried out first. The moisture content of a material of high electrical conductivity cannot be measured with a capacitance technique. A narrow particle-size distribution combined with a small average particle size is more favorable for infrared reflection than a wide distribution combined with a large average particle size (e.g., larger than 1 mm).

The second step is a laboratory assessment of available measurement techniques, with the successful methods identified. Online calibration trials come next, and finally, a system is commissioned.

Electrical Capacitance

The method relies on a change of the capacitance of a capacitor when the water level of the process flow changes. The method is unsuitable for materials that can conduct electricity. The capacitance of a capacitor is directly proportional to the dielectric constant of the material between the plates. The material between the plates is the process material. Up to a frequency of 10^9 s^{-1} (Hz), the dielectric constant of water is approximately 80. The dielectric constants of most other materials are in the range 1 to 10 and are thus much lower. Sand, for example, has a dielectric constant between 3 and 4. The method uses the difference between the dielectric constant of water and the dielectric constant of other materials to assess a material's water content. The dielectric constant of the material measured is a function of its water content. The explanation for water's anomalous behavior is that water molecules are dipoles that, on applying an electric field, align with the field that is thus reinforced. Ice's dielectric constant is approximately 3 because the water molecules in the crystal lattice cannot orient themselves in the electric field. By the same token, bound water molecules cannot orient themselves as easily as free water molecules can. This fact explains the phenomenon that the online moisture measurement based on capacitance is less effective for bound water than for free water.

The (small) conductivity of the material itself can, in principle, disturb the result to some degree. By employing sophisticated electronics, it is possible to eliminate this effect completely. In practice, the effect can also be removed by using frequencies exceeding 10^6 Hz.

An advantage of this method is that it presents an overall picture (i.e., local water-level fluctuations do not interfere). In many instances, the material to be analyzed moves over a probe's surface. Instead of being opposite to each other, the two plates of the capacitor are in line. A minimum layer thickness of 10 mm must be observed.

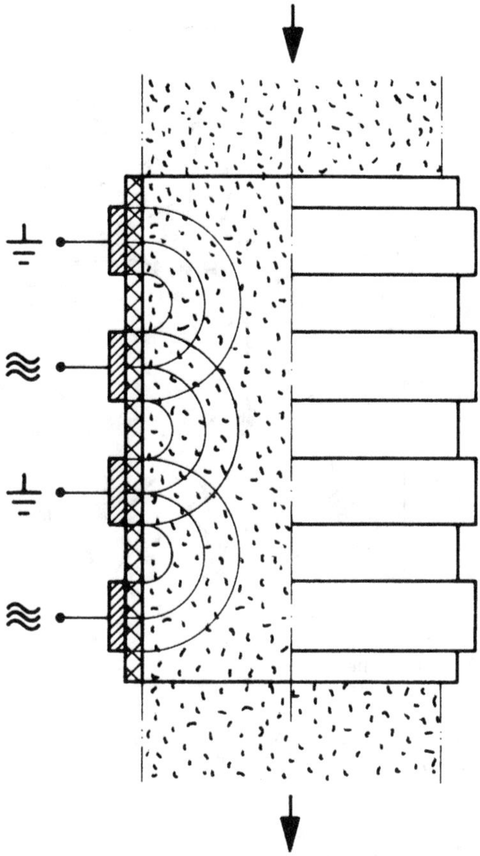

Figure 15.1 Schematic of a continuous online dc moisture measurement cell.

However, a direct physical contact is not always necessary (e.g., a measuring head can be located under a moving conveyor belt). The distance between the head and the material shall not exceed several millimeters and must remain constant. A further possibility is to install a measurement cell as depicted in Figure 15.1. The process flow is interrupted and the measurement space is filled with material. The field is indicated schematically by the curved lines.

The technique has found many applications for agricultural products (coffee, grain) and for such materials as sand and fertilizers. If the bulk density of the analyzed materials varies, it is recommended that the variable be measured simultaneously by means of an appropriate technique. The combination of the two signals may still give a good indication of the moisture content.

Besides the variables discussed, further variables that influence the instrument's output are the temperature (effect on material properties) and the presence of impurities (small amounts of impurities can have pronounced effects).

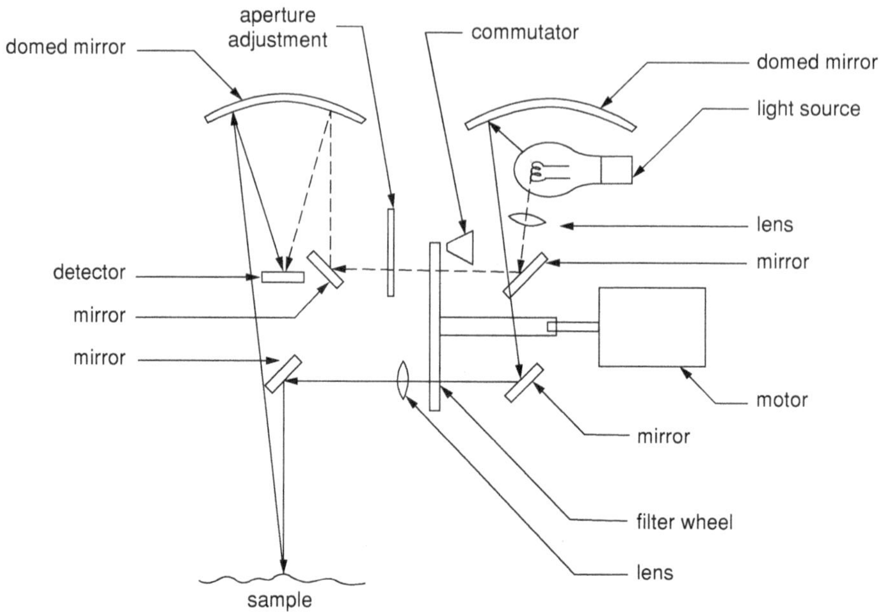

Notes:
1. Solid lines - main channel
2. Dashed lines - prime channel

Figure 15.2 Layout of the optical system of an IR reflection instrument. (Courtesy of Process Sensors Corp., Milford, MA.)

Infrared Reflection

The method is based on the ability of water to absorb infrared light. The light is cast upon the material to be analyzed and part of it is reflected (see Fig. 15.2). The reflected light's intensity is a measure of the water content. Infrared reflection is a contactless method that gives information of the material's surface only. It is substantially independent from the bulk density.

The method is to be distinguished from infrared transmission. There, the emission source and the detector are located at different sides of the material. IR transmission also relies on the absorption of radiation and can be applied to web and sheet material.

Background The IR light that is used has wavelengths between 1 and 3 μm (near infrared). Infrared light can interact with the vibration of the O—H bonds in the water molecule. However, interactions with O—H bonds in other molecules (e.g., alcohols and sugars) are also possible. Molecules different from water should, when present, cause a constant absorption of energy.

Frequently, the wavelengths employed are 1.45, 1.94, and 2.95 μm, with 1.45 and 1.94 μm being used more often than 2.95 μm. Of the two, the 1.94-μm radiation is the more powerful one. Hence, a wavelength of 1.45 μm is often used for materials that have a high moisture content (e.g., paper and textiles containing 80% water by weight). To cancel the influence of variables other than the water level (e.g., the color of the material), a reference wavelength is chosen near to the water absorption band (the word *band* is used because the NIR (near infrared) spectrum of free water exhibits rather broad maxima caused by the influence of H bonding). A typical reference wavelength is 1.8 μm. The usual setup contains a scanning disk with the two interference filters (e.g., 1.94 and 1.8 μm). These select sequentially the chosen narrow bands from the tungsten radiation source. A lead sulfide photoelectric cell serves as a detector.

The installation of a combined source–detector over a conveying belt is, in continuous plants, the usual method. Often, special provisions (plough, comb) ensure that a fresh surface is exposed. Calibration of the instrument occurs by means of a comparison of the instrument's reading with the result of an absolute water analysis (e.g., Karl Fischer). By the same token, regular checking of the readings is advisable.

Practical Aspects Variables that influence an instrument's output are:

1. *Drying-up:* The surface of a warm and wet material can dry up on a conveying belt. Just before the measuring cell, a bar or plough mounted over the conveyor can expose a fresh surface.
2. *Surface structure:* A bar mounted over the conveyor can often assist to smooth the surface. This remedy, however, cannot compensate for intrinsic surface differences.
3. *Particle-size distribution:* A narrow particle-size distribution combined with a small average particle size is more favorable than a wide distribution combined with a large average particle size (e.g., larger than 1 mm).
4. *Chemical composition:* Small amounts of impurities or additives, (e.g., 1% by weight) can influence the reading markedly.
5. *Color:* An aspect to be kept in mind, although its influence is generally ruled out by means of the reference beam. Black materials need extra attention.
6. *Temperature:* Cooling the source and detector is advisable when the instrument would otherwise be exposed to temperatures exceeding 40°C.
7. *Daylight:* A changing environment might influence the performance. A cylinder through which the light is passed onto the material can be a remedy.
8. *Vibrations:* These might adversely affect the functioning of the instrument.
9. *Light intensity:* An instrument compensating for the effect of the light intensity decreasing with time is discussed below.
10. *Optics:* See "Light intensity" above.
11. *Distance head/bed:* Generally, the distance should be kept constant. Altering the distance by several centimeters can influence the instrument's reading by as much as 1% by weight.

Figure 15.3 Installed continuous moisture measurement system employing IR reflection. (Courtesy of Process Sensors Corp., Milford, MA.)

Measuring System A typical measuring system is shown schematically in Figure 15.3. The signal from the reference wavelength and the signal from the measuring wavelength are combined to form a ratio (reference divided by measure). This ratio is directly proportional to the moisture level. The Quadra Beam Analyzer uses two additional beams to stabilize the instrument and to eliminate drift. (Drift is the change in an instrument reading caused by factors other than moisture.) The beams have the same wavelengths as the measure and the reference wavelengths and are passed directly to the detector. Their path is indicated by means of dashed lines. The four signals are then put through a calculation:

$$\frac{R}{M} \cdot \frac{M'}{R'}$$

The R and M signals are reflected from the sample whereas the R' and M′ signals pass internally from the source to the detector. A change in light intensity at the measuring wavelength will affect both the M and M' values proportionally. The same applies for the R and R' values.

The infrared reflection method is a standard method for the measurement of the moisture content of tobacco. It also works successfully for measurement of the moisture content of rayon, vacuum-pan salt, and detergents.

Microwaves

Microwaves are electromagnetic waves such as radio waves or light. The length of the microwave varies from 1 to 10 cm, and the frequency varies correspondingly from approximately $3 \cdot 10^{10}$ to $3 \cdot 10^9$ s^{-1}. The idea that microwaves could play a role in the measurement of moisture was inspired by the observation that rain attenuates radar signals.

Two independent physical properties can be utilized for measuring the moisture content by means of microwaves: the material's potential to cause dielectric losses and the material's dielectric constant. Microwaves can be used in three different ways: in transmission, by resonance, and in reflection. Almost all commercially available instruments measure dielectric losses (absorption) by means of transmission. The measurement of absorption by means of transmission is contactless. Resonance cavities can be employed for foils and threads.

Background Water molecules rotate as a whole, and the energy associated with this rotation is quantized. Microwaves can interact with this movement and thus cause transitions between defined energy levels. Since the mobility of bound water molecules is restricted, the method is more effective for free water than for bound water. Commonly used microwave frequencies are the S band ($2.6 \cdot 10^9$ to $3.95 \cdot 10^9$ s^{-1}) and the X band ($8.2 \cdot 10^9$ to $12.4 \cdot 10^9$ s^{-1}). The result of the measurement is strongly dependent on the bulk density and the temperature. Therefore, process instruments measure the packing density and temperature simultaneously. The three signals are put through a calculation. The bulk density measurement can proceed by means of a radioactive source. The relationship is

$$I = I_0 \cdot \exp(-\mu \cdot \rho \cdot D)$$

where

I = signal at the detector
I_0 = signal at the detector with air or water in the chamber (reference)
μ = radiation absorption coefficient, $m^2 \cdot kg^{-1}$
ρ_b = bulk density, $kg \cdot m^{-3}$
D = line diameter, m

Water's absorption potential increases with increasing temperature. The presence of electrolytes affects the measurement less than it does in the capacitance technique. The size of the particulate material should not exceed one-fourth of the wavelength of the microwaves to prevent undue scattering of the incident beam.

Practical Aspects Microwave instruments are applied extensively in the field of food and feed: cereals, grain, nuts, animal feeds, and dried pet foods. Here, too much moisture adversely affects the shelf life and the nutritional value. Too little moisture may make the product less enjoyable and perhaps even lead to valuable nutrient loss that is removed with the water. The right moisture content must be adhered to, and this justifies a sophisticated instrument. A further advantage is that it is often possible to narrow the specification limits leading to, on average, a slightly higher water content. Cases have occurred in which the investment was recovered in three months. The method is also used for paper.

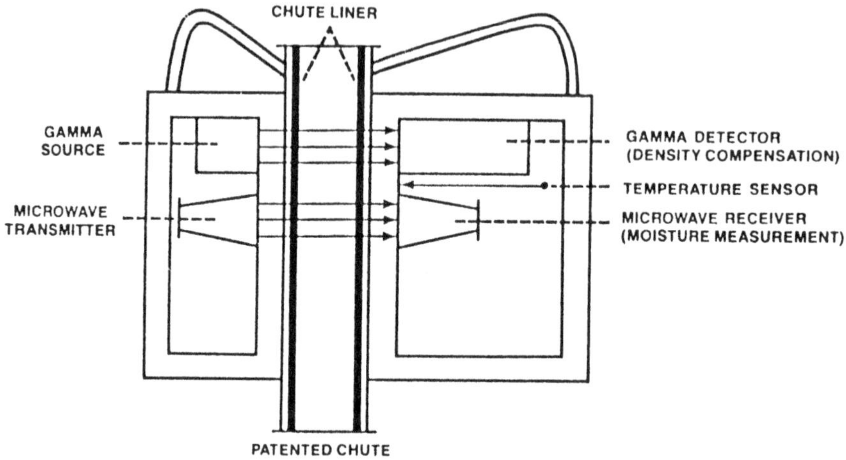

Figure 15.4 Online, noncontacting moisture measurement by means of microwaves.

Figure 15.4 exhibits a flow chute. This chute is filled with process material during the measurement. The walls are lined with special microwave energy-absorbent material to avoid microwave reflections in the chute. Because of leakage, the sample cross-sectional area must exceed considerably the area through which the beam is sent. Figure 15.5 shows an installed system. The two microwave antennas are located at different sides of the process stream on the conveyor belt. The process stream on

Figure 15.5 Continuous measurement of the moisture content of wood chips by means of microwaves. (Courtesy of Berthold Technologies GmbH & Co. KG, Bad Wildbad, Germany.)

the belt must have a constant thickness. The bulk density is measured by means of a radioactive source.

Temperature Measurement Temperature measurements can be used to control the moisture content of the dried product, the attraction being that temperature-measuring systems are reliable, cheap, and require minimal maintenance. Consider a batch fluid-bed dryer. This dryer is typical for the group of convective dryers. Free water (surface water) is evaporated in the first drying phase. The drying-gas temperature drops to the adiabatic saturation temperature as long as the product's moisture content exceeds the critical moisture content. The product temperature equals the adiabatic saturation temperature in that phase, which leads immediately to an important conclusion: temperature measurements cannot be used to establish moisture contents greater than the critical moisture content. Assume that drying proceeds beyond the critical moisture content. The exit-gas temperature (equal to the product temperature) starts to rise. It was established empirically that in many instances there exists a relationship between the difference (actual exit-gas temperature – adiabatic saturation temperature) and the moisture content. So, the adiabatic temperature booked in phase 1 can be stored electronically and used as a reference point to terminate the drying in the second phase. This principle can be simplified in many applications:

1. The continuous fluid-bed drying of vacuum pan salt is controlled by means of the highest product–gas temperature.
2. The flash drying of sodium sulfate is controlled by the exit-gas temperature.
3. The spray drying of rubber accelerators is checked by means of the exit-gas temperature.

The three examples given concern cross-flow and concurrent flow. To assess the dried product's moisture content, temperature measurements can also be used for countercurrent flow.

It is possible in most instances to use this principle to control the final moisture content at convective drying. At contact drying, the dried product's temperature is used for performance control.

15.2 CONTINUOUS MOISTURE-MEASUREMENT METHODS FOR GASES

Sometimes it is necessary to monitor the moisture content of the exhaust gas of a dryer. A high moisture content can reflect a satisfactory drying efficiency, and vice versa. Furthermore, the product quality may call for a certain humidity of the gas with which the product was in contact. A great variety of hygrometers are on the market. The heart of the system is the sensor. A complication is that the exhaust gases of a dryer are usually contaminated to some degree. It is possible to use sampling systems filtering the gas and, if necessary, cooling the gas. These two steps should not change the gas moisture content. A gas pretreatment is never perfect and complicates the measuring system.

Four different methods can be used satisfactorily in drying systems: (1) psychrometry, (2) chilled mirror hygrometry, (3) zirconium oxide potentiometry, and (4) electrical capacitance. The first two methods are absolute methods, whereas the other two methods are inferential methods.

Psychrometry

A dry-bulb thermometer measures the air temperature. A wet-bulb thermometer measures the temperature of a wetted wick in flowing air. The difference between the two measurements is due to evaporation. Figure 15.6 exhibits an instrument that is slightly more sophisticated but is still based on this principle. The gas enters at 1 and the flow is measured by venturi 2. Next, the dry-bulb temperature of the gas flow is

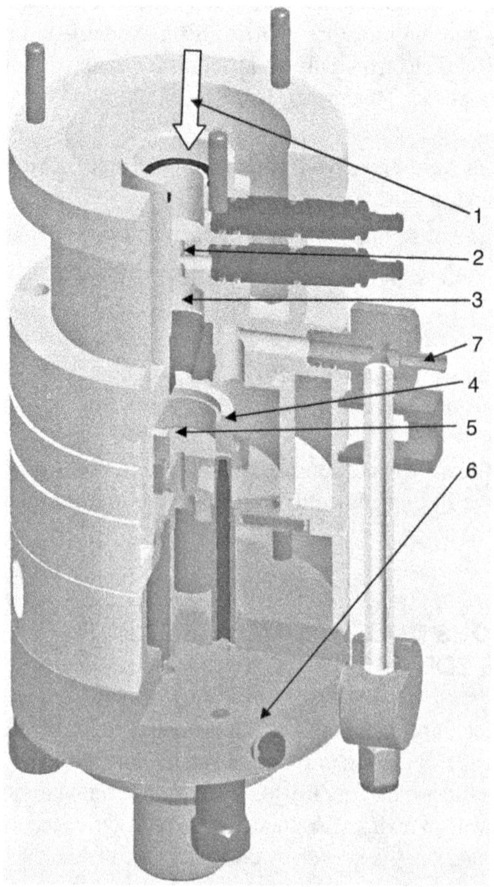

Figure 15.6 Process psychrometer. (Courtesy of Bartec GmbH, Gotteszell, Germany.)

measured at 3, whereafter the gas flow impinges on the surface of water-filled cylinder 4. The temperature of the water is measured by means of wet-bulb thermometer 5. The difference between the temperatures of the two thermometers is a measure for the water content of the gas flow. For temperatures between 0 and 100°C, this can be understood as follows. If the relative humidity of the gas flow is 100%, no water evaporation from the cylinder will occur and the two temperatures will be identical. If the relative humidity of the gas flow is 0%, the driving force for water evaporation from the cylinder will be maximum, which means that the temperature difference will also be maximum. Other temperature differences are a function of the moisture content of the gas. The method can also be used to measure the moisture content of air that has a temperature higher than 100°C. Water is pumped to the cylinder and flows over the rim to be recycled. The gas removes water from the instrument and the water depletion is automatically made-up. The gas is removed by means of the ejector 6, and the pressure in the measuring space is measured at 7.

The instrument is typically applied at the drying of paper. It is also applied at the drying of fertilizers and detergents.

Chilled-Mirror Hygrometer

A chilled-mirror hygrometer provides continuous dew-point measurements (see Fig. 15.7). The dew-point temperature is a function of the amount of water vapor in air or other gases. The chilled-mirror hygrometer is widely used as a reference instrument for the calibration of humidity sensors and instruments. The gas flow to be analyzed flows from left to right through the instrument. The heart of the instrument is a mirror cooled by a heat pump. General Electric's hygrometers are discussed as an example. The chilled mirror consists of a small polished hexagonal rhodium or platinum mirror. Platinum mirrors are often used in mildly corrosive environments.

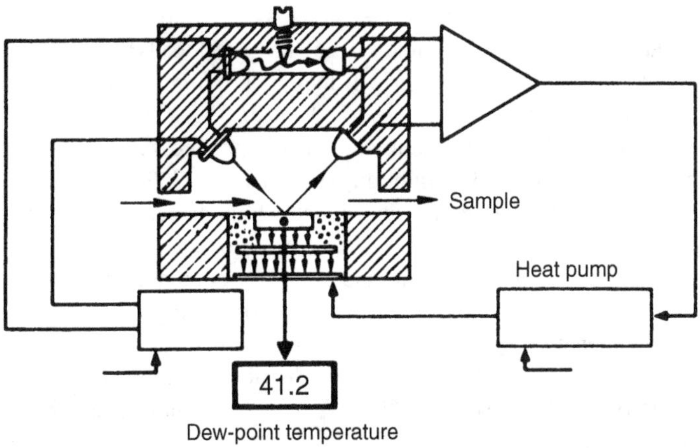

Figure 15.7 Condensation dew-point hygrometer. (Courtesy of GE General Eastern Instruments, Wilmington, MA.)

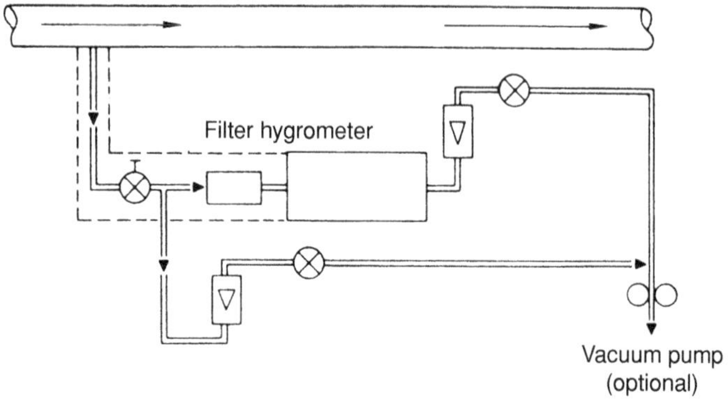

Figure 15.8 Hygrometer sampling system. (Courtesy of GE General Eastern Instruments, Wilmington, MA.)

When the mirror has the dew- or frost-point temperature, a dew or ice layer forms on the surface and the mirror's light-reflecting capability changes. This can be detected by a photoelectric cell receiving the light from a light-emitting diode via the mirror. The light used is infrared light. A separate light-emitting diode–photodetector pair provides a reference light measurement. The mirror surface temperature is controlled automatically and continuously at the dew- or frost-point temperature of the sample gas. A precision platinum resistance thermometer, which is embedded just beneath the mirror surface, is used to accurately measure the mirror temperature. GE's instrument may be used with five interchangeable chilled mirror sensors to provide a measurement range from –80°C to 85°C frost- or dew-point temperature with 0.2 K accuracy. The instrument responds quickly, and the long-term stability is excellent.

Unlike temperature or pressure sensors, a condensation hygrometer must be in intimate contact with the gas to be analyzed. Thus, they are subject to contamination. Often, a sampling system is recommended for applications involving considerable contamination (see Fig. 15.8). The setup is also used to cool sample gas below 100°C. A contamination with a soluble salt such as NaCl leads to a systematic error. The dew-point temperature indication will then be too high. This event can readily be explained, since at temperatures between 0 and 100°C, a saturated brine solution exerts approximately 75% of the saturated water vapor pressure. It also means that dry salt gets wet when the relative humidity is higher than 75%. So a mirror contaminated with NaCl exhibits a scattering of reflected light at a temperature higher than the true dew point.

GE introduced a system to alleviate the effects of contamination. Intermittently, the mirror is cooled to a temperature well below the prevailing dew point. This causes an excess of condensing water. The salt is dissolved. On heating the mirror above the dew point, the water evaporates, leaving the redistributed salt in clusters. This effect is illustrated in Figure 15.9. Most of the surface is clean again. The effect is that the time between mirror cleanings becomes 10 to 100 times longer.

The frequency of the automatic intermediate cleanings is once per 2, 6, 12, or 24 h. A typical cleaning cycle lasts 3 min. The time intervals between cleaning could

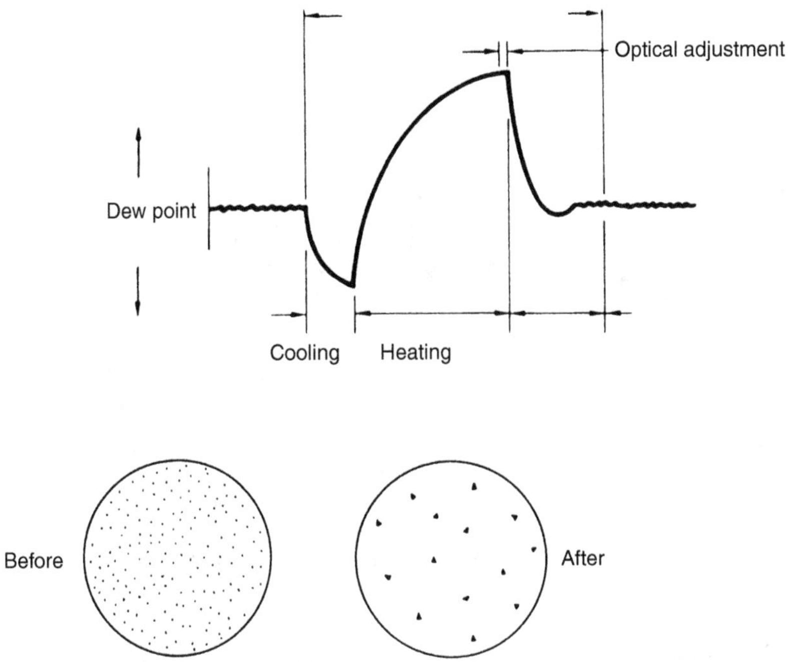

Figure 15.9 Pacer cycle for a condensation hygrometer. (Courtesy of GE General Eastern Instruments, Wilmington, MA.)

vary widely. In heavily contaminated areas, the mirror may have to be cleaned every day, whereas in clean environments the mirror could be operated without service for several weeks or months. In typical industrial applications with relatively clean air, the mirror is cleaned once or twice a month. Chilled-mirror dew-point sensors are generally not suitable for operation in very corrosive environments.

Example 15.1 In a plant for PVC stabilizers, the product was dried by means of a flash dryer. The gas flow leaving the dryer was transported through a duct. The duct was located near an outer plant wall. In wintertime, the gas flow was cooled, and this sometimes caused water condensation in the duct. To prevent this, it was decided to check the dew-point temperature of the leaving gas flow continuously by means of a chilled-mirror hygrometer. The dew-point temperature should always be below a prescribed temperature. It was considered that this robust instrument does not need calibration and that drift does not occur. The installation was a success.

Zirconium Oxide Potentiometry

Zirconium oxide potentiometry can measure the water content of gas flows having a temperature up to 300°C. The instrument employs an inferential method. In the

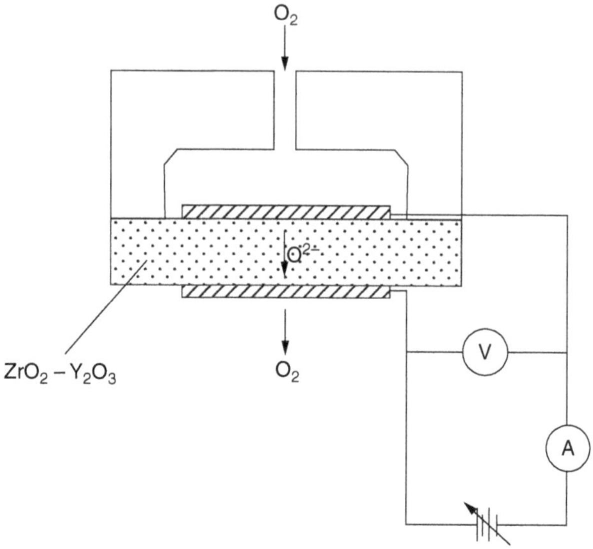

Figure 15.10 Zirconium oxide potentiometric sensor. (Courtesy of Bartec GmbH, Gotteszell, Germany.)

original version of the instrument, the water content is derived from the oxygen content of the gas flow. In this version it is assumed that the gas flow consists of nitrogen, oxygen, and water only. Because of the fixed ratio between the first two gases, the water vapor content follows unambiguously from the oxygen content. The electrolytic cell functions as an oxygen pump. Gaseous oxygen molecules dissolve into the cell, are transported through the cell, and are resorbed at the other cell side (see Fig. 15.10). As the tension is raised, the current increases meaning that, initially, the pump capacity increases. When the tension reaches and surpasses a certain value, the current no longer increases. This current plateau is characteristic for the oxygen level in the gas flow. The explanation is as follows. There is a diffusion hole with a circular cross-section and a diameter of 20 μm between the gas flow and the cell. The capacity to transport oxygen molecules through this hole is dependent on the concentration difference between the gas flow and the cell. The oxygen concentration at the cell's surface is nil when the current plateau is reached. Then the oxygen transport capacity of the diffusion hole is a function of the oxygen concentration in the gaseous flow only.

In the second version of the instrument, it is possible to determine the water content of the gaseous flow when additional permanent gases, (e.g., carbon dioxide) are present. A first plateau current indicates the oxygen content of the gas flow. On continuing the tension increase, the current starts to increase again because a decomposition at which oxygen is produced from water vapor is occurring. A second plateau indicates the sum of the oxygen level produced by the water vapor decomposition and the oxygen content of the gas flow. Subtracting the first plateau from the second plateau leads to the water level in the gaseous flow.

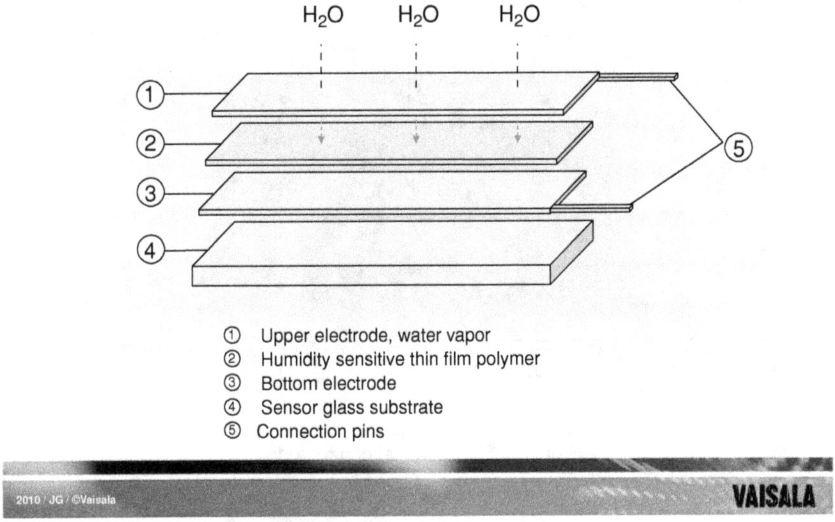

Figure 15.11 Capacitive thin-film polymer sensor. (Courtesy of Vaisala Oyj, Helsinki, Finland.)

The second version of the instrument can be used up to 300°C. Its accuracy is better than 1%. A typical application is the measurement of the moisture content of the exit gas at the drying of gypsum slabs.

Electrical Capacitance

The thin-film polymer in the sensor either absorbs or releases water vapor when the relative humidity of the air rises or falls (see Fig. 15.11). The dielectric properties of the polymer film depend on the amount of water it contains. The instrument measures the capacitance of the sensor and converts it into a humidity reading. The instrument can be used from –40°C up to 180°C. Its accuracy is approximately 2% relative humidity in the range –40 to 100°C. A typical response time to an instantaneous change in relative humidity from 10% to 95% is 3 min. The sensor is immune to particulate contamination and many chemicals. The instrument is relatively cheap. The measurement of the moisture content of the exit gas at wood drying is a typical example.

15.3 DRYER PROCESS CONTROL

The production of dried material at minimal cost and maximum throughput is the objective of many drying operations. There is no need for automatic control as long as the mass balance and the process conditions do not change. However, changes do occur, and it is possible to compensate for their effect by means of process

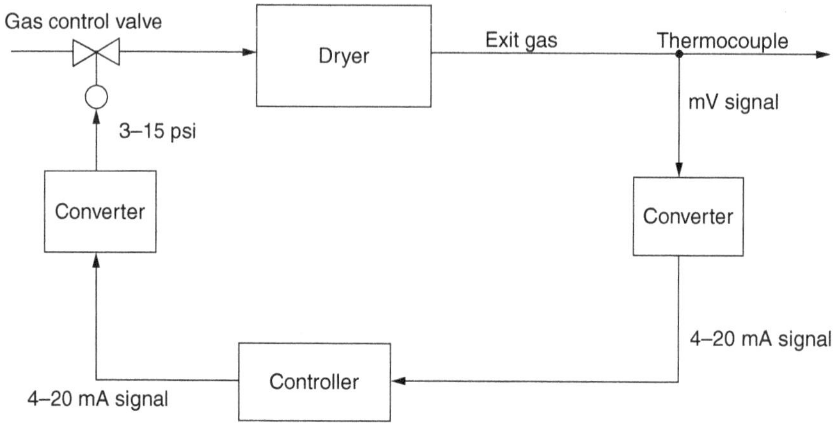

Figure 15.12 Typical control loop.

control, which comprises the measurement of a process variable and the subsequent automatic adjustment of a different process variable. At continuous convective drying, the temperature of the exit gas is often measured and used as an input to a control loop. Often, the moisture content of the dried product can be inferred from this measurement. Furthermore, measuring temperatures is simple, accurate, reliable, and cheap. In some instances it is also customary to measure the moisture content of the dried material in an inferential way (e.g., a textile dryer can be controlled, employing the conductivity of the exiting material as the input). Another example is the measurement of the dried product moisture content in food processing by means of microwave absorption and using the signal for process control. Various online continuous product moisture measurement methods were treated in Section 15.1. In principle, the analog signals of these instruments can be used for process control.

If water is evaporated, small underpressures in the drying chambers are preferred to avoid dust emissions. If a solvent is evaporated, small overpressures in the drying chambers are preferred to avoid air ingress.

A rule of thumb for process control systems is that the response time of the process control system should be smaller than, or approximately equal to, the product's residence time.

Many processes proceed in a hazardous area (electrical classification), and this is one reason to locate the control equipment in a control-room environment or "safe area." Consider the physical realization of a control loop measuring the exit-gas temperature of a continuous convective dryer and adjusting the natural gas flow to the burner (see Fig. 15.12): (1) measurement of the exit-gas temperature by means of a thermocouple, (2) converting the mV signal into a 4 to 20-mA signal, (3) processing the signal by means of an analog-to-digital converter, which is part of a microprocessor (controller), and (4) sending an output signal of 4 to 20 mA to a device converting the electronic signal into a pneumatic signal (3 to 15 psi) operating the control valve.

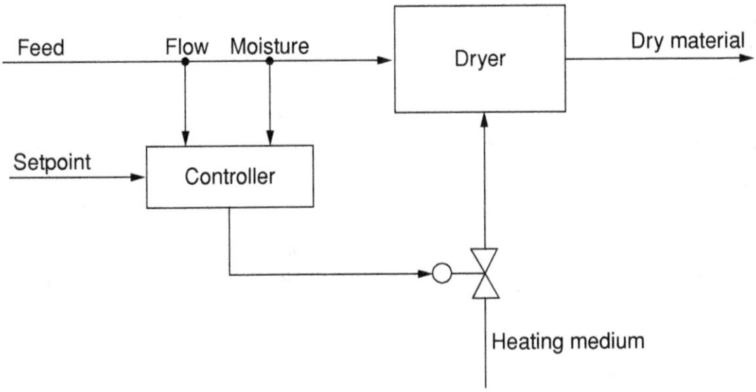

Figure 15.13 Feedforward dryer control.

Control Systems

Two different control systems, feedforward and feedback control, can be distinguished.

Feedforward Control The principle is shown in Figure 15.13. Wet material enters the dryer continuously, and when the flow or the moisture content changes, a different heating-medium flow is required. The controller takes care of adjustments. Feedforward control prevents an upset of the dryer from occurring; however, the actual result of the control action is not checked. Thus, small gradual shifts are often not noticed.

Feedback Control Figure 15.14 exhibits the principle. A temperature measurement serves to check the proper drying action. A change in this temperature initiates a controller action; for example, when the temperature decreases, more heat is sent to the dryer. This sounds simple; however, there are complications. By the time

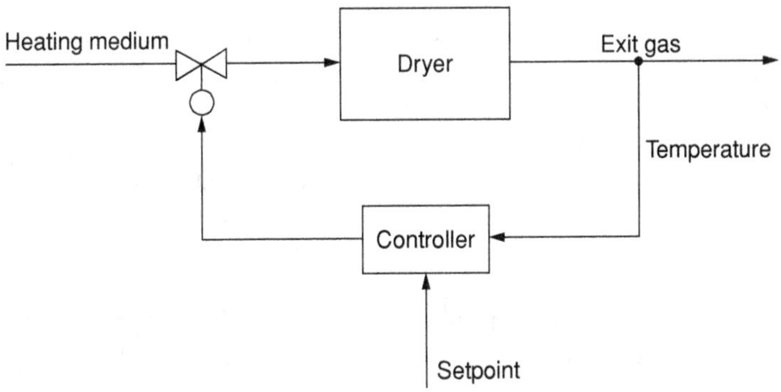

Figure 15.14 Feedback dryer control.

the temperature decrease is detected, some wetter material has already entered the dryer (delay time). Furthermore, at the moment the controller sends more steam to a heater, the mass of the heater begins to heat up. It takes some time to heat the metal and hence to notice the effect of the controller on an increased drying-air temperature (resistance–capacitance lag). The latter effect is not present if the temperature measurement acts on the gas flow to a burner.

One way to deal with the effects of delay time and resistance–capacitance lag is to tune the control loops under operating conditions. Ziegler and Nichol's rule is a practical means to attain this goal. A different avenue is the construction of a process model.

Advantages are sometimes found if feedforward and feedback controls are combined. Microprocessors and other process computers offer options to realize this. Three basic modes of continuous control exist: proportional action, integral action, and derivative action (collectively, PID actions). Proportional action implies a corrective action that is proportional to the deviation of the value measured from the value desired (setpoint). The bandwidth of the proportional action is an important variable. Beginning with the situation that the value desired equals the setpoint, the actuator's position is defined. A deviation occurs (e.g., a temperature drop of the exit gas because of wetter feed), and the actuator (e.g., a valve) admits a larger heating-medium flow.

In the new situation, the value desired cannot equal the setpoint because the original valve stem position would then be resumed. The difference is called the *offset*. The integral action comprises a corrective action to undo the offset. It sees to it that the valve's stem gets a new position. The integral time is a variable.

The derivative action implies a control mode proportional to the rate of change of error. When a deviation occurs rapidly, fast action is required, and vice versa. The derivative time is the variable here. Derivative action enables anticipation.

A three-term controller combines the three modes. In the past, three-term controllers often consisted of small boxes combined with a recorder. In conventional hardwired control rooms, many of these PID boxes can be found. Today's practice implies the use of microprocessors. A typical setup is the combination of eight control loops in one microprocessor.

The use of process computers does allow (if required) the use of more sophisticated (often tailormade) algorithms than the PID actions. Furthermore, the combination of control loops (as found in cascade control systems) is easier than with conventional control equipment.

A typical aspect of process computers is the process data-acquisition capability. Temperature, flow, and pressure data can be stored, displayed on visual display units or monitors, plotted (a hard copy can be obtained), and passed on to a supervisory process computer.

Control valves are activated pneumatically almost exclusively. A time lag of several seconds would prevail when the pneumatic lines are long. Electric signals are intermediate between the control room and the local pneumatic valve activation system.

Continuous Convective Dryers

Three flows enter the drying system: the feed, the air, and the heating medium. The possibility to select feedforward control was discussed earlier. However, the majority of the continuous convective dryers have feedback control, which implies a check of the drying function and acting upon one of the flows entering the system. In many cases, the process variable acted on is the heat supply or the feed to the dryer. Control of the airflow is hardly ever used. The background is that the air is sometimes important for the transport function (e.g., flash drying) and that the air ducts are wide; hence, this would necessitate large and expensive control valves. The air flows through continuous convective dryers are usually adjusted manually (by means of butterfly valves).

A general remark for all continuous convective dryers equipped with a steam heater is that the steam flow to the heater can be varied by varying the steam pressure. A higher pressure leads immediately to a higher condensation temperature, and vice versa.

Flash Dryers The residence times of both the solids and the gas are short, several seconds. The holdup of material is quite small. In practice, a situation often met is that the flash dryer is fed by a centrifuge or filter. A centrifuged product's moisture content can gradually increase from 3% to 5% by weight between wash-outs. The usual approach is to connect the exit-gas temperature to the heating-medium flow (see Fig. 15.15). The exit-gas temperature signal can be used alternatively to alter the setpoint of the device controlling the heater exit temperature. The latter setup has a short response time. Control by means of the feed is attractive because it implies a small response time. However, quite often the flash dryer must accept what the preceding equipment produces. Furthermore, controlling a flow of wet solids is not too easy, although it can, in principle, be done by means of a buffer and a rotary lock with variable speed.

When the drying system is equipped with a steam heater, the resistance–capacitance lag due to the heater is considerable. Rapid changes in the feed quantity or moisture content can lead to "hunting" by the control system if the exit-air temperature acts on the steam flow. A connection of the temperature measurement to the feed flow may be necessary.

Spray Dryers Here, too, the residence times are short. Connecting the exit-gas temperature to the feed flow is quite common (see Fig. 15.16). The background is that this process control method implies a small response time and that it is relatively easy to control a liquid flow. It must be realized that, in principle, varying the feed flow influences the particle-size distribution. The control method does not function if the feed line gets choked up. The spray drying system must, in that case, be stopped by an alarm. The drawback of that fallback provision is that proper functioning of the alarm is not continuously apparent. If the alarm fails, hot air can flow directly into a filter in which the filter bags can melt or ignite. It is also possible to connect the

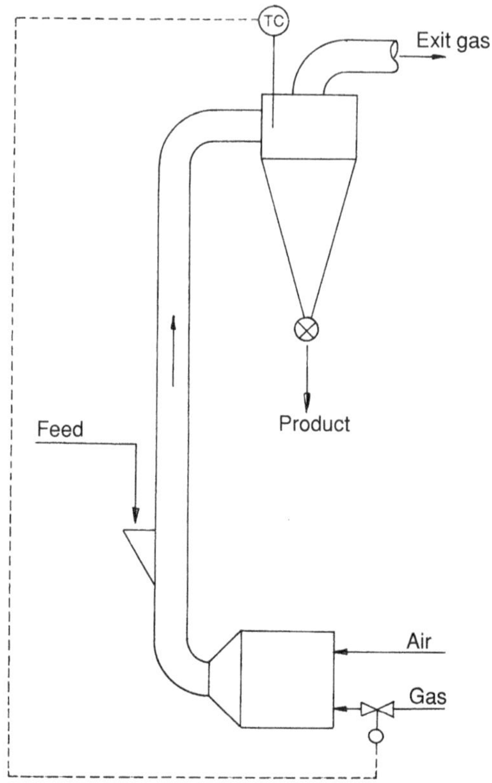

Figure 15.15 Flash dryer process control.

exit-gas temperature to the supply of the heating medium. This method does not carry the drawback of the previous process control method but has a longer response time.

The remarks made concerning a steam heater in a flash drying system also apply for a spray dryer.

Fluid-Bed Dryers Both circular and rectangular dryers use relatively long residence times for the solids and short residence times for the gas. Rapid changes in the feed's quality or quantity will not readily upset the dryers because of the buffer capacity with respect to the solids. The usual control method for the rectangular fluid-bed dryer is to measure the gas temperature in the fluid bed just ahead of the cooling section and connect it to the heating-medium supply (see Fig. 15.17).

Usually, the dryer processes whatever the plant produces. Circular fluid-bed dryers are commonly controlled by connecting the exit-air temperature to the burner fuel supply. The time delay is nil here, and changes occur very gradually because of the bed's buffer capacity.

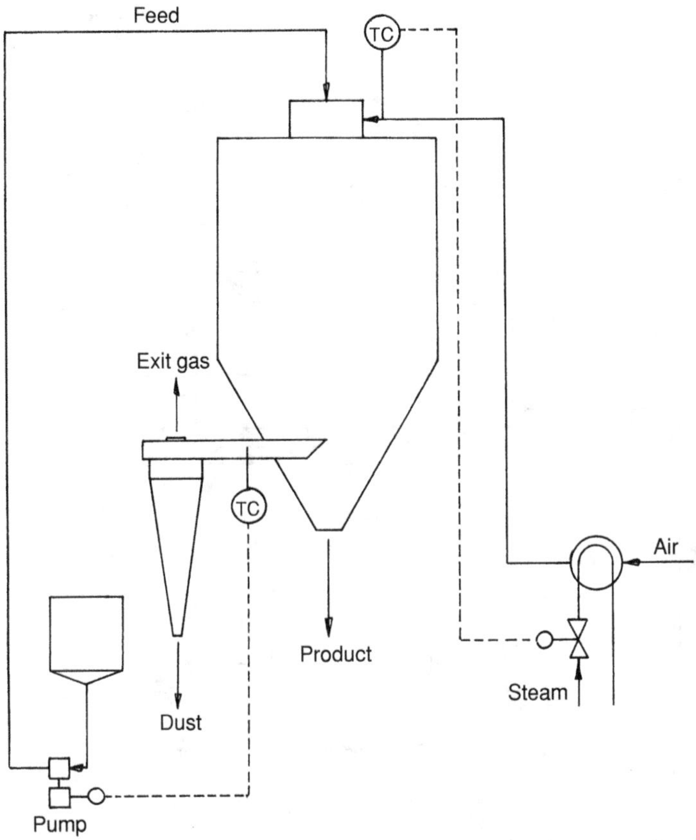

Figure 15.16 Spray dryer process control.

Rotary Dryers This dryer type can be controlled via measurement of the exit-gas temperature, which is then connected to the heating medium.

Conveyor Dryers Often, conveyor dryers are equipped with a steam heater that heats the makeup air indirectly. The temperature of the air circulating through the product layer is measured, and this value is used to adjust the steam control valve. The relative humidity of the recycled air is controlled manually by means of a butterfly valve in the exit-air line.

Miscellaneous Continuous Convective Dryers Control of these dryers can be obtained by application of the methods that were discussed in this section. For example, the Hazemag Rapid Dryer and the Anhydro Spin Flash Dryer can be compared to the flash dryer.

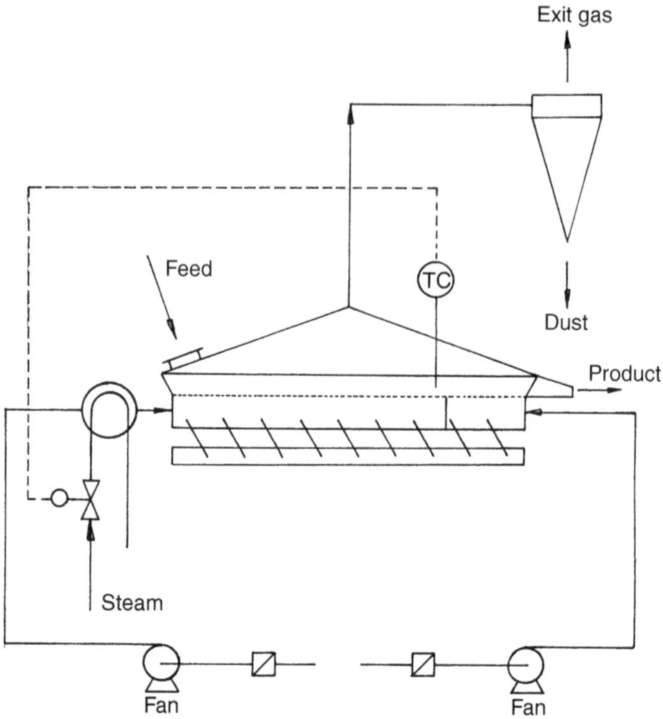

Figure 15.17 Fluid-bed dryer process control.

Continuous Contact Dryers

The general approach is to measure the dried product's temperature and to use this signal to increase or decrease the heating medium's temperature. For example, the pressure of the steam flowing to a jacket is increased if the temperature of the dried product falls. However, too high a medium temperature can, in some instances, lead to caking or product degradation. The airflow or nitrogen flow (to remove the evaporated water or solvent) is adjusted manually.

Batch Dryers

The control practice most commonly encountered is to fix the heating-medium temperature and to measure the product temperature. The latter measurement can be used to terminate the drying step. Measurement of the product's temperature can be difficult. It is then often possible to infer the product's moisture content from other process conditions, such as the vacuum. The vacuum in a dryer increases (the pressure in the dryer becomes lower) when the bulk of the moisture content has been evaporated. A different possibility is to check the amount of liquid condensed overhead. It is sometimes also possible to stop drying automatically when a certain amount of time has elapsed.

15.4 ENERGY RECOVERY

Drying alone accounts for 8% of all UK industrial energy demand, summed over all energy-consuming buildings and processes, all forms of energy, and all industrial sectors. The water evaporated in total, $2 \cdot 10^{10}$ kg, is equivalent to the contents of a reservoir of size $2000 \cdot 1000 \cdot 10$ m^3 [1]. The energy consumption of the drying operation has been reviewed in Chapter 1.

At steam drying, energy is recovered (see Chapter 12). Furthermore, heat can be recovered from two conventional dryer process flows: the product and the exit gas. Heat recovery from the exhaust gas is met more often than heat recovery from the dried product. Hence, the discussion will be restricted to the former type of efficiency improvement, which is met at convective drying. However, three aspects cause the heat exchangers to become large:

1. The energy to be recovered is spread out in space (the exhaust gas is rather bulky).
2. The gases are not particularly hot.
3. The heat transfer coefficient gas/metal is low; a value of 100 $W \cdot m^{-2} \cdot K^{-1}$ is typical.

The possible heat recovery methods are (1) recycle of part of the exhaust air to the air inlet line, (2) indirect gas–gas heat exchange, (3) indirect gas–liquid heat exchange, (4) scrubbing, and (5) using a heat pump. It is possible to combine several methods. It is necessary to balance the advantages against the investments required.

Exhaust-Gas Recycle

Part of the exhaust gas is recycled to the air-inlet line, however, condensation may not occur there. The recycle leads to a higher moisture level in the exhaust gas. Often, the capacity can be maintained only if the exhaust-gas temperature is increased. But such a measure counteracts the first step. Furthermore, for various reasons, it is necessary to rigorously remove dust particles from the recycle gas flow.

If the temperature of the gas leaving a dryer is high (e.g., 150°C), it is possible to recycle part of this gas and mix it with the hot gases from a burner.

Indirect Gas–Gas Heat Exchange

Indirect gas–gas heat exchange is depicted for a spray dryer in Figure 15.18. Usually, only sensible heat is removed. The heat exchanger can become contaminated on the exhaust-gas side. Consequently, the steps to counteract this are (1) air filtering, (2) installing a glass heat exchanger (note: glass is fragile), or (3) periodic heat-exchanger cleaning.

Indirect Gas–Liquid Heat Exchange

Figure 15.19 exhibits a runaround coil installed to recover heat from the exhaust of a spray dryer. It can offer advantages over the former type of heat exchange because it

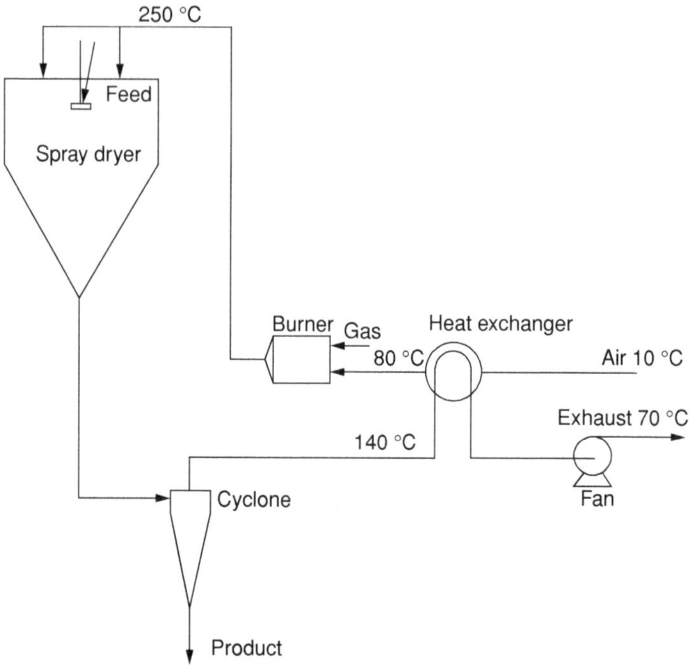

Figure 15.18 Heat recovery from a spray dryer using a gas–gas heat exchanger.

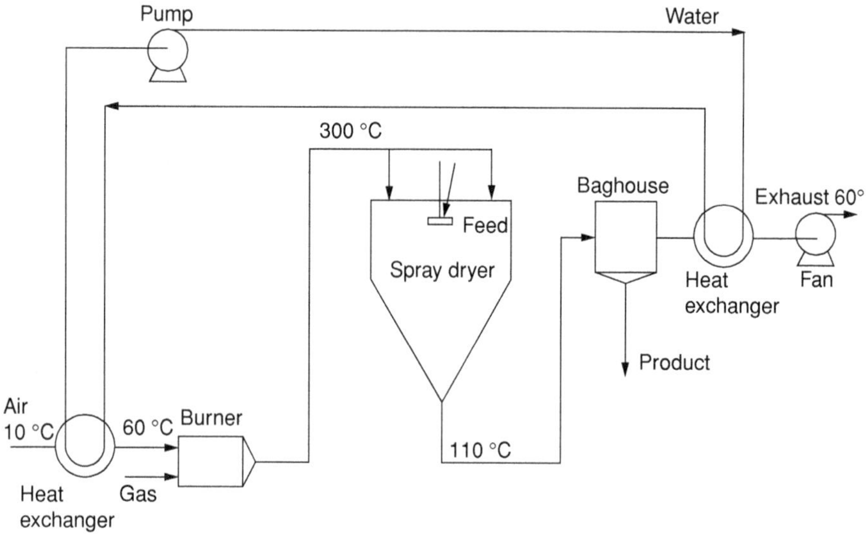

Figure 15.19 Heat recovery from a spray dryer by means of a runaround coil.

is now not necessary to construct (so to speak) a flyover crossing of roads (the heat exchanger).

Another possibility for recovering heat concerns the installation of a heat pipe. The warm gas causes a thermal fluid to evaporate by indirect heat exchange. The vapor condenses in a different part of the conduit and transfers heat to the cold gas. It stands to reason that the distance between the two parts cannot be too large.

Scrubbing

The exhaust gas is scrubbed with cold water, and warm water is obtained. This method can also recover part of the latent heat in the exhaust-gas flow. It is necessary to have an outlet for the warm water (use it as, e.g., process water). Thorough cleaning of the exhaust gas is a bonus of this method.

Heat Pump

A heat pump can accept thermal energy at a low level (both sensible and latent heat) and deliver the energy at a higher level. This is accomplished by means of mechanical energy. A thermal fluid acts as an intermediate. The thermal fluids in use at present cannot get warmer than approximately 100°C because of their thermal stability.

REFERENCE

[1] Bahu, R., Kemp. I. (1994). Chapter 6 (Drying) in *Separation Technology: The Next Ten Years*, edited by Garside, J., IChemE, Rugby, UK.

16

GAS–SOLID SEPARATION METHODS

The movement of gases is inherent to drying operations. The gases may entrain particulate material, and a gas–solid separation step is usually required. Convective drying implies intimate contact between the air used for drying and the material being dried; this is not the case for contact drying. Some dryers have a preseparation of the gas and the solids in the dryer itself (e.g., fluid-bed and rotary dryers). Preseparation does not occur in a flash dryer, whereas spray dryers can operate in both modes. In conveyor dryers, the solid remains stationary, and usually only low levels of entrainment are found. The pieces of equipment for the separation of entrained solid particles from gases treated in this chapter are cyclones, filters, scrubbers, and electrostatic precipitators.

Cyclones are capable of operating with high inlet dust levels (e.g., 200 $g{\cdot}m^{-3}$), and it is common to install them upstream of such other collectors as bag filters or scrubbers. However, they are not suitable for the efficient collection of particles smaller than 5 to 10 μm.

Because of their positive nature, fabric filters will generally ensure a high collection efficiency (greater than 99%) even on submicrometer-sized dusts. Their main limitation is the heat sensitivity of the filter media. Natural fibers cannot be used at temperatures exceeding approximately 80°C, whereas synthetic fibers can be used up to 150°C. Glass fiber will withstand continuous operation up to 260°C.

A low-energy scrubber can collect fine soluble particles effectively, and fine insoluble particles can be collected using high-energy scrubbers; however, the energy requirement increases exponentially in the submicrometer range as the particle size

Drying in the Process Industry, First Edition. C.M. van 't Land.
© 2012 John Wiley & Sons, Inc. Published 2012 by John Wiley & Sons, Inc.

becomes smaller. On using a scrubber to separate insoluble particles, a gas–solid separation task is converted into a solid–liquid separation task.

Electrostatic precipitators are used for the separation of inorganic, incombustible particles from large gas flows. An example is the application at the drying of marl, a raw material used in cement manufacture. Marl is dried convectively using high air temperatures, in the range 400 to 500°C. Installing filters is a risk, as the hot air might come into direct contact with the fabric. The performance of electrostatic precipitators is almost as good as the performance of filters.

Cyclones, filters, scrubbers, and electrostatic precipitators are widely used in convective drying plants and are available in packaged assemblies. Contact dryers are frequently equipped with a small fabric filter, which is suitable to retain the small particles that are entrained with the low gas velocities in contact dryers.

It is important to avoid condensation in dry collectors. Cyclones can become plugged by incrustations, and filter fabrics are easily clogged by caking material and are then difficult to clean. Condensation in dry electrofilters should also be avoided. (Wet electrofilters exist as well, but, they are not used in drying plants.)

At gas–solid separation, the solids content in milligrams per normal cubic meter of the cleaned gas is an important criterion. The demand in the spray drying of coffee concentrate at Sara Lee's plant in Joure, The Netherlands is a maximum of 10 $mg{\cdot}nm^{-3}$ in the filtered exiting gases. The daily practice is 3.5 $mg{\cdot}nm^{-3}$.

16.1 CYCLONES

A cyclone is a device that accomplishes a solid–gas separation by imparting a spinning motion to the two-phase system. The particles move outward to the wall of the cyclone because of the centrifugal force and subsequently travel downward to enter a discharge hopper. The gas that is freed from solids leaves at the top through a central outlet. Figure 16.1 shows a Stairmand cyclone with all the necessary dimensions [1]. The Stairmand cyclone is a high-efficiency, medium-throughput cyclone. The dust-containing gas enters the cyclone body tangentially and moves downward in a vortex motion. The flow direction reverses at the apex of the cone, and the cleaned gas travels upward in a second vortex.

Sometimes it is stated that a cyclone is a cheap centrifuge, and certainly an element of truth is contained in this statement. However, it must be recognized that cyclones generate much lower radial acceleration than that of centrifuges. For example, a cyclone with a diameter of 1 m and an inlet velocity of 20 $m{\cdot}s^{-1}$ has a radial acceleration (expressed as a multiple of the acceleration due to gravity) of $20^2/(0.5{\cdot}9.81) = 81.5$. This acceleration is considerable but it is not spectacular, since peeler centrifuges operate with an acceleration figure in the range 500 to 1000. A cyclone with a diameter of 2 m and the same inlet velocity has a radial acceleration figure of 40.75. Consequently, large cyclones are less efficient than small ones. However, the pressure drop is, in principle, equal for cyclones if their geometric ratios and the inlet and outlet velocities are equal. These factors provide an incentive to arrange several cyclones in parallel operation rather than using only one large cyclone.

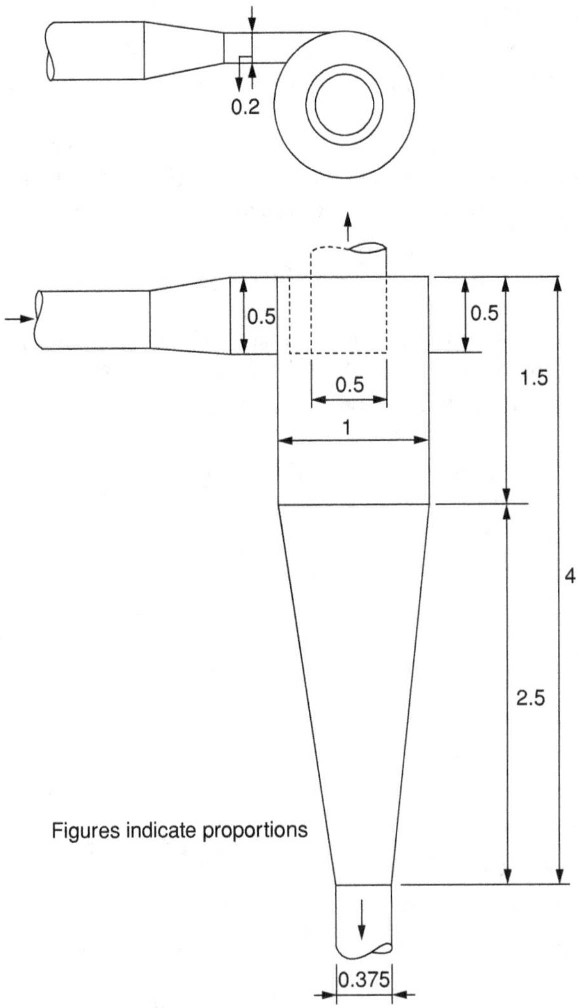

Figure 16.1 Stairmand cyclone for gas-solids separations. Figures indicate proportions.

Cyclones are cheap devices for dust collection, but their main limitation is their poor efficiency in collecting particles smaller than about 5 μm. High dust loadings can be handled, and hence cyclones can be used to separate flash-dried particles from the gas leaving the dryer.

Sizing and Process Data

Cyclones typically have diameters in the range 0.1 to 2 m. Cyclones can be sized using the rule of thumb that the normal flow rate in $m^3{\cdot}s^{-1}$ equals $1.5{\cdot}D_c^2$ for a high-efficiency, medium-throughput pattern. The factor 1.5 becomes 4.5 for a medium-efficiency, high-throughput pattern. The entrance velocity is approximately 15 $m{\cdot}s^{-1}$ in both types. An entrance velocity in excess of 30 $m{\cdot}s^{-1}$ causes turbulence, which, in turn, leads to bypassing and reentrainment of separated particles and hence a decrease in efficiency [2].

Pressure drops of between 50 and 150 mm in a water gauge are normal. Koch and Licht [1] recommend Shepherd and Lapple's approach to estimate the pressure drop in a cyclone. First, the friction loss factor is calculated:

$$N_H = K(a{\cdot}b/D_e^2) \tag{16.1}$$

K is 7.5 if a neutral inlet vane is present and 16 if an inlet vane is absent. a and b are, respectively, the height and width of the inlet duct, and D_e is the diameter of the outlet duct. Second, the pressure drop is calculated:

$$\Delta p = N_H{\cdot}0.5{\cdot}\rho_g{\cdot}v_i^2 \qquad \mathrm{N{\cdot}m^{-2}} \tag{16.2}$$

Example 16.1

Geometrical ratios cyclone

$a = 0.5{\cdot}D_c$

$b = 0.2{\cdot}D_c$

$D_e = 0.5{\cdot}D_c$

$$\frac{a{\cdot}b}{D_e^2} = \frac{0.5{\cdot}0.2}{0.5^2} = 0.4$$

$K = 16$ (no inlet vane)

$N_H = 16{\cdot}0.4 = 6.4$

Take $\rho_g = 1.0$ $kg{\cdot}m^{-3}$ and $v_i = 15$ $m{\cdot}s^{-1}$

$\Delta p = 6.4{\cdot}0.5{\cdot}1.0{\cdot}15^2 = 720$ $N{\cdot}m^{-2}$(73.5 mm water gauge)

Koch and Licht[1] recommend relationships for the calculation of collection efficiency. A potential source for poor cyclone performance is the air ingress into the dust outlet of the cyclone. In Figure 16.1, the ratio between the cyclone height and the diameter of the cylindrical part is 4. If this ratio increases to 6, for example, the collection efficiency increases. However, the capital cost increases as well. The maximum solids content of the gas treated by a cyclone is generally 20 $mg{\cdot}nm^{-3}$.

Cyclones are applied frequently in convective drying plants. Their construction is simple and the energy requirements are low. They can withstand high temperatures and be cleaned efficiently. Cyclones can become plugged with particulate material and the supporting structure must be able to support the filled cyclone.

Explosion relief vents, which are located on the roof of the cyclone, are used when processing explosive dusts. Whether it is possible to achieve the required relief area without affecting performance should be checked. Units should be grounded to avoid electrostatic charge buildup.

16.2 FABRIC FILTERS

Fabric filters are devices that perform gas–solid separation by means of a fabric, with the gas passing through a cloth and the particulate material being retained. Fabric filters can separate particles as small as 0.1 μm, and the treated gas usually contains less than 10 mg·nm^{-3} of solids. Today's trend is to even less than 5 mg·nm^{-3}. The structural enclosure containing the bag-shaped filters is termed a *baghouse*. For large convective dryers the reverse-pulse filter today is state of the art. Figure 16.2 shows such a filter schematically, and Figure 16.3 shows such a filter as part of a spray-drying installation. This filter type will be described. The dust-laden air enters via an inlet pipe and flows to the bags, where the dust accumulates outside the bags and the gas passes through the dust layer and the cloth and leaves the filter through the outlet pipe. The bags are supported by wire cages that have a cylindrical cross-section. Venturi nozzles are located in the clean-gas outlets of the bags. For cleaning, jets of high-velocity air are directed through the venturi nozzles and into the bags. The pulses of air expand the bags and dislodge the collected dust.

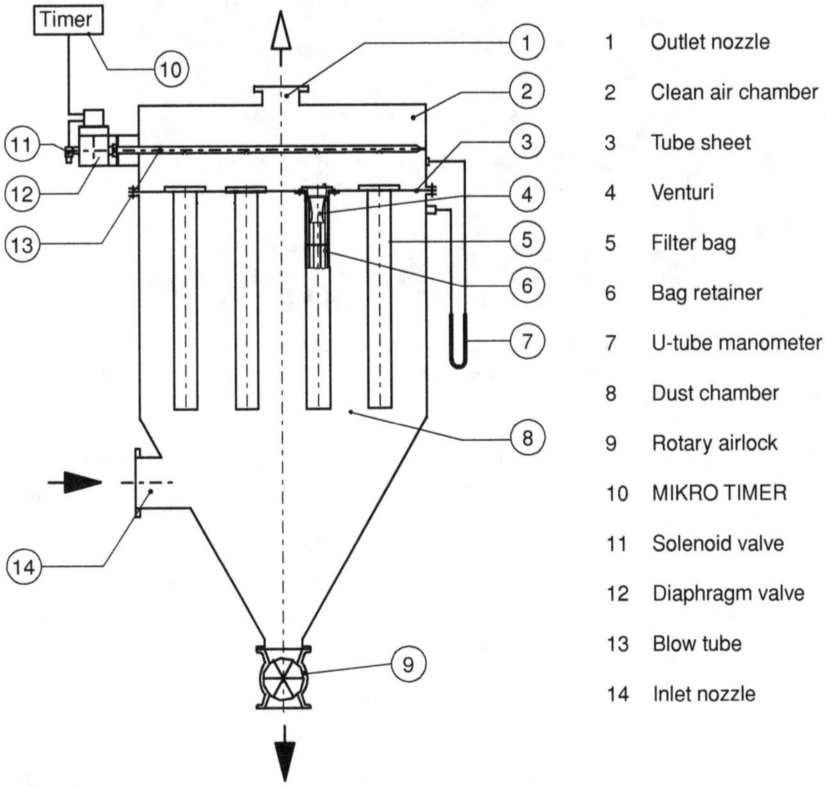

Figure 16.2 Sectional view of a reverse-pulse fabric filter. (Courtesy of Mikropul GmbH, Cologne, Germany.)

Figure 16.3 Reverse-pulse filter to separate silica particles from the gases leaving a spray dryer, having an area of 1370 m^2. (Courtesy of Mikropul GmbH, Cologne, Germany.)

The gas velocity is reduced from the value in the inlet pipe (e.g., 15 $m \cdot s^{-1}$) to the low fabric approach velocity (e.g., 1 $m \cdot min^{-1}$). Subsequently, the gas velocity increases to, for example, 15 $m \cdot s^{-1}$ in the outlet pipe. Often, filters are provided with internals that assist in an orderly decrease of the velocity in the inlet pipe to the fabric approach velocity. With abrasive dusts, this is important to avoid fabric wear. The fabric approach velocity is low, to enable bag cleaning and prevent bag plugging. The consequence is a large filtration area.

Fabrics

The dust layer may be completely removed from the surface of the fabric due to the cleaning action of the pulse. Hence, the fabric itself must serve as the principal filter medium for at least a substantial part of the filtration cycle. Woven fabrics are unsuitable for such service, and felts of various types must be used. The felt ensures that a good collection efficiency is also maintained until a dust layer has formed.

Felted fabrics consist of fibers that are compressed under high pressure. The commonly used range of filtration velocities for felted fabrics is 1.2 to 2.5 m·min^{-1} [3].

As an alternative for felt, it is possible to use a polyester fabric coated with a Teflon membrane. The membrane allows passage of the gas and retains the solid material at the surface.

Baghouse Construction and Operation

The cross-section of the baghouse can be circular or rectangular. For relatively heavy dusts, the gas enters the filter in the lower part to enable direct settling of part of the dust. For relatively light dusts, the gas enters the filter in the upper part to enable uniform fabric charging. Typical filter bag diameters are 70, 120, and 160 mm. The maximum length of the bags is 8 m, as the jets of high-velocity air must also be able to reach the bottom parts of the bags. The filter bags are mounted in the supporting plate by means of a bayonet locking.

The cleaning air pulses typically last 80 ms. It is common to clean a row of bags at a time. Each row (or pair of rows) of filter elements is cleaned in turn while the gas or air continues to flow through the remaining elements. A typical time between cleaning by means of air pulses is 15 s. The air pressure is usually 6 bar, but it is possible, if need be, to use lower pressures. Normal filter pressure losses are in the range 80 to 150 mm water gauge. Initially, the cloth is the filter medium; however, the surface layer generally becomes the dominating filter medium once it has formed. The gas flow through both the cloth and the dust layer is laminar and hence D'Arcy's equation is applicable. This means that the pressure drop across the fabric itself is proportional to the velocity, and the pressure drop that is due to flow through the dust deposit is proportional to both the velocity and the dust layer thickness. Despite the air pulses, the filter pressure loss increases with time, and eventually it is necessary to wash the filter bags. Ultimately, it will be necessary to replace the filter bags.

Table 16.1 contains the maximum recommended operating temperatures for common fabrics. Glass can be used at a relatively high temperature; however, the finish

Table 16.1 Maximum Recommended Operating Temperatures for Fabrics

Fabric	Temperature (°C)
Cotton	80
Wool	90
Nylon	90
Polypropylene	90
Polyacrylonitrile	125
Polyester/Teflon	140
Nomex (*m*-aramid)	230
Glass fiber	260
Ceramic fiber and sintered metal	650

limits the maximum temperature range. Fabric filters are not recommended for large convective drying operations using high air-inlet temperatures. Fabric filters are used primarily in drying systems for organic chemicals, where drying air-inlet temperatures generally do not exceed 200°C and the exit gases are usually not above 100°C. However, the possibility of noncooled drying air passing directly into the baghouse must be evaluated. If the dust can cause a dust explosion, baghouses must be safeguarded against them. Thorough bonding is necessary, and metal- or carbon-coated fibers are often included into the fabrics. Explosion panels that relieve directly or through short lines to the atmosphere are normal features of baghouses. A dust explosion can be followed by a fire because considerable dust holdup in the baghouse can occur.

The resistance against acids and alkalis of the filter fabric should be checked. *Perry's Handbook* [3] provides much information concerning the properties (including mechanical) of fibers for fabrics. To avoid condensation, it is important that the exit gas be maintained above its dew point. Insulation of the filter is recommended to avoid condensation.

16.3 SCRUBBERS

A scrubber is a device that uses a liquid (normally, water or an aqueous solution) to clean a gas. The material that is removed from the gas can be in either a liquid phase (droplets) or a solid phase (particles). In drying plants, scrubbers remove solids and are often preceded by cyclones.

Scrubbers are used in the chemical industry for several reasons, one of which is that the gaseous flows in that industry are usually relatively modest (e.g., 50,000 $m^3 \cdot h^{-1}$). Also, there is considerable processing of liquids and suspensions in the industry.

Good contact must be obtained between the liquid and the gaseous phase; to this end, the scrubbing liquid is dispersed in a jet or spray or spread in a film over the internal surfaces of the scrubber. The three relevant basic particle collection mechanisms in all scrubbers are centrifugal deposition, inertial impaction, and the Brownian movement.

The centrifugal deposition mechanism comprises spinning out of particles by centrifugal force caused by a change in gas flow direction. This is the mechanism prevailing in cyclone separators and is effective for particles down to approximately 5 μm.

Inertial impaction occurs when a gas stream flows around a small object and the particles suspended in the gas stream continue to move toward the object and eventually are trapped. This mechanism is effective for particles as small as 0.5 μm.

Actually, there is no fundamental difference between these two collection mechanisms. The first is concerned with large-scale flow-direction changes, and the second concerns small-scale changes. The two collection methods are only slightly dependent on the residence time. Hence, multiple-stage contacting has no real advantage over single-stage contacting for the first two collection methods.

Brownian diffusion becomes important for particles that are smaller than 0.3 μm. It comprises the movement at random of particles due to collisions with gas molecules. This mechanism becomes quite effective for very small particles, and the performance depends on the residence time.

For a given particulate, there is a relationship between the collection efficiency of a scrubber and the energy dissipated in the gas–liquid contacting process [4]. It appears that the efficiency is relatively independent of scrubber geometry and depends very little on the way the power is applied.

The gaseous flow entering a scrubber contains particles as small as 0.5 μm for example. The gaseous flow leaving the scrubber often contains droplets of 100 μm, for example. The separation of these droplets from the gaseous flow is relatively simple and can be achieved by a cyclone or a chevron demister (see Fig. 16.4).

There is no risk of a dust explosion or a fire if water or an aqueous solution is used as the scrubbing liquid. If the solid particles are insoluble in the scrubbing liquid, the solid–gas separation task is replaced by a solid–liquid separation task.

A distinction is often made between low-, medium-, and high-energy scrubbers. High-energy scrubbers have pressure drops exceeding 500 mm water gauge and can collect particles as small as 0.5 μm; however, their running costs are high. Low-energy types have pressure drops of up to 100 mm water gauge [5].

It is not clear whether wettable particles are more readily collected than non-wettable particles or whether the use of wetting agents promotes collection. Normal

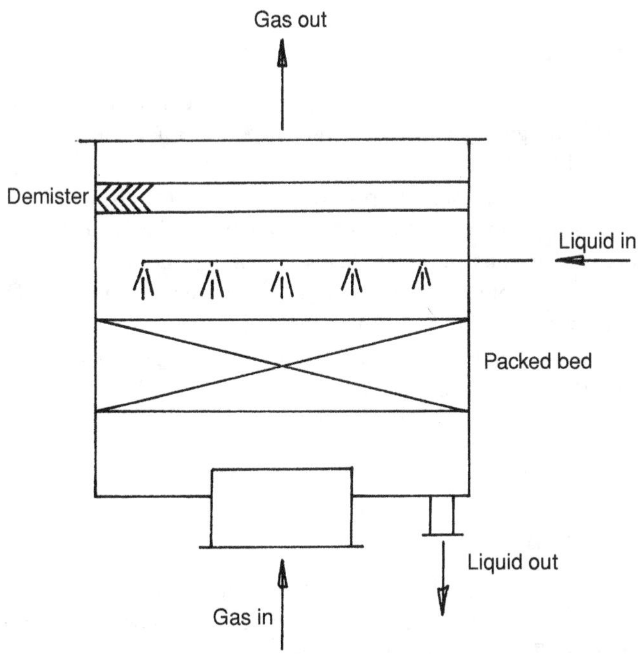

Figure 16.4 Packed-bed scrubber.

liquid/gas ratios are between 0.1 and 10 m^3 of liquid per 1000 m^3 of gas. Solid concentrations in the recycle scrubbing liquid can be as high as 10% by weight [5].

The scrubber types most frequently used for drying systems are packed bed, venturi, and mechanical. These three scrubber types are considered next in some detail.

Packed-Bed Scrubbers

Figure 16.4 is the schematic of a tower scrubber. Normally, packed-bed scrubbers are used for the removal of soluble particles since high loadings of an insoluble dust can cause plugging. The scrubbers may be packed with, for example, Raschig rings or Berl saddles, and gas–liquid contacting is performed in a countercurrent manner. Typical superficial gas velocities are on the order of 1 $m{\cdot}s^{-1}$, and droplet entrainment is prevented by means of a chevron demister.

The removal efficiency depends on the kinetic energy possessed by the particle, which is $0.5{\cdot}m{\cdot}v^2$. Thus, the gas velocity must be increased by a factor of 2.15 if a particle of 3 μm must be captured with the same efficiency as a particle of 5 μm. Because pressure drop is proportional to the square of the gas velocity, the scrubber pressure drop to capture the 5-μm particle must be multiplied by a factor of 4.62 (2.15^2) to be able to capture the 3-μm particle. In a countercurrent packed scrubber, the power applied (as reflected by pressure drop) is limited by hydraulic flooding of the packing. As a rule of thumb, for a 3-μm particle, 75 to 80% removal efficiency can be obtained [6].

Venturi Scrubbers

A venturi scrubber (Fig. 16.5) in its simplest form consists of a constriction in the duct carrying the gas that increases the velocity to a value in the range 60 to 100 $m{\cdot}s^{-1}$. The width of the constriction is adjustable by means of a servomotor, to enable adjustment of the pressure drop. Scrubbing liquid is introduced at or upstream of the constriction. The liquid is atomized, and the high relative velocity between the liquid droplets and the dust particles leads to efficient collection of even the fine particles. The liquid droplets are then separated from the gas in the cyclonic separator. This high-energy scrubber is the most efficient type of scrubber if the gas flow through the venturi scrubber can be maintained. The pressure drop is in the range 300 to 2000 mm water gauge. Particles as small as 0.1 μm can be removed with approximately 50% efficiency. Figure 16.6 shows a venturi scrubber installed in a chemical plant.

Mechanical Scrubbers

Figure 16.7 shows a mechanical scrubber schematically. This scrubber type incorporates one or two rotors in the wet collection zone. The rotors produce a fine spray that contacts the gas. The liquid droplets are then separated from the gas in a cyclonic separator. The circumferential velocity of the rotors is in the range 60 to 70 $m{\cdot}s^{-1}$.

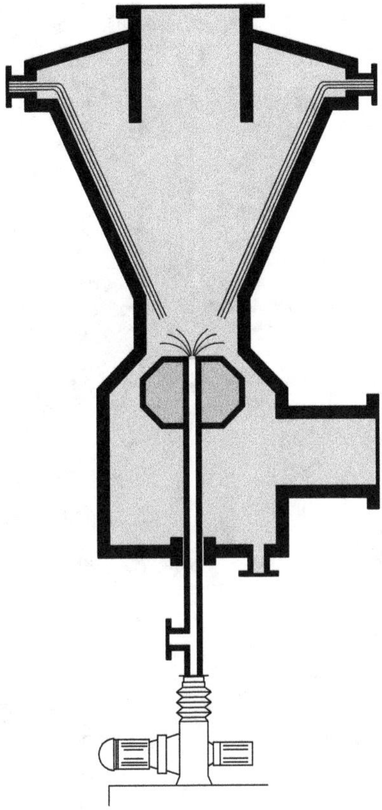

Figure 16.5 Venturi scrubber (seen schematically). (Courtesy of TREMA Verfahrenstechnik GmbH, Kemnath, Germany.)

This scrubber can remove particles of 0.2 μm with approximately 50% efficiency. The investment is higher than for a venturi scrubber, but the efficiency is not dependent on the gas velocity. The pressure loss is in the range 40 to 100 mm water gauge. However, the energy consumption per 1000 m^3 gas is comparable to the energy consumption of the venturi scrubber because of the rotor energy consumption. Figure 16.8 shows a mechanical scrubber installed in a plant.

16.4 ELECTROSTATIC PRECIPITATORS

Electrostatic precipitators can achieve a separation between particulates and a gas stream. To this end, the gas stream is directed through a number of grounded channels in which high-voltage discharge electrodes are centrally positioned at regular intervals. The process is illustrated in Figure 16.9. Uncharged particles entering the channels are charged by the negative discharge electrodes or ionized gas and then move toward the grounded walls of the channel. The term *electrostatic* is not entirely

Figure 16.6 Venturi scrubber in a chemical plant. (Courtesy of TREMA Verfahrenstechnik GmbH, Kemnath, Germany.)

correct, as a direct current is always flowing from the collecting plates, through the channel filled with gas and to the electrodes.

The pressure drop to move the gas through the separator is very low, which compensates partially for the relatively high capital cost of the equipment.

Electrostatic precipitators, also called *electrofilters*, are suitable for very large gas flows (e.g., 10^6 $m^3 \cdot h^{-1}$). They effectively remove particles as small as 1 μm and can clean gases down to 20 $mg \cdot nm^{-3}$ and, in some cases, even down to 10 $mg \cdot nm^{-3}$. Electrofilters are unsuitable for combustible materials. They are sometimes employed in convective drying systems in which high (e.g., up to 600°C) drying-air temperatures are used. Filters are then less suitable, as there is a risk that the hot gas enters the filter and melts or ignites the fabric. A typical average tension of the discharge electrode is −55 kV; the tension varies sinusoidally from −47 to − 63 kV.

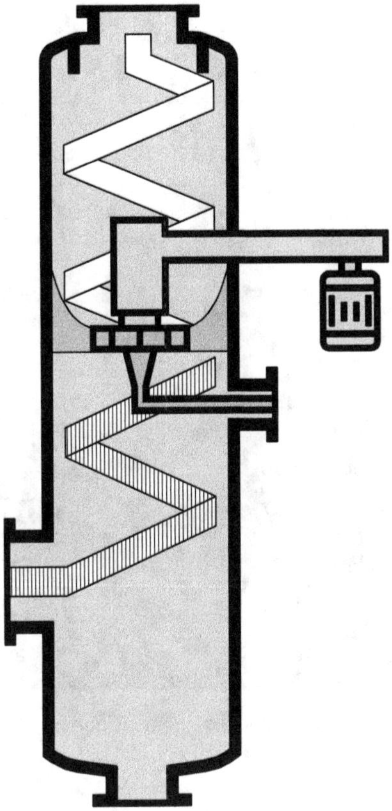

Figure 16.7 Mechanical scrubber (seen schematically). (Courtesy of TREMA Verfahrenstechnik GmbH, Kemnath, Germany.)

Principle of Operation

Figure 16.9 shows a perspective view of an electrostatic precipitator. The gases pass horizontally through a number of ducts. Two parallel rows of vertically mounted collecting plates form a duct in which a number of discharge electrodes are suspended vertically between the plates. A strong electrical field between the discharge electrodes and the grounded collecting plates is created by the high negative tension applied to the discharge electrodes. The field strength is maximum near the discharge electrodes. Electrical breakdown of the gas takes place on raising the tension. The breakdown, called *corona*, appears as a bluish glow around the discharge electrode. The precipitation mechanism is as follows:

1. The gas always contains free electrons. These electrons move rapidly away from the discharge electrode, collide with gaseous molecules, and remove additional electrons from these molecules.

Figure 16.8 Mechanical scrubber in a chemical plant. (Courtesy of TREMA Verfahrenstechnik GmbH, Kemnath, Germany.)

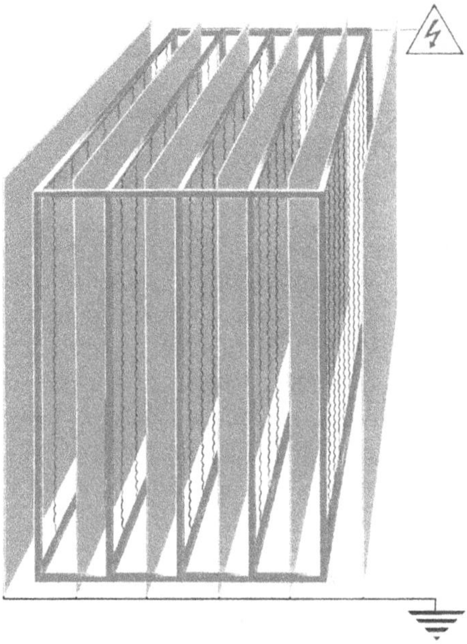

Figure 16.9 Electrostatic precipitator (seen schematically). (Courtesy of FLSmidth A/S, Valby, Denmark.)

2. At a certain distance from the electrode, the free electrons are captured by gaseous molecules which become negatively charged and also move away from the discharge electrode.
3. The next step is that the negatively charged gaseous molecules (ions) are captured by dust particles, which become negatively charged and move toward the collecting plates. These particles form a dust layer on the plate surface which is dislodged by rapping.
4. The dust particles are discharged by the collecting plates and the charge flows to ground.
5. The positively charged gaseous molecules (ions) move toward the discharge electrode and are discharged. The discharge electrode is rapped as well.

Process Data

The velocity of the gases passing through the ducts is typically 1 $m \cdot s^{-1}$. The maximum temperature of the separation process is in the range 400 to 500°C. The process loses efficiency at higher temperatures.

A typical dust resistivity is 10^9 Ω·cm. When the resistivity is relatively low, particles reaching the collecting plates lose their charge relatively easily and the dust can be reentrained. At high levels of dust resistivity, an insulating layer is formed on the collecting plates which prevents normal corona discharge, and this can lead to electrical instability and decreased collection efficiency.

A transformer–rectifier set converts the available low-voltage alternating-current supply to a high-voltage direct-current supply for the discharge electrodes. If the primary tension into the transformer is gradually increased, the high tension also increases until a maximum is achieved. A further increase in the primary tension also increases the sparkover, and the high tension falls. The automatic control system keeps the primary tension continually at the optimum value.

Scaling-up experiments are usually carried out on a 1/16 to 1/8 scale [7].

Construction

Sometimes, the precipitator unit is divided into two or more separate chambers to allow one chamber to be off-line for inspection and maintenance. Each chamber consists of several (e.g., three) fields in series. Each field operates independently, with a separately optimized energy supply. The reason is that the gas flows in plug flow through the fields and the dust content decreases rapidly.

Typical collecting plate spacings are 300 and 400 mm. A typical field length is 3.5 m, and a typical plate height is 12 m. The plates are generally made of mild steel, but stainless steel can be used where corrosion or abrasion resistance is required. Figure 16.10 shows an electrostatic precipitator installed in a plant.

Figure 16.10 Electrostatic precipitator in a chemical plant. (Courtesy of FLSmidth A/S, Valby, Denmark.)

Example 16.2 This example describes the use of electrostatic precipitators in the drying of marl. Dried marl is a raw material used to make cement. It is a mineral obtained by means of mining and typically contains 10% water by weight. The solid material consists of 90 to 95% by weight calcium carbonate and 10 to 5% by weight sand or clay. Marl is dried in large direct-heat rotary dryers by using hot air having a temperature of 450°C, for example. The separation of entrained particles from the gases leaving the dryer is achieved by means of electrostatic precipitators. This equipment is preferred over fabric filters because fabric cannot resist high temperatures.

Example 16.3 This example concerns the use of electrofilters in waste incineration plants to clean gases leaving a spray dryer. In such plants, waste is incinerated and the gases leaving the incinerator are cleaned carefully before they are passed on to the atmosphere. An electrostatic precipitator in a typical waste incineration plant will be described briefly. First, the gases are used to heat water in heat exchangers. Steam is raised from the hot water and the steam is used to generate electricity. The first step in cleaning the gases, now having a temperature of 245°C, is an electrostatic precipitator to separate fly ash. The second step is the cooling of these gases in a spray dryer by spray drying a suspension of gypsum particles in an aqueous solution of calcium chloride. The suspension originates from a downstream cleaning step of the gases. On spray drying the suspension, fine dry gypsum particles and dry calcium chloride particles are produced. Ten percent by weight of the powder production falls into the spray dryer cone and 90% by weight is entrained by the gases. The gases, having a temperature of 170°C, pass through a second electrofilter, separating the particulate material from the gaseous flow. Again, electrofilters are selected for this application because of their ability to process hot gases and their efficiency.

REFERENCES

[1] Koch, W.H., Licht, W. (1977). New design approach boosts cyclone efficiency. *Chemical Engineering*, *84*, 80–88.

[2] Muir, D.M. (1985). *A User Guide to Dust and Fume Control*, The Institution of Chemical Engineers, Rugby, UK, pp. 37–48.

[3] Perry, R.H., Green, D.W. (2008). *Perry's Chemical Engineers' Handbook*, McGraw-Hill, New York, pp. 17-46 to 17-51.

[4] Holzer, K. (1979). Wet scrubbing of particles and aerosols. *Chem. Ing. Tech.*, *51*, 200–207 (in German).

[5] Reference [2], pp. 49–60.

[6] Strigle, R.F., Jr. (1987). *Random Packings and Packed Towers*, Gulf Publishing, Houston, TX, pp. 31, 32.

[7] Reference [2], pp. 61–72.

17

DRYER FEEDING EQUIPMENT

Feeders are defined as devices that introduce a variety of materials directly into dryers. Typically, the feeder is located at the interface between material-handling equipment and the dryer. Some feeds can be pumped; for example, all spray-dryer feeds are pumped through a nozzle or to a rotary atomizer. Many wet cakes can be made to flow and can thus be pumped to a vertical thin-film dryer or a drum dryer. But in the majority of cases, the feed cannot be pumped, so specialized equipment for the handling of bulk solids is used for storing, transporting, dosing, and controlling dryer feeds.

Often, large storage containers are required because the feed is transported over long distances and is then stored in a bin or hopper. The design of the storage container must be based on the flow properties of the material being handled. If the dryer feed is received directly from equipment that operates batchwise, an intermediate storage container is required. These containers are deliberately not oversized, to reduce the likelihood of setting and caking. However, if the dryer feed is received continuously from a processing plant, then in principle, a buffer is not required, even though process control factors may necessitate its use. Dryers with small product holdup (e.g., flash dryers and vigorously agitated contact dryers) can be seriously affected by feed variation and interruption.

Intermediate storage containers are often combined with feeding or dosing equipment such as a bottom-mounted screw in a hopper. Transport devices in drying systems are identical to those that are used for the transport of bulk solids; but a selection must be made because not all transport equipment is suitable for handling wet solids. The most commonly used systems are screw, belt, and vibrating conveyors. Each of these systems can be utilized for dosing duties, but belt weighers and loss-in-weight feeders are the most reliable and accurate systems.

Drying in the Process Industry, First Edition. C.M. van 't Land.
© 2012 John Wiley & Sons, Inc. Published 2012 by John Wiley & Sons, Inc.

A loss-in-weight feeder can be a bin with a bottom-mounted twin screw. The bin's weight is measured by use of strain gauges, and their signal is used to adjust the rotational speed of the twin screw. Thus, variations in bulk density of the feed can be compensated. The steady flow of material to the screws is promoted by a slowly rotating agitator.

Screws, rotary locks, and vibrating conveyors are also used for dosing; the rotational speed (of screws and rotary locks) and frequency (of vibrating conveyors) are related to the throughput. However, if the bulk density varies, a recalibration of the system should be undertaken, which is often impractical because of the frequency with which changes occur. Thus, masswise, in those cases, the exact feed rate is frequently not known. Nevertheless, such systems can be used; for example, the exit-gas temperature in a flash dryer can be used to control the rotational speed of a rotary lock so that a steady exit temperature is maintained.

Often, it is important to prevent gas from flowing through feeders since ingress of hot gases from the dryer may give rise to incrustations and lead to losses, whereas cold air flowing through the feeder to the dryer might interfere with the control system if the gas temperature is used for process control. If a screw feeder is being considered, it is usually advantageous that one and a half pitches of the screw flight be left off at the discharge end of the feed screw. The choke formed by material moving as a plug will often minimize air leakage. Many feeders become hot during operation, which is sometimes acceptable, but in other instances it may be necessary to provide indirect cooling.

In Sections 17.1 through 17.10 we deal with feeders to various dryer types, with the important convective dryers covered first.

17.1 FLUID-BED DRYERS

In circular fluid-bed dryers, distribution and acceptance of the feed are relatively easy because the bed is deep and well mixed. But it is important that the feed not be introduced near a wall (i.e., a screw must protrude into the freeboard over the bed and then release the feed). Normally, the feeder is not subject to heating because the gas that flows past it has been cooled by the bed.

Gas leakage can be either into or out of the dryer. Leakage into the dryer does not influence the thermal efficiency because the vented gas is not used for drying purposes; however, it can depress the gas-vent temperature, and if this is used for process control, the process can be disturbed. Leakage out of the dryer generates a dust problem and may introduce incrustations in the feeder.

Usually, screw and rotary lock feeders are used in circular fluid-bed dryers.

In rectangular fluid-bed dryers, if the bed is relatively deep (e.g., 0.5 m), distribution and acceptance of the feed are comparatively easy. However, if the dryer has a shallow bed, the feed must be distributed over the full width of the dryer. In these circumstances, it is customary to receive the output from the feeder on a chute that has triangular baffles (like those of a sample splitter). Vibrating fluid-bed dryers have shallow beds so that the amount of material being vibrated is minimized; also, the vibrating action prevents incrustations in the air distributor. However, sometimes special measures to facilitate feed acceptance are necessary: More air needs to be passed through the feed section or a relatively low temperature must be used for the drying gas to prevent incrustations. The information given above on feeder temperature and gas leakages for circular fluid-bed dryers is equally applicable to rectangular fluid-bed dryers.

Two examples of feeder systems that are used with fluid-bed dryers are as follows:

1. Sodium chlorate crystals containing 3% water by weight and coming from pusher centrifuges are fed to a circular fluid-bed dryer by means of a screw [1]. The gas-vent temperature is kept at 250°F (121°C) by adjusting the drying-gas temperature within the range 400 to 450°F (204 to 232°C).
2. Vacuum-pan salt containing 3% water by weight and coming from pusher centrifuges continuously is transported to a rectangular fluid-bed dryer-cooler by means of conveyor belts. Feeding occurs by means of a vibrating feeder, the feed entering the dryer via a chute with triangular baffles. Process control is achieved by keeping the highest bed temperature constant. As this temperature decreases, the pressure of the steam flow to the indirect heat exchanger, which heats the drying air, is increased automatically, and vice versa. There is a small underpressure in the freeboard, and air leaks into the dryer at the feeding point. Figure 17.1 depicts a vibrating feeder driven by a 1-hp motor that employes the natural-frequency principle. It is 2 ft wide and 7 ft 6 in. in length.

Figure 17.1 Vibrating feeder. (Courtesy of Carrier Vibrating Equipment, Inc., Louisville, KY.)

17.2 DIRECT-HEAT ROTARY DRYERS

Three feeders are frequently encountered: chute, screw, and vibratory. If the dryer operates countercurrently, the heat load of the feeder is limited. Although ingress of air does not reduce the dryer's economy, it does depress the vented air temperature, and this can interfere with operation of the control system. Air leaking out at the feeding point creates a dust problem. If the dryer operates concurrently, the feeder may become overheated, and a special construction or indirect cooling with water may be required. Ingress air then reduces the heat economy, and the reverse gives rise to dust problems.

17.3 FLASH DRYERS

Flash dryers are used for free-flowing powders, granular and crystalline solids, slurries, and pastes. The latter two feeds can be accepted only if the dry product is recycled to make the feed acceptable for handling. The maximum particle size should not exceed 1 to 2 mm because the particles must be transported vertically and larger particles often require longer residence times than can be afforded by flash dryers.

The distance between the feed stream and the wall is small at the point at which the feed enters, and the entering feed is not mixed with material already present in the feed section. These two aspects distinguish flash dryers from the other types of convective dryers and call for careful consideration of the properties of the feed entering the dryer.

The combination of drying and vertical transport is an advantage of the flash dryer. There are no mechanical means for transport. However, air takes care of the vertical transport in a direction opposed to gravity. Often, special devices must be integrated into the flash-dryer system to ensure proper feed acceptance and processing. The main devices used are mixers, mills, slings, and classifiers. A mixer can mix feed and dry product to improve the flow characteristics of the feed (see Fig. 7.1). Figure 17.2 depicts a double-paddle mixer suitable for this procedure, which is often

Figure 17.2 Double-paddle mixer. (Courtesy of Alstom Power, Inc., Warrenville, IL.)

termed *backmixing*. A mill can take the physical form of a sling, a relatively simple mill, or a standard mill (refer again to Fig. 7.1, which shows a cage mill). A mill is often installed at the lowest point of the drying system, since any coarse material fed falls down, is reduced in size, and can then be entrained by the airstream. A classifier is located at the top of the dryer tube, where it separates coarse material that is subsequently recycled to the feed via an in-line mounted mill.

So although a flash dryer looks simple, a set of auxiliary devices is often required to guarantee solid handling. The drying of filter cakes, for example, usually requires a mixer and a cage mill.

The operation of a flash dryer is concurrent, and this implies that the feeding system can become overheated. Hence, special solutions must sometimes be used, such as indirect shaft cooling of a mill.

Ingress air at the feed point affects the heat economy adversely, and hot gas leaking out of the dryer may create a dust problem and lead to incrustations of equipment.

Two examples of feed systems in flash dryers are as follows:

1. Anhydrous sodium sulfate crystals containing approximately 5% water by weight are fed to a flash dryer by means of a screw in which the feed is not mixed with previously dried material. A sling is installed at the lowest point of the drying system to disintegrate larger material coming down. Coarse material is recycled from the dryer head into the feed screws via a mill mounted in line.
2. A filter cake containing approximately 25% water by weight is mixed with previously dried material and fed by means of a screw to a sling at the lowest point. Recycling of the milled material is carried out.

17.4 SPRAY DRYERS

Generally, the design practices used for the storage and transport of solutions and slurries are applied. It should be noted that on introducing the feed, the distance between the feed and the walls is rather large, and in this respect, spray dryers differ from the other important convective dryers. The following examples concern feeder systems that are used with spray dryers:

1. Two process vessels are used to prepare and store a slurry, which then passes through a colloid mill and is then pumped to a rotary atomizer. The colloid mill protects the atomizer against blockages.
2. A series of two pumps transport a solution to a single-fluid nozzle. The first is a low-pressure pump, and the second pump generates the pressure required for atomization.

17.5 CONVEYOR DRYERS

Two different types of conveyor dryer feeds may be distinguished: the first comprises rather large particles (e.g., bread crumbs, nuts, and tobacco), whereas the second

Figure 17.3 Oscillating spreader. (Courtesy of CPM Wolverine Proctor Ltd, Glasgow, UK.)

comprises feeds that must be physically formed. The latter procedure is often carried out in the chemical industry when filter cakes are processed using conveyor dryers.

Various types of equipment can be used for the first type of feed, such as the oscillating spreader (Fig. 17.3) and the wiper feed (Fig. 17.4), both of which spread the material to be dried evenly across the full width of the moving conveyor at the feed end.

A simple feeder is a hopper, but its application is limited to hard spherical objects that flow easily. Special feeders have been developed for tobacco and cereals.

Several techniques can be used for the second type of feed, which needs preforming. A short review of the methods that are employed is: (1) granulation (i.e., particle size enlargement by agglomeration), (2) extrusion to form spaghetti-like pieces about 6 mm in diameter and several centimeters long, (3) preforming on a steam-heated finned drum, (4) preforming of thixotropic filter cakes by scoring with knives on the filter, and (5) briquetting [2].

Rolling extruders are in quite common use in the chemical industry to preform filter cakes. The material must remain sufficiently fluid after dewatering for this technique to be successful. Figure 17.5 illustrates a rolling extruder. The width of the trough corresponds to the width of the conveyor dryer and the filter (e.g., a rotary vacuum filter). The arms containing the rolls reciprocate and the rolls roll over the perforated trough wall while the paste is pushed through the holes forming the extrudates. If the extrusion pressure is too low, the extrudates are mechanically weak,

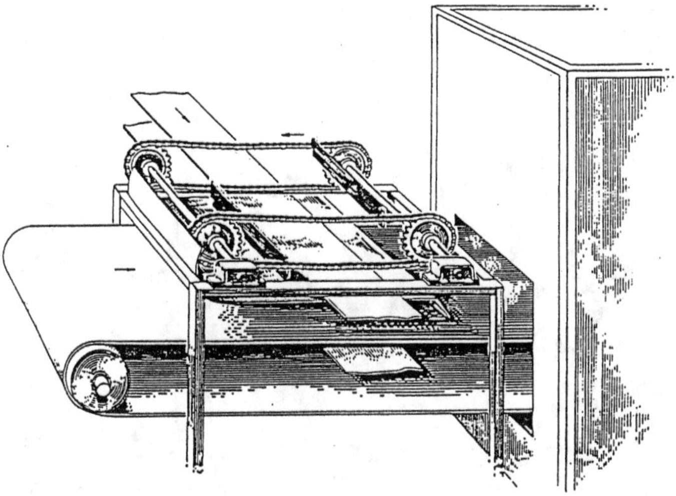

Figure 17.4 Wiper feed for a conveyor dryer (old version). (Courtesy of CPM Wolverine Proctor Ltd, Glasgow, UK.)

which can lead to dust formation. An extrusion pressure that is too high results in extrudates having poor dispersion properties or even dewatering in the forming equipment, which may cause jamming. The pressure is influenced by the water content of the cake, the cake properties, and the hole sizes, along with the percentage of free area.

17.6 HAZEMAG RAPID DRYER

The majority of Hazemag Rapid Dryers are equipped with a double-pendulum self-sealing feed and dry product connection. Figure 17.6 depicts a single-chamber pendulum lock. The chamber is situated between the upper pair of flaps and the lower pair of flaps. Sealing the dryer is necessary as there is much turbulence inside the

Figure 17.5 Rolling extruder.

Figure 17.6 Single-chamber pendulum lock (hydraulically controlled). (Courtesy of Hazemag & EPR GmbH, Dülmen, Germany.)

drying chamber. How the feeder works can be seen by referring to Figure 9.9. Two pairs of flaps are mounted above each other; one pair opens while the other is closed. Thus, the material travels through the lock and enters the dryer. Normally, the pressure difference across the feeder should not exceed 100 mm water gauge [3].

Flap gates cannot be used as extraction devices under bins or hoppers since the flaps would then have to close against the weight of the column of material resting on them.

It is also possible to install a pendulum lock having two chambers, which are formed by three pairs of flaps. The two-chambers design is used when it is important to minimize the airflow leaking into the dryer. For a two-chamber pendulum lock, hydraulic operation is used.

The feed and the drying air pass through the dryer concurrently, and thus, to some extent, the feeder is exposed to elevated temperatures. However, cooling by the feed stream also occurs.

Pivoted flap gates are especially suitable for handling sticky, caking, or lumpy materials. The gates are made in sizes of up to $2000 \cdot 1000\ \text{mm}^2$ internal dimensions.

17.7 ANHYDRO SPIN FLASH DRYER

The Anhydro Spin Flash Dryer is shown in Figure 9.8. Normally, feeds for this dryer type can be characterized as dry pastes, filter cakes, and in some instances, sludges. The feed enters a feed vessel in which an agitator rotates at a low speed. The agitator elements have a small wall clearance. The vessel also contains horizontal flow breakers. The agitator mixes the material and transports it axially to the vessel bottom. A single- or double-screw conveyor is installed excentrically at the vessel bottom. A single-screw conveyor is used for relatively small capacities and a double-screw conveyor is used for large capacities. The feed flows handled range from 100 $kg{\cdot}h^{-1}$ to 40 $t{\cdot}h^{-1}$. The conveyor transports the material to be dried to the drying chamber. The rotational speeds of both the agitator and the conveyor are continuously variable.

17.8 PLATE DRYERS

Plate dryers can operate under atmospheric or reduced pressure; in the latter case, special vacuum-tight rotary locks are used for the feed introduction and product extraction. Generally, rotary locks, kibblers, granulating screens, and table feeders are used for the introduction of the feed, although it is possible that other pieces of equipment may be preferred in some instances. In a way, granulating screens perform the same function as a roller extruder for conveyor dryers; they can form a cake into a type of extrudate that can be processed. Kibblers disintegrate the feed, to enable processing. Figure 17.7 shows a table feeder suitable for vacuum operation. The rotating table is present in the lower part of the feeder and is bottom-driven. The feed is distributed to the dryer through a peripheral hole. Baffles mounted on the table convey the feed to the chute.

17.9 VIGOROUSLY AGITATED CONTACT DRYERS

Buss-SMS-Canzler in Butzbach, Germany manufactures vigorously agitated horizontal contact dryers equipped with an auger feed blade in the feed section. The system resembles that of their vertical thin-film dryer. Horizontal contact dryers can accept wet, free-flowing powders and pastes that can be pumped. Figure 17.8 shows a feed system suitable for nonpumpable feeds that can be used in continuous vacuum drying. The output of a batch centrifuge is received in a vibrating grid feeder and the feed enters the dryer via a pendulum lock. Feed shaping is not required since the dryer takes care of the formation of a thin film. (*Note:* Some filter cakes containing up to 60 to 70% water by weight can be made to flow under shear.)

17.10 VERTICAL THIN-FILM AND DRUM DRYERS

The feed for a vertical thin-film dryer has a low viscosity. It is customary to employ gear or Moyno pumps since these pumps achieve smooth dosing. The feed enters the

Figure 17.7 Table feeder suitable for vacuum operation. (Courtesy of Andritz KMPT GmbH, Vierkirchen, Germany.)

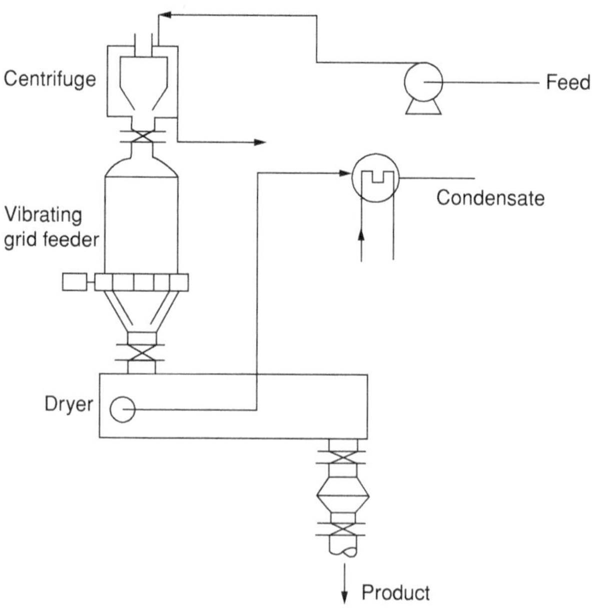

Figure 17.8 Vibrating grid feeder. (Courtesy of Buss-SMS-Canzler GmbH, Butzbach, Germany.)

top part of the dryer and is distributed by means of the agitator. The solution or slurry flows downward.

The feed to a drum dryer can be pumped into the space between the two rolls, or by means of a top- or bottom-application roll. The feed can also be pumped into a trough and applied by dipping, splashing, or spraying.

REFERENCES

[1] Vreeland, R., Baccheti, J.H. (1982). Equipment plugging eliminated with fluid bed dryer. *Chemical Processing*, *45*(*12*), 24–25.

[2] Perry, R.H., Green, D.W. (2008). *Perry's Chemical Engineers' Handbook*, McGraw-Hill, New York, p. 12–64.

[3] Motek, H. (1983). Possibilities for using pivoted flap gates in bulk material preparation. *Journal for Preparation and Processing*, *24*, 439–443.

NOTATION

A	Constant in equation (4.1)	—
	Constant in equation (4.6)	$kg \cdot kg^{-1}$
	Heat transfer area	m^2
	Moisture content	$kg \cdot kg^{-1}$
	Product area	m^2
	Sample area	m^2
A_1	Area of body 1	m^2
	Feed moisture content	% by weight
	Nozzle area	m^2
A_2	Area of body 2	m^2
	Product moisture content	% by weight
A_C	Corrected nozzle area	m^2
A_{crit}	Critical moisture content	$kg \cdot kg^{-1}$
Airflow_1	Airflow taken in by the dryer (10°C)	$m^3 \cdot h^{-1}$
Airflow_2	Airflow leaving the dryer (TA_{out})	$m^3 \cdot h^{-1}$
Ar	Archimedes number	—
a	Constant in $TA_{out} = a \cdot TA_{in} + b$	°C
	Cyclone inlet height	m
	Exponent in equation (8.3)	—
	Heat diffusion number	—
	Proportionality constant in equation (13.3)	m^{-1}
	Product area per m^3 dryer volume	$m^2 \cdot m^{-3}$
	Gas area per m^3 of liquid	$m^2 \cdot m^{-3}$
B	Bed width	m
	Constant in equation (4.1)	—
	Constant in equation (4.6)	K^{-1}
	Rotary dryer parameter	$\mu m^{-0.5}$
Bi	Biot number	—
b	Constant in $TA_{out} = a \cdot TA_{in} + b$	°C
	Cyclone inlet width	m
	Exponent in equation (8.3)	—

Drying in the Process Industry, First Edition. C.M. van 't Land.
© 2012 John Wiley & Sons, Inc. Published 2012 by John Wiley & Sons, Inc.

Symbol	Meaning	Unit
C	Constant in equation (4.1)	—
	Constant in equation (8.1)	—
	Constant in a Nusselt equation and in a Sherwood equation	—
C_C	Nozzle area correction factor	—
C_K	Constant in equation (8.5)	—
Cap	Product mass flow	$kg \cdot h^{-1}$
c	Constant in expressions for Q_2 and Q_3	K
	Exponent in Equation (8.3)	—
c_g	Water vapor concentration	$kmol \cdot m^{-3}$
c_g^*	Water vapor concentration at saturation	$kmol \cdot m^{-3}$
c_p	Air specific heat at constant pressure	$J \cdot kg^{-1} \cdot K^{-1}$
c_s	Solid specific heat	$J \cdot kg^{-1} \cdot K^{-1}$
c_v	Air specific heat at constant volume	$J \cdot kg^{-1} \cdot K^{-1}$
c_w	Wien's constant (2897)	$\mu m \cdot K$
D	Dryer diameter	m
	Line diameter	m
	Nozzle diameter	m
D_c	Cyclone diameter	m
D_e	Cyclone gas outlet diameter	m
d	Agitator diameter	mm
	Constant in expression for Q_{tot_2}	—
	Exponent in equation (8.3)	—
	Nozzle diameter	m
d_1	Atomizer wheel diameter	m
d_{50}	Weight-average droplet size	μm
	Weight-average particle size	μm
d_{95}	Droplet size below which 95% by weight of a sample can be found	μm
d_d	Distance between the feeding point and the axis of an atomizing wheel	m
d_p	Particle size	μm
$\bar{d}_p$	Sauter diameter (Chapters 5 and 8)	μm
	Average particle size (Chapter 6)	μm
E	Electric field strength	$V \cdot m^{-1}$
$E(x)$	Electric field strength at depth x	$V \cdot m^{-1}$
Evap	Dryer water evaporation load	$kg \cdot h^{-1}$
		$kg \cdot s^{-1}$
F	Specific product mass flow	$kg \cdot m^{-2} \cdot h^{-1}$
Fanpower	Fan power	kW
Fr	Froude number	—

f	Constant in expression for $Airflow_2$	—
	Electromagnetic field frequency	s^{-1}
G	Specific dry airmass flow	$kg{\cdot}m^{-2}{\cdot}h^{-1}$
G_{air}	Dry airmass	kg
G_{H_2O}	Evaporated water mass	kg
G_{mf}	Minimum fluidization mass velocity	$kg{\cdot}m^{-2}{\cdot}s^{-1}$
G_s	Solid mass	kg
$Gasflow_2$	Gas flow leaving the dryer	$m^3{\cdot}h^{-1}$
g	Acceleration due to gravity	$m{\cdot}s^{-2}$
H	Enthalpy per kilogram of dry air	$J{\cdot}kg^{-1}$
		$kJ{\cdot}kg^{-1}$
	Flash dryer height	m
	Spray dryer chamber cylindrical height	m
ΔH	Heat of evaporation	$J{\cdot}kmol^{-1}$
		$J{\cdot}kg^{-1}$
		$kJ{\cdot}kg^{-1}$
h	Atomizer wheel vane height	m
	Height	m
	Planck's constant ($6.6{\cdot}10^{-34}$)	J·s
	Settled bed height	m
h_{exp}	Expanded bed height	m
I	Microwave detector signal	—
I_0	Microwave reference detector signal	—
K	Coefficient in equation (6.4)	—
	Constant in equation (8.3)	—
	Cyclone pressure-loss constant	—
K_{st}	Deflagration index	$bar{\cdot}m{\cdot}s^{-1}$
k	Mass transfer coefficient	$m{\cdot}s^{-1}$
k_l	Partial mass transfer coefficient (liquid-side)	$m{\cdot}s^{-1}$
L	Dryer length	m
	Line length	m
Le	Lewis number	—
L_1	Fluid-bed dryer length for free-water evaporation	m
L_2	Fluid-bed dryer length for cross-flow heating	m
L_3	Fluid-bed dryer length for cross-flow cooling	m
L_e	Height of a transfer unit	m
M	Atomizer wheel feed	$kg{\cdot}h^{-1}$
	Mass	kg
	Molecular weight	$kg{\cdot}kmol^{-1}$

Symbol	Description	Units
m	Exponent of the Reynolds number	—
	Mass	kg
N	Agitator rotational speed	min^{-1}
	Atomizer wheel rotational speed	min^{-1}
	Rotary dryer rotational speed	min^{-1}
	Theoretical energy for the adiabatic compression of 1 kg of air	$J \cdot kg^{-1}$
Nu	Nusselt number	—
Nu_{lam}	Nusselt number for laminar flow	—
Nu_{turb}	Nusselt number for turbulent flow	—
N_H	Number of velocity heads	—
N_t	Number of transfer units	—
n	Exponent in equation (6.4)	—
	Number	—
	Number of vanes in an atomizer wheel	—
P	Power consumption	kW
		W
	Pressure	bar absolute
	Specific power generation	$W \cdot m^{-3}$
P_{air}	Air partial pressure	$N \cdot m^{-2}$
P_{max}	Maximum explosion pressure	bar absolute
P_{red}	Maximum pressure on venting	bar gauge
P_{sat}	Saturated water vapor pressure	$N \cdot m^{-2}$
P_{stat}	Pressure at which a relief vent opens	bar absolute
P_{tot}	System pressure	$N \cdot m^{-2}$
$Power_{G1}$	Fan power required for $Airflow_1$	kW
$Power_{G2}$	Fan power required for $Gasflow_2$	kW
$Power_{rot}$	Rotary dryer motive power	kW
Pr	Prandtl number	—
p	pressure	bar
		mbar
p_1	Air pressure before compression	$N \cdot m^{-2}$
	Fluid pressure upstream of a nozzle	$N \cdot m^{-2}$
p_2	Air pressure after compression	$N \cdot m^{-2}$
	Fluid pressure downstream of a nozzle	$N \cdot m^{-2}$
p_c	Capillary water vapor pressure	$N \cdot m^{-2}$
p_g	Water vapor partial pressure	$N \cdot m^{-2}$
p_g^*	Saturated water vapor pressure	bar absolute
		$N \cdot m^{-2}$
Δp	Pressure loss	$N \cdot m^{-2}$
Q	Radiation energy density	$W \cdot m^{-2}$
Q_1	Enthalpy flow for the evaporated water	$kJ \cdot h^{-1}$
Q_2	Enthalpy flow for the dry solids	$kJ \cdot h^{-1}$

Q_3	Enthalpy flow for the residual water	$kJ{\cdot}h^{-1}$
Q_{tot_1}	Net heat	$kJ{\cdot}h^{-1}$
Q_{tot_2}	Steady-state dryer-heat requirement	$kJ{\cdot}h^{-1}$
R	Contact dryer radius	m
	Universal gas constant (8314)	$J{\cdot}kmol{\cdot}K^{-1}$
R_A	Air specific mass	$kg{\cdot}m^{-3}$
R_D	Universal gas constant for water	$J{\cdot}kg^{-1}{\cdot}K^{-1}$
R_W	Water vapor specific mass	$kg{\cdot}m^{-3}$
Re	Reynolds number	—
r_c	Caplillary radius	m
S	Dryer area	m^2
	Product area per meter of dryer length	$m^2{\cdot}m^{-1}$
	Rotary dryer slope	—
Sc	Schmidt number	—
Sh	Sherwood number	—
Sol	Dry solids flow	$kg{\cdot}h^{-1}$
T	Temperature	K °C
T_1	Initial air temperature	K
T_f	Feed temperature	°C
T_g	Gas temperature	K
T_i	Heating medium inlet temperature	°C
$T_p(t)$	Product temperature at time t	°C
$T_p(x)$	Product temperature at position x	°C
T_s	Adiabatic saturation temperature	K
T_W	Wet-bulb temperature	K
TA_{in}	Drying air temperature	°C
TA_{out}	Spent drying-gas temperature	°C
TP_{in}	Feed temperature	°C
TP_{out}	Product exit temperature	°C
ΔT	Temperature difference	K
$(\Delta T)_m$	Logarithmic mean temperature difference	K
t	Time	s
t_1	Duration of the first drying period	s
t_2	Duration of the second drying period	s
t_3	Duration of the cooling period	s
U	Heat transfer coefficient	$kJ{\cdot}m^{-2}{\cdot}h^{-1}{\cdot}K^{-1}$ $W{\cdot}m^{-2}{\cdot}K^{-1}$
V	Air volume for batch fluid-bed drying	m^3
	Dryer volume	m^3

	Sample volume	m^3
	Volume	m^3
v	Atomizer wheel circumferential velocity	$m \cdot s^{-1}$
	Circumferential velocity	$m \cdot s^{-1}$
	Contact dryer agitator circumferential velocity	$m \cdot s^{-1}$
	Electromagnetic field propagation velocity	$m \cdot s^{-1}$
	Particle terminal velocity	$m \cdot s^{-1}$
	Velocity	$m \cdot s^{-1}$
v'	Water specific volume	$m^3 \cdot kg^{-1}$
v_1	Fluid velocity upstream of a nozzle	$m \cdot s^{-1}$
v_2	Fluid velocity downstream of a nozzle	$m \cdot s^{-1}$
v_F	Fluidization velocity	$m \cdot s^{-1}$
v_i	Cyclone inlet velocity	$m \cdot s^{-1}$
v_{mf}	Minimum fluidization velocity	$m \cdot s^{-1}$
v_{rel}	Gas–liquid relative velocity	$m \cdot s^{-1}$
W	Rotary dryer mass including product	kg
W_{in}	Water flow to the dryer	$kg \cdot h^{-1}$
W_{out}	Water flow leaving the dryer	$kg \cdot h^{-1}$
WA_{flow}	Evaporated water flow	$m^3 \cdot h^{-1}$
w	Rotary dryer holdup	kg
X	Mass fraction	—
X_3	Size range mass fraction	—
X_e	Equilibrium solid moisture content per kg of dry matter	$kg \cdot kg^{-1}$
x	Gas water content in kg per kg of dry air	$kg \cdot kg^{-1}$
	Kilogram of water vapor raised per kg of steam fed	$kg \cdot kg^{-1}$
	Length or depth	m
x_i	Entering air water content	$kg \cdot kg^{-1}$
x_o	Leaving air water content	$kg \cdot kg^{-1}$
α	Heat transfer coefficient	$W \cdot m^{-2} \cdot K^{-1}$
	Radiation absorption coefficient	—
α_o	Heat transfer coefficient for fluidized particle	$W \cdot m^{-2} \cdot K^{-1}$
γ	$\gamma = c_p/c_v$	—
δ	Penetration depth of electromagnetic waves	m
ε	Radiation emission coefficient	—
ε_0	Dielectric permittivity of free space ($8.854 \cdot 10^{-12}$)	$F \cdot m^{-1}$
ε_1	Radiation emission coefficient of body 1	—
ε_2	Radiation emission coefficient of body 2	—

ε'_r	Relative dielectric constant	—
ε''_r	Loss factor	—
η	Efficiency	—
	Gas–liquid ratio	—
η_{ad}	Adiabatic compression efficiency	—
λ	Electromagnetic field wavelength	m
λ_G	Gas thermal conductivity	$W \cdot m^{-1} \cdot K^{-1}$
λ_{max}	Wavelength of maximum radiation energy density	μm
λ_S	Solid thermal conductivity	$W \cdot m^{-1} \cdot K^{-1}$
μ	Fluid dynamic viscosity	$N \cdot s \cdot m^{-2}$
	Fluidization gas dynamic viscosity	$N \cdot s \cdot m^{-2}$
	Microwave radiation absorption coefficient	$m^2 \cdot kg$
μ_l	Liquid dynamic viscosity	$N \cdot s \cdot m^{-2}$
υ	Fluidization gas kinematic viscosity	$m^2 \cdot s^{-1}$
ρ	Specific mass	$kg \cdot m^{-3}$
ρ_b	Bulk density particulate material	$kg \cdot m^{-3}$
ρ_F	Fluidization gas specific mass	$kg \cdot m^{-3}$
ρ_g	Gas specific mass	$kg \cdot m^{-3}$
ρ_l	Liquid specific mass	$kg \cdot m^{-3}$
ρ_p	Particle specific mass	$kg \cdot m^{-3}$
ρ_S	Solid specific mass	$kg \cdot m^{-3}$
$\Delta\rho$	Specific mass difference	$kg \cdot m^{-3}$
σ	Air–water surface tension	$N \cdot m^{-1}$
	Stefan–Boltzmann's constant ($5.675 \cdot 10^{-8}$)	$W \cdot m^{-2} \cdot K^{-4}$
τ	Drying time	h
	Residence time in dryer	min
ϕ''_h	Heat flow	$W \cdot m^{-2}$
ϕ_{H_2O}	Water-vapor flow rate	$kg \cdot s^{-1}$
ϕ_l	Water flow rate through a nozzle	$L \cdot min^{-1}$
ϕ_m	Airmass flow	$kg \cdot s^{-1}$
ϕ''_{mol}	Mass flow	$kmol \cdot m^{-2} \cdot s^{-1}$
ϕ_p	Product flow rate	$kg \cdot s^{-1}$
ϕ_υ	Gas flow	$m^3 \cdot h^{-1}$
ϕ	Loss angle	—
Ψ	Air relative humidity	—

INDEX

Drying in the Process Industry, First Edition. C.M. van 't Land.
© 2012 John Wiley & Sons, Inc. Published 2012 by John Wiley & Sons, Inc.

www.ingramcontent.com/pod-product-compliance
Lightning Source LLC
LaVergne TN
LVHW011529200926
848484LV00001B/207